"十三五"高等教育规划教材

高等院校电气信息类专业"互联网+"创新规划教材

Android 开发工程师案例教程

（第 2 版）

倪红军　编著

内 容 简 介

本书以 Android 开发基础、Android 开发提高和 Android 高级开发为主线，通过开发实例和项目案例，由浅入深、循序渐进地介绍 Android 应用开发的主要技术。本书的整体编写思路是"案例项目诠释理论基础、理论基础拓展项目创新"，即通过典型的项目案例既要告诉读者该项目的实现过程和步骤，又要告诉读者该项目实现时所需要的理论知识。读者在掌握理论知识后还要会灵活运用，并在新的项目开发中不断拓展，真正能够实现"教、学、做"的有机融合，实现学生从项目模仿到应用创新的递进式培养。

本书所举的开发案例步骤清晰详细，项目案例典型实用，操作步骤图文并茂，案例视频、微课、习题等资源丰富，以便读者更好地学习和掌握 Android 开发技术，提高实际开发水平，快速成为一名合格的 Android 开发工程师。

图书在版编目(CIP)数据

Android 开发工程师案例教程/倪红军编著．—2 版．—北京：北京大学出版社，2019.9
高等院校电气信息类专业"互联网+"创新规划教材
ISBN 978-7-301-30627-7

Ⅰ.①A… Ⅱ.①倪… Ⅲ.①移动终端—应用程序—程序设计—高等学校—教材 Ⅳ.①TN929.53

中国版本图书馆 CIP 数据核字(2019)第 168721 号

书　　　名	Android 开发工程师案例教程（第 2 版） Android KAIFA GONGCHENGSHI ANLI JIAOCHENG （DI-ER BAN）
著作责任者	倪红军　编著
策 划 编 辑	郑　双
责 任 编 辑	李娉婷　郑　双
数 字 编 辑	刘　蓉
标 准 书 号	ISBN 978-7-301-30627-7
出 版 发 行	北京大学出版社
地　　　址	北京市海淀区成府路 205 号　100871
网　　　址	http://www.pup.cn　新浪微博:@北京大学出版社
电 子 信 箱	pup_6@163.com
电　　　话	邮购部 010-62752015　发行部 010-62750672　编辑部 010-62750667
印 刷 者	北京虎彩文化传播有限公司
经 销 者	新华书店
	787 毫米 ×1092 毫米　16 开本　28.75 印张　680 千字 2014 年 7 月第 1 版 2019 年 9 月第 2 版　2022 年 1 月第 2 次印刷
定　　　价	69.00 元

未经许可，不得以任何方式复制或抄袭本书之部分或全部内容。
版权所有，侵权必究
举报电话：010-62752024　电子信箱：fd@pup.pku.edu.cn
图书如有印装质量问题，请与出版部联系，电话：010-62756370

第 2 版前言

近年来，移动互联网技术发展非常迅速，全世界大部分的 IT 公司都将业务重心转移到移动互联网，移动互联网业务也成为业内最大的利润增长点。Android 凭借其开源性、优异的用户体验和极为方便的开发方式而广泛应用于手机、平板电脑等移动设备上，并且赢得了广大用户和开发者的青睐，目前已经发展成为市场占有率最高的智能终端设备操作系统。

2014 年，编者根据应用型本科人才培养要求，结合计算机类、电子信息类专业的课程体系，在多年 Android 项目开发积累和教学实践基础上编写了《Android 开发工程师案例教程》，并获得教育部高等教育司产学合作协同育人项目——Google 中国教育合作部 2014 年移动应用技术人才培养课程建设项目资助。随着 Android 新技术的不断涌现、Android Studio 的日趋成熟和 Google 提出停止对其他开发环境的支持，编者对《Android 开发工程师案例教程》进行修订，编写了本书。本书保持了《Android 开发工程师案例教程》的写作风格：内容编写上注重由简到繁、由浅入深和循序渐进；选择典型且实用的项目案例阐述 Android 应用开发新技术；知识结构条理清晰，源码注释通俗易懂，让初学者看得懂、学得会、做得出，让有一定基础的 Android 应用开发者强基础、拓思路、升能力。

本书共 10 章，分为 4 个部分。第一部分介绍 Android 的开发环境和应用程序结构，由第 1～2 章组成；第二部分介绍 Android 应用开发中的基本界面组件与布局、高级界面组件与布局优化、菜单和对话框的设计与实现，由第 3～5 章组成；第三部分介绍服务、消息广播、数据存储与访问和多媒体应用开发，由第 6～8 章组成；第四部分介绍 Android 的网络应用、传感器与位置服务应用开发，由第 9～10 章组成。

与同类教材相比，本书具有以下特色。

1. 新技术、新环境

本书根据 Google 发布的 Android 技术进行编写，除介绍常用界面组件和开发技术外，还增加了 ViewPager、ViewFlipper、RecyclerView 等目前比较流行的高级界面组件，第三方网络请求框架 OkHttp，以及传感器与位置服务等 Android 应用开发的新技术。书中的案例源码全部在 Android Studio 开发环境下实现，在 Android 6.0 以上版本平台测试运行。

2. 重理论、强实践

本书按照编者近年来参与的实际工程项目和教学实践来安排各章节内容，便于读者掌握和理解。每章均以案例项目的开发过程为主线，介绍每个案例设计与实现时所涉及的 Android 开发技术。每一个案例项目都是 Android 项目开发中的常见需求，读者既可以直接将这些案例的解决方案应用于自己的开发项目中，也可以通过案例的实现过程巩固理论知识、强化实践能力、提高学习效率。

3. 多资料、便学习

本书为"互联网+"教材，既提供了教学课件、教学大纲、课后习题和程序源代码等传统的教学资源，还提供 27 个精彩项目案例的 61 个微课视频，并且已经在腾讯课程上开设免费课程，读者可以在腾讯课程上搜索"Android 开发工程师案例教程"关键字查找，也可以直接输入网址进行学习：https://ke.qq.com/course/349479。读者也可以随时随地扫码观看重点、难点内容的讲解，以便更好地学习和掌握 Android 开发技术，提高实际开发水平，快速成为一名合格的 Android 开发工程师。

本书由南京师范大学泰州学院倪红军编写。南京师范大学泰州学院李霞、周巧扣老师在习题的编写和整理等方面做了许多工作，北京大学出版社郑双对本书的编写工作提供了帮助和指导，编者在此一并表示感谢。

由于编者理论水平和实践经验有限，书中疏漏和不足之处在所难免，恳请广大读者提出宝贵的意见和建议，相关问题请联系 tznkf@163.com。

编　者

2019 年 2 月

【资源索引】

目 录

第1章 Android 开发环境 1
1.1 Android 的发展与现状 2
1.2 Android 的基本架构与特性 4
    1.2.1 Android 平台基本架构 4
    1.2.2 Android 的特性 6
1.3 Android 开发环境与常用工具 6
    1.3.1 安装 JDK 6
    1.3.2 安装 Android Studio 9
本章小结 18
习题 18

第2章 Android App 结构 20
2.1 创建 App 21
2.2 剖析 Android App 24
    2.2.1 Android 应用目录剖析 24
    2.2.2 资源的使用 26
    2.2.3 AndroidManifest.xml 文件的
           结构 30
2.3 Android 的四大组件 31
    2.3.1 Activity 32
    2.3.2 BroadcastReceiver 32
    2.3.3 Service 32
    2.3.4 Contentprovider 33
本章小结 33
习题 33

第3章 基本界面组件与布局 35
3.1 用户界面基础 36
    3.1.1 组件和布局管理器 36
    3.1.2 View 类和 ViewGroup 类 ... 37
3.2 计算器的设计与实现 38
    3.2.1 预备知识 38
    3.2.2 计算器的实现 46
3.3 高仿 QQ 登录界面的设计与
    实现 49
    3.3.1 预备知识 50
    3.3.2 高仿 QQ 登录界面的
           实现 56
3.4 注册界面的设计与实现 60
    3.4.1 预备知识 61
    3.4.2 注册界面的实现 72
3.5 考试系统界面的设计与实现 81
    3.5.1 预备知识 81
    3.5.2 考试系统界面的实现 86
3.6 打老鼠游戏的设计与实现 91
    3.6.1 预备知识 91
    3.6.2 打老鼠游戏的实现 100
3.7 猜扑克游戏的设计与实现 105
    3.7.1 预备知识 106
    3.7.2 猜扑克游戏的实现 112
本章小结 116
习题 117

第4章 高级界面组件与布局优化 119
4.1 Adapter 120
    4.1.1 基本概念 120
    4.1.2 Adapter 的常用子类 120
4.2 通讯录的设计与实现 122
    4.2.1 预备知识 122
    4.2.2 通讯录的实现 133
4.3 仿微信主界面的设计与实现 139
    4.3.1 预备知识 140
    4.3.2 仿微信主界面的实现 143
4.4 仿今日头条主界面的
    设计与实现 148

| 4.4.1　预备知识 …………… 148
       4.4.2　仿今日头条主界面的
              实现 ……………………… 154
   4.5　轮播效果的设计与实现………… 157
       4.5.1　预备知识 …………… 157
       4.5.2　左右轮播的实现 …… 158
       4.5.3　上下轮播的实现 …… 160
   4.6　商品列表布局切换效果的
        设计与实现 …………………… 162
       4.6.1　预备知识 …………… 163
       4.6.2　商品列表布局切换效果的
              实现 ……………………… 167
   本章小结 ……………………………… 172
   习题 …………………………………… 172

第5章　菜单和对话框 …………… 175
   5.1　概述………………………………… 176
       5.1.1　菜单（Menu）………… 176
       5.1.2　对话框（Dialog）…… 177
   5.2　满意度调查表的设计与实现…… 177
       5.2.1　预备知识 …………… 178
       5.2.2　满意度调查表的实现 …… 190
   5.3　其他常用菜单 …………………… 197
       5.3.1　选项菜单
              （OptionsMenu）………… 197
       5.3.2　子菜单（SubMenu）…… 199
       5.3.3　活动栏（ActionBar）…… 200
       5.3.4　弹出式菜单
              （PopupMenu）………… 201
   5.4　宾馆预订App界面的设计与
        实现…………………………………… 202
       5.4.1　预备知识 …………… 202
       5.4.2　宾馆预订App界面的
              实现 ……………………… 207
   本章小结 ……………………………… 213
   习题 …………………………………… 213

第6章　服务和消息广播 ………… 215
   6.1　概述………………………………… 216
   6.2　电话监听器的设计与实现……… 217

6.2.1　预备知识 …………… 217
       6.2.2　电话监听器的实现 …… 227
   6.3　短信拦截器的设计与实现……… 230
       6.3.1　预备知识 …………… 231
       6.3.2　短信拦截器的实现 …… 236
   6.4　闹钟的设计与实现 ……………… 239
       6.4.1　预备知识 …………… 239
       6.4.2　闹钟的实现 …………… 241
   6.5　定时短信发送器的设计与实现 … 245
       6.5.1　预备知识 …………… 245
       6.5.2　定时短信发送器的
              实现 ……………………… 248
   本章小结 ……………………………… 251
   习题 …………………………………… 251

第7章　数据存储与访问 ………… 253
   7.1　概述………………………………… 254
   7.2　幸运抽奖器的设计与实现……… 255
       7.2.1　预备知识 …………… 255
       7.2.2　幸运抽奖器的实现 …… 272
   7.3　实验室安全测试系统的
        设计与实现 …………………… 280
       7.3.1　预备知识 …………… 280
       7.3.2　实验室安全测试系统的
              实现 ……………………… 289
   7.4　应用程序间的数据共享 ………… 299
       7.4.1　预备知识 …………… 300
       7.4.2　学生信息共享应用的
              实现 ……………………… 302
       7.4.3　使用Android提供的
              ContentProvider ……… 310
   本章小结 ……………………………… 313
   习题 …………………………………… 313

第8章　多媒体应用开发 ………… 317
   8.1　概述………………………………… 318
   8.2　音乐播放器的设计与实现……… 318
       8.2.1　预备知识 …………… 318
       8.2.2　音乐播放器的实现 …… 324

8.3 视频播放器的设计与实现 ········ 333
    8.3.1 预备知识 ················ 333
    8.3.2 视频播放器的实现········ 334
8.4 录音机的设计与实现 ··········· 338
    8.4.1 预备知识 ················ 338
    8.4.2 录音机的实现 ··········· 339
8.5 照相机的设计与实现 ··········· 341
    8.5.1 预备知识 ················ 341
    8.5.2 照相机的实现 ··········· 345
本章小结 ··························· 351
习题 ······························· 352

第9章 网络应用开发 ············ 354

9.1 概述 ························· 355
9.2 在线中英文互译工具的
    设计与实现 ··················· 357
    9.2.1 预备知识 ················ 357
    9.2.2 在线中英文互译工具的
        实现 ···················· 366
9.3 股票即时查询工具的设计与
    实现 ························· 371
    9.3.1 预备知识 ················ 371
    9.3.2 股票即时查询工具的
        实现 ···················· 378
9.4 快递单查询工具的设计与实现 ··· 388

    9.4.1 预备知识 ················ 388
    9.4.2 快递单查询工具的
        实现 ···················· 393
9.5 WebView 的应用 ·············· 396
    9.5.1 预备知识 ················ 397
    9.5.2 WebView 定制浏览器 ······ 400
    9.5.3 Android 与 JavaScript
        交互 ···················· 405
    9.5.4 在 HTML 页面显示 Android
        通讯录联系人信息········ 412
本章小结 ··························· 416
习题 ······························· 416

第10章 传感器与位置服务
        应用开发 ··············· 419

10.1 概述 ······················· 420
10.2 指南针的设计与实现 ·········· 422
    10.2.1 预备知识················ 422
    10.2.2 指南针的实现 ··········· 426
10.3 百度地图在 Android 中的
    开发应用 ···················· 430
    10.3.1 预备知识················ 430
    10.3.2 百度地图应用实现 ······· 435
本章小结 ··························· 450
习题 ······························· 450

第 1 章 Android 开发环境

从 2007 年苹果发布第 1 代 iPhone 引发智能手机革命之后,移动互联网这个全新的市场就此打开。在我国互联网的发展过程中,PC 互联网已日趋饱和,移动互联网却呈现井喷式发展。伴随着移动终端价格的下降及 WiFi 的广泛铺设,移动网民呈现爆发趋势,同时 Android App 开发和 iOS App 开发也成为移动应用开发的主体,近年来 Android App 已经超越了手机产业,迅速扩张到更多相关的智能领域。Android "N" 的到来,让 Android 开发技术应用得更加广泛,Android App 开发拥有比较好的发展前景。

教学目标

了解 Android 的发展历史与现状。
熟悉 Android 的基本架构与特性。
掌握 Android 开发环境的搭建步骤与常用工具的使用。

教学要求

知识要点	能力要求	相关知识
Android 的发展与现状	(1) 了解 Android 的发展历程 (2) 了解 Android 各个版本的特点及现状	Android 名字的由来
Android 平台的基本架构与特性	(1) 掌握 Android 的平台架构及其每一层功能 (2) 了解 Android 平台的特性	Linur 内核
Android 的开发环境与常用工具	(1) 掌握 Windows 平台下 Android 开发环境的搭建步骤 (2) 掌握创建模拟器的方法及其他工具的使用	系统环境变量设置

1.1　Android 的发展与现状

随着 Android 和移动互联技术的迅猛发展，Android 已经从最初的智能手机操作系统发展为应用于平板电脑、可穿戴设备、车载导航、家电等移动终端设备上的具有广泛影响力的操作系统。

Android 这一词最早出现在法国作家利尔·亚当 1886 年发表的科幻小说《未来的夏娃》中，作者将外表像人类的机器起名为 Android。Android 的 Logo 最初是由伊琳娜·布洛克设计的，由于 Google 希望将 Android 平台应用于移动设备，伊琳娜·布洛克和她的设计团队被要求设计一个能够很容易被消费者记住的机器人标志，后来伊琳娜·布洛克从厕所门上经常出现的男人和女人的形象中得到灵感，画了一个有锡罐形的躯干、头上有天线的简易机器人作为 Android 的 Logo，如图 1.1 所示。

图 1.1　Android 的 Logo

Android 最初由安迪·罗宾等人于 2003 年 10 月创建的 Android 公司开发，2005 年 8 月由 Google 注资收购，并聘用安迪·罗宾为 Google 公司的工程部副总裁，继续负责 Android 项目的研发工作。

2007 年 11 月，Google 正式向外展示 Android，并且宣布与其他 30 多家手机厂商（包括摩托罗拉、华为、HTC、三星、LG 等）、软件开发商、芯片制造商和电信运营商联合组成开放手机联盟（OHA），这一联盟将支持 Google 发布的手机操作系统及应用软件。同时，Google 以 Apache 免费开源许可证的授权方式，发布了 Android 的源代码。

2008 年 5 月，帕特里克·布雷迪在 Google I/O 大会上提出了 Android HAL（硬件抽象层）架构图；同年，Android 获得了美国联邦通信委员会（FCC）的批准；同年 9 月，Google 正式发布 Android 1.0 和由 HTC 代工制造、运营商 T-Mobile 定制的手机——T-Mobile G1（图 1.2）。这款手机当时定价为 179 美元，采用了 3.17in（1in＝2.54cm）、分辨率为 480×320 的屏幕，手机内置 528MHz 处理器，拥有 192MB RAM 及 256MB ROM。当时智能手机领域还是诺基亚的天下，Symbian 在智能手机市场中占有绝对优势，所以 Google 发布的 Android 1.0 并没有被外界看好，甚至有言论称最多一年 Google 就会放弃 Android。

2009 年 4 月，Android 1.5 正式发布。从 Android 1.5 开始，Google 开始将 Android 的版本以甜品的名字命名，Android 1.5 命名为 Cupcake（纸杯蛋糕）。随后每隔一段时间 Google 就会发布 Android 的新版本或更新版本，见表 1-1。

图 1.2　T-Mobile G1

表 1-1　Android 发布一览表

版　本　号	名　　称	API 级别	发　布　日　期
Android 1.0		1	2008 年 9 月
Android 1.1		2	2009 年 2 月
Android 1.5	Cupcake（纸杯蛋糕）	3	2009 年 4 月
Android 1.6	Donut（甜甜圈）	4	2009 年 9 月
Android 2.0	Eclair（泡芙）	5	2009 年 10 月
Android 2.01	Eclair（泡芙）	6	2009 年 12 月
Android 2.1	Eclair（泡芙）	7	2010 年 1 月
Android 2.2-2.2.3	Froyo（冻酸奶）	8	2010 年 5 月—11 月
Android 2.3-2.3.2	Gingerbread（姜饼）	9	2010 年 12 月—2011 年 1 月
Android 2.3.3-2.3.7	Gingerbread（姜饼）	10	2011 年 2 月—9 月
Android 3.0（平板电脑）	Honeycomb（蜂巢）	11	2011 年 2 月
Android 3.1（平板电脑）	Honeycomb（蜂巢）	12	2011 年 5 月
Android 3.2（平板电脑）	Honeycomb（蜂巢）	13	2011 年 7 月
Android 4.0-4.0.2	Ice Cream Sandwich（冰激凌三明治）	14	2011 年 4 月—11 月
Android 4.0.3-4.0.4	Ice Cream Sandwich（冰激凌三明治）	15	2011 年 12 月—2012 年 2 月
Android 4.1（平板电脑）	Jelly Bean（果冻豆）	16	2012 年 7 月
Android 4.2（平板电脑）	Jelly Bean（果冻豆）	17	2012 年 10 月
Android 4.3（平板电脑）	Jelly Bean（果冻豆）	18	2013 年 7 月
Android 4.4	KitKat（奇巧）	19	2013 年 9 月
Android 4.4（手表）	KitKat（奇巧）	20	2014 年 6 月

(续)

版 本 号	名 称	API 级别	发 布 日 期
Android 5.0	Lollipop（棒棒糖）	21	2014 年 10 月
Android 5.1	Lollipop（棒棒糖）	22	2015 年 3 月
Android 6.0	Marshmallow（棉花糖）	23	2015 年 5 月
Android 7.0	Nougat（牛轧糖）	24	2016 年 5 月
Android 7.1	Nougat（牛轧糖）	25	2016 年 12 月
Android 8.0	Oreo（奥利奥）	26	2017 年 8 月
Android 8.1	Oreo（奥利奥）	27	2017 年 12 月
Android 9.0	Pie（馅饼）	28	2018 年 8 月

近几年，基于 Android 平台的硬件设备进入了一个爆炸式增长的状态。最新数据显示，在中国智能手机操作系统分布中，Android 的占比超过了 80%，基于 Android 平台的移动软件开发、系统 ROM 定制、机顶盒开发、可穿戴设备开发及智能家居开发等风起云涌。由此可以看出，Android 在移动互联网中占有绝对的主导地位，Android 开发拥有广阔的发展前景。

1.2　Android 的基本架构与特性

1.2.1　Android 平台基本架构

Android 并不是传统的 Linux 风格的规范或分发版本，也不是一系列可重用的组件集成，而是基于 Linux 内核的软件平台和操作系统。它采用了分层的架构，从高层到低层分别是应用层、应用框架层、系统运行库层和 Linux 内核层。Android 具体架构如图 1.3 所示。

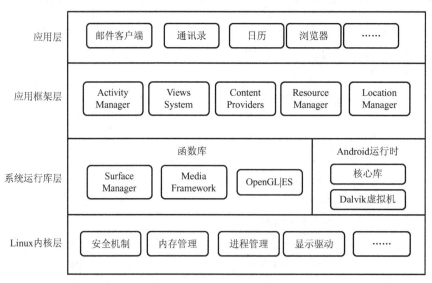

图 1.3　Android 具体架构

1. 应用层

Google 发行的 Android SDK 包自带了一个核心 App 集合，这些程序是用 Java 语言编写的运行在虚拟机上的，如 E-mail 客户端、SMS 短消息程序、浏览器、地图、联系人管理程序等，如图 1.3 中最上层部分所示。用户也可以用 Java 语言开发更加丰富的 App 在该层上运行。

2. 应用框架层

应用框架层是 Android 平台为 App 的开发而提供的 API 框架，它提供了 Android 平台基本的管理功能和组件重用机制。这一机制允许开发人员替换组件来开发自己的 App。API 框架中的所有组件和服务都可以被用户的应用重复利用。每个 App 有可能会使用到的应用框架如下。

（1）丰富的、可扩展的视图集合（Views）。该视图集合中包括列表框（ListView）、编辑框（EditText）、按钮（Button）、网格（GridView）甚至内嵌的网页浏览器等，可以用来设计 App 视图部分，也就是用户界面（UI）。

（2）内容提供器（Content Providers）。内容提供器提供了一种共享私有数据，实现跨进程的数据访问机制，使得 App 能访问其他 App（如通讯录）的数据或共享自己的数据。

（3）资源管理器（Resource Manager）。资源管理器可以用于对本地化字符串、图片、涉及布局的 XML 文件等非代码资源进行访问。

（4）活动管理器（Activity Manager）。活动管理器管理着 App 的生命周期，并且提供常用的导航回退机制。

（5）位置管理器（Location Manager）。位置管理器用来管理与地图相关的服务功能，提供一系列方法用来解决地理位置相关的问题。

3. 系统运行库层

系统运行库层已经涉及底层，和 App 关系不是很密切。Android 包含一些 C/C++ 库，有 C 语言标准库（libc）、多媒体库（Media Framework）、3D 效果支持（OpenGL ES）、关系数据库（SQLite）、Web 浏览器引擎（WebKit）等，这些库能被 Android 中的不同组件使用。该层的核心库与进程运行相关，它是应用框架的支撑，也是连接应用框架层与 Linux 内核层的重要纽带。

Android 包括一个核心库，该核心库提供了 Java 语言 API 中的大多数功能，同时也包含了 Android 的一些核心 API，如 android.os、android.net、android.media 等。每一个用 Android 开发的 App 都有独立的 Dalvik 虚拟机为它提供运行环境，让它在自己的进程执行。Dalvik 虚拟机只执行 .dex 的可执行文件，Java 源代码编译成 CLASS 后再由 SDK 中的 dx 工具转化成 .dex 格式后才能在 Dalvik 虚拟机上执行。

4. Linux 内核层

Android 的底层基于 Linux 2.6 内核，其核心系统服务如安全性、内存管理、进程管理、网络协议及驱动模型都依赖于 Linux 2.6 内核。Linux 2.6 内核也作为硬件和软件之间的抽象层，它隐藏具体硬件细节为上层提供统一的服务。

1.2.2 Android 的特性

随着 Android 的发展和用户体验好评的逐年上升，Android 的市场占有率也已经取代 iOS 成为全球第一大的智能系统，存在极大的市场潜力。

1. 优点

Android 有以下几方面的优点。

（1）开放特性，得到众多厂商的支持。开放的 Android 平台允许任何厂商加入 Android 联盟中来，每个厂商可以推出千奇百怪、功能迥异的 Android 平台设备以满足各种不同的需求。

（2）开发成本不高，App 发展迅速。由于 Android 平台的设备性价比高，为第三方开发商提供了十分宽泛、自由的环境，Android Market 对 App 也不做特别严格的控制，所以伴随着 Android 的发展，Android 平台软件越来越多，且有很多免费软件。

（3）无缝结合 Google 应用。Android 由 Google 主导研发，Google 服务如地图、邮件、搜索等一应俱全，应用方面拥有其他系统无可比拟的优势。用户在使用 Android 时，可以实现 Google 服务的完全同步。

2. 面临的问题

当然，除了上述优点外，Android 发展过程中也面临一些问题。

（1）版本过多，升级过快；系统升级滞后。由于 Android 的开放性，很多厂商推出了各自定制的界面，如小米的 MIUI、华为的 EMUI、OPPO 的 Color OS、vivo 的 Funtouch OS 等，在提供给用户丰富选择的同时，也造成版本过多、升级较慢的缺点。因为 Android 版本升级速度很快，而厂商要推出新固件需要经过深度的研发，这就造成升级滞后的问题。

（2）适配难，质量不能保证。由于 Android 在不同的厂商、不同的配置下均有不同的机型，所以造成有些 App 在机型适配上会有一些问题，如运行缓慢、卡顿、异常等问题，这导致用户体验满意度急剧下降。

1.3 Android 开发环境与常用工具

由于 Android 版本发展较快，开发环境的配置也随着版本的更新不断变化，Android Studio 作为 Google 主推的 Android 集成开发环境逐渐取代了传统的 Eclipse + ADT。根据 Google 官方建议硬件配置需要内存最低 4GB，推荐 8GB；硬盘容量最低 128GB，推荐使用固态硬盘；CPU 推荐 Intel i3 以上。硬件环境的配置可以根据实际情况进行选择，软件环境的配置以开发者普遍采用的 Android Studio 集成开发环境为例进行介绍。

【开发平台搭建、常用工具使用】

1.3.1 安装 JDK

1. 下载 JDK

登录 http://www.oracle.com/technetwork/java/javase/downloads/index.html 下载 Java SE 安装包，下载页面如图 1.4 所示（因网网页更新等情况，读者打

第1章 Android开发环境

开的网页可能与书中介绍的略有不同，本书汲及的网址均有这种可能，后边不再赘述）。单击 Downloads 下载 JDK，以 Java SE 9.0.1 为例。开发者也可以根据不同的操作系统选择相应版本的 JDK，如图 1.5 所示。

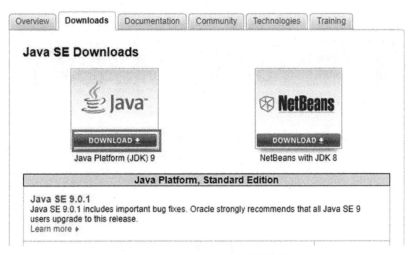

图 1.4 Java SE 安装包下载页面

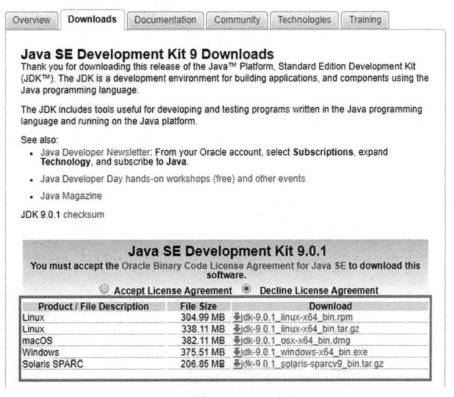

图 1.5 其他操作系统 JDK 下载页面

2. 安装JDK

一般情况下，JDK 安装位置保持默认设置即可，此处默认安装在 C 盘，用户也可以根据自己的需要选择安装位置，如图 1.6 所示。

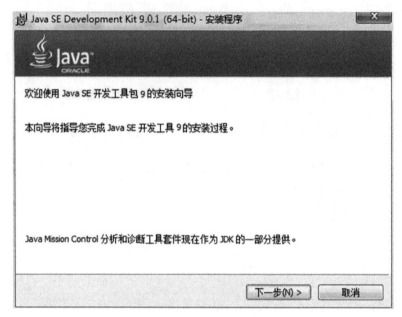

图1.6　安装 JDK

3. 配置环境变量

安装完 JDK 后，右击桌面上"我的电脑"图标，在弹出的快捷菜单中选择"属性"选项，在"系统属性"对话框中选择"高级"选项卡，单击"环境变量"按钮，新建环境变量见表 1-2。

表1-2　新建环境变量

变量名	变量值	说　　明
JAVA_HOME	C:\Program Files\Java\jdk1.9.0.1	JDK 安装目录
PATH	%JAVA_HOME%\bin;%JAVA_HOME%\jre\bin	若系统有，直接将变量值加到原有值前面，但用";"连接
CLASSPATH	%JAVA_HOME%\lib;%JAVA_HOME%\lib\tools.jar	指明.class 文件的目录，可以省略

至此，在命令窗口中输入 java 或 javac 命令。运行 javac 后显示如图 1.7 所示界面，则表示配置成功。如果使用集成开发环境（IDE）编写 Java 代码，环境变量一般不用显式设置，IDE 会根据系统配置情况自动完成。

第1章 Android开发环境

图1.7 运行 javac 后的界面

1.3.2 安装 Android Studio

Android 开发环境除了需要 JDK 环境外（Java 开发包），还需要 Android SDK 组件和 Android Studio，所以搭建 Android 开发环境时，需要下载这两个软件包。Google 为了方便开发者，在其提供的软件包中包含了这两个软件，下面将详细介绍它的配置过程。

1. 下载 Android Studio

登录 https://developer.android.google.cn/studio/index.html 下载 Android Studio 开发包，下载页面如图1.8所示。单击下载 ANDROID STUDIO 2.3.3 FOR WINDOWS 按钮后开始下载 android-studio-bundle-162.4069837-windows.exe 安装文件。该安装文件包含了 Android 开发必要的 Android SDK 组件和 Android Studio 集成开发环境。

图1.8 Android Studio 开发包下载页面

2. 安装 Android Studio 文件

双击下载完成的 android-studio-bundle-162.4069837-windows.exe 文件开始安装 Android Studio，如图 1.9 所示。单击 Next 按钮，显示如图 1.10 所示窗口。在该窗口中用户可以选择需要安装的组件 Android SDK（Android 软件开发包，默认选中）和 Android Virtual Device（Android 模拟设备，默认选中），单击 Next 按钮同意安装协议，显示如图 1.11 所示窗口。在该窗口中选择 Android Studio 和 Android SDK 的安装位置（笔者安装位置如图 1.11 所示），单击 Next 按钮，直至安装完毕。安装完毕后，目标位置包含 Android Studio 文件夹（文件夹下内容如图 1.12 所示）和 sdk 文件夹（文件夹下内容如图 1.13 所示）。

图 1.9　安装 Android Studio

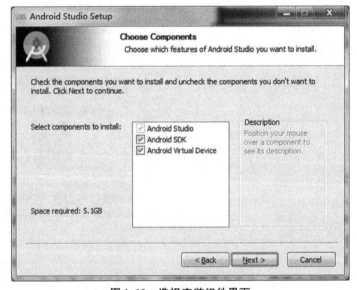

图 1.10　选择安装组件界面

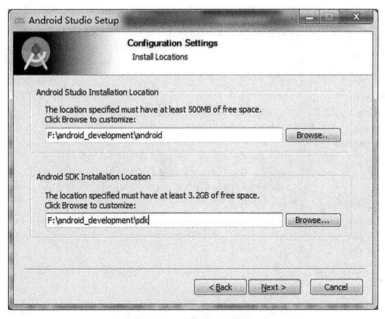

图1.11　安装位置设置

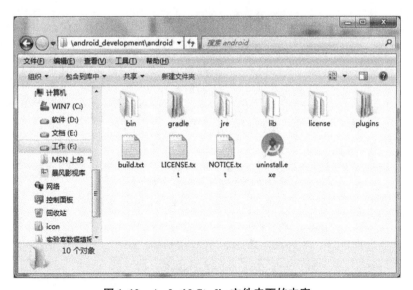

图1.12　Android Studio 文件夹下的内容

3. 运行 Android Studio

安装完毕后，第 1 次启动 Android Studio 需要等待一段时间下载最新版本的 Android SDK，默认保存在"C:\Users\Auser\AppData\Local\Android\Sdk"位置，然后出现如图 1.14 所示窗口，在该窗口中有如下选项。

（1）Start a new Android Studio project：开始创建一个新的 Android 项目。

（2）Open an existing Android Studio project：打开一个原有的 Android 项目。

图 1.13　sdk 文件夹下的内容

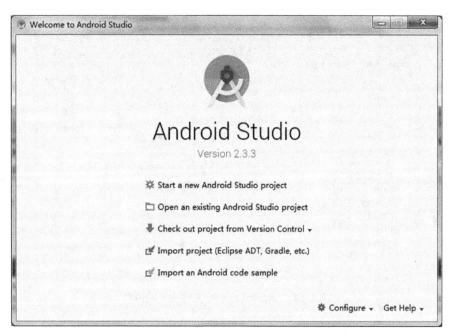

图 1.14　Android Studio 启动界面

（3）Check out project from Version Control：通过版本控制器检查项目。

（4）Import project（Eclipse ADT，Gradle，etc.）：导入 Eclipse ADT、Gradle 等创建的项目。

（5）Import an Android code sample：导入一个 Android 代码例子。

（6）Configure：配置开发环境，单击后弹出如图1.15所示下拉菜单，单击 SDK Manager 进入 Default Settings 对话框，如图1.16所示。默认状态下 Android SDK Location 设置值为第1次启动 Android Studio 时下载的 SDK 的存放位置（笔者的机器存放位置：C:\Users\Auser\AppData\Local\Android\Sdk），此时可以单击 Edit 按钮修改 Android SDK 存放位置，笔者修改后指向安装 Android Studio 时的目标位置"F:\android_development\sdk"。

图1.15 Configure 下拉菜单

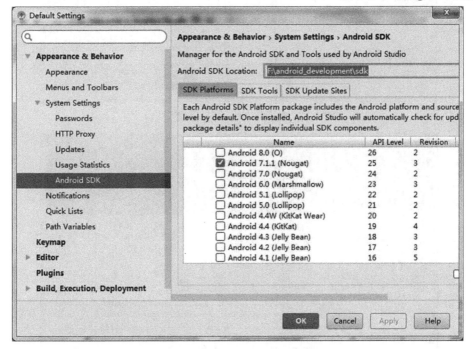

图1.16 Default Settings 对话框

4. 创建 Android Studio 项目

单击图1.14上的 Start a new Android Studio project 选项，出现如图1.17所示窗口，在该窗口下有如下选项。

- Application name：项目名和首个 App 名。
- Company domain：开发者域名，用来合成包名。
- Project location：项目和首个 App 存放位置。

在图1.17所示窗口中进行相应设置后，单击 Next 按钮，出现如图1.18所示窗口，在该窗口下有如下选项。

- Phone and Tablet：App 运行于 Android 手机与平板设备。
- Wear：App 运行于 Android 穿戴设备。
- TV：App 运行于 Android 电视。

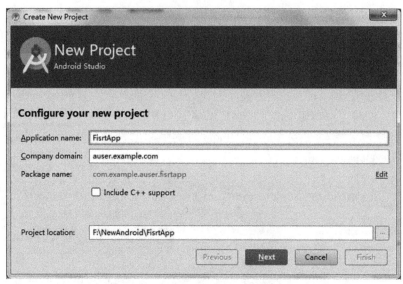

图 1.17 Create New Project 窗口(1)

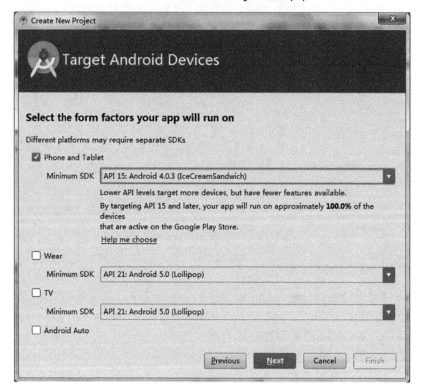

图 1.18 Create New Project 窗口(2)

- Android Auto：App 运行于 Android 汽车。
- Minimum SDK：最低兼容到的 Android 版本。

在图 1.18 所示窗口中进行相应设置后，单击 Next 按钮，出现如图 1.19 所示窗口，选择 Empty Activity 选项，单击 Next 按钮后出现如图 1.20 所示窗口，在该窗口下有如下选项。

- Activity Name：App 的第 1 个 Activity 的名称。
- Layout Name：App 的第 1 个 Activity 对应的布局文件（用户界面）的名称。

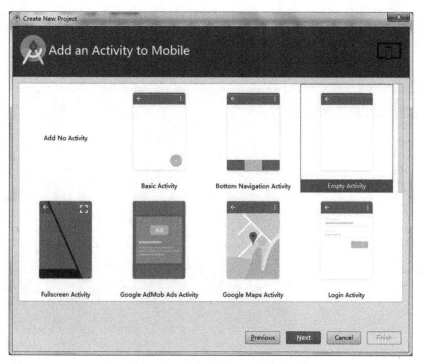

图 1.19　Create New Project 窗口(3)

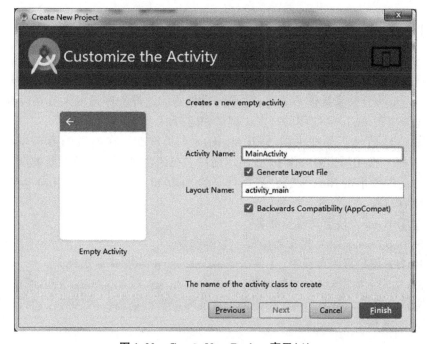

图 1.20　Create New Project 窗口(4)

最后单击 Finish 按钮，集成开发环境根据前面的设置自动创建第 1 个 Android Studio 项目和第 1 个 App 程序，创建完成后的窗口如图 1.21 所示。

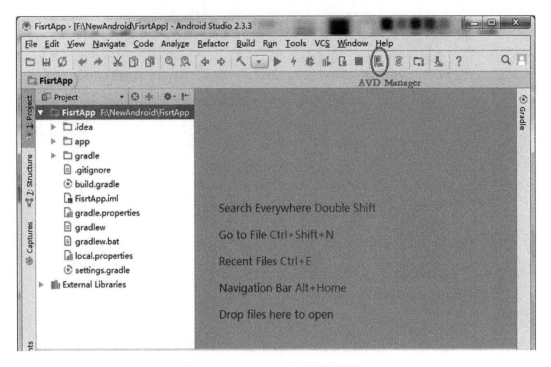

图 1.21　Android Studio 开发环境窗口

5. 创建 Android 模拟器

在进行 Android 应用开发时需要调试程序代码和运行程序来展现开发的 App 的运行效果，而要运行 App 通常需要一个 Android 平台的设备。为了方便开发人员进行开发、调试和仿真，Google 为开发者提供了 Android 平台模拟器。这样程序开发人员在没有实际设备的情况下也能实现 Android 应用的开发、调试和运行。

单击图 1.21 所示工具栏中的 AVD Manager 工具按钮，在出现的窗口中单击 Create Virtual Device... 按钮后出现如图 1.22 所示窗口。在该窗口中选择模拟器类别、屏幕大小和分辨率等；继续单击 Next 按钮，选择对应的 Android Target 后单击 Next 按钮，出现如图 1.23 所示窗口。

（1）AVD Name：自定义模拟器名。

（2）Startup orientation：模拟器横屏、竖屏选择。

（3）RAM：模拟器内存设置。（单击 show advanced setlings 后出现）

（4）SD Card：模拟器 SD 卡的设置。（单击 show advanced setlings 后出现）

完成屏幕设置后，单击 Finish 按钮即可完成模拟器的创建，然后可以运行模拟器，运行效果如图 1.24 所示。

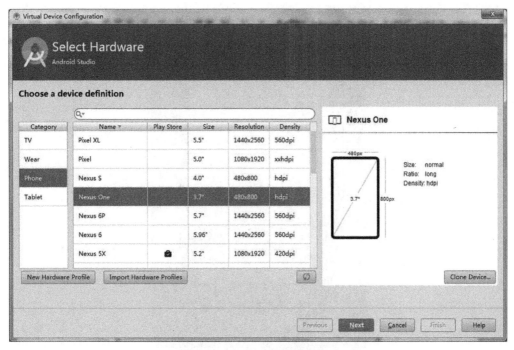

图1.22 Android 模拟器创建(1)

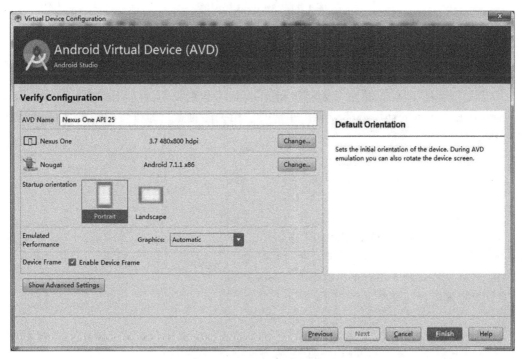

图1.23 Android 模拟器创建(2)

图 1.24 Android 模拟器

本 章 小 结

本章简要介绍了 Android 的发展历程和 Android 平台的架构特性，详细列出了 Windows 下 Android 开发平台的环境搭建步骤，并指明了相关注意点，为读者后续的 Android 应用软件开发学习打下基础。

习　　题

一、选择题

1. 以下关于 Android 模拟器的说法错误的是（　　）。
A. 两个模拟器之间能模拟电话呼叫
B. 创建模拟器时需要设置模拟器的 Android SDK 版本
C. 模拟器不支持模拟手机 SD 卡
D. 模拟器不支持模拟 USB 连接和设备耳机

2. 在 Android 程序调试过程中，可以使用 Log 类进行写日志文件，使用不同的函数可以写入不同等级的信息，如果写入调试信息需要调用（　　）。
　　A. Log. v( )　　　　　　　　　B. Log. d( )
　　C. Log. e( )　　　　　　　　　D. Log. w( )

3. 下面（　　）不是 Android 的特点。
   A. 开源，得到众多厂商支持　　　　B. 无缝结合 Google 应用
   C. App 发展迅速　　　　　　　　　D. Andorid 中默认浏览器是 safari

4. 下列不是手机操作系统的是（　　）。
   A. Android　　　　　　　　　　　　B. Window Mobile
   C. Apple IPhone iOS　　　　　　　D. Windows Vista

5. 在 Android 程序调试过程中，可以使用 Log 类进行写日记文件，使用不同的函数可以写入不同等级的信息，如果写入警告信息需要调用（　　）。
   A. Log. v( )　　　　　　　　　　　B. Log. d( )
   C. Log. e( )　　　　　　　　　　　D. Log. w( )

6. Android 的具体架构分为四层，分别是应用层、应用框架层、系统运行库层和 Linux 内核层，其中 Dalvik 虚拟机在（　　）层。
   A. 应用层　　　　　　　　　　　　B. 系统运行库层
   C. 应用框架层　　　　　　　　　　D. Linux 内核层

二、填空题

1. Android 平台架构分为四层，从高到低分为应用层、应用框架层、系统运行库层及_____。

2. 安装 Android 开发环境时，首先需要安装_____，然后安装集成开发环境 Android Studio。

3. Android 是基于_____内核的软件平台和操作系统。

4. Android App 的安装文件的后缀名为_____。

5. 在 Android 程序调试过程中，可以使用 Log 类进行写日志文件，使用不同的函数可以写入不同等级的信息，如果写入错误信息需要调用_____。

6. Android 是_____公司推出的手机操作系统。

三、判断题

1. Android 中软件发展速度很快。　　　　　　　　　　　　　　　　　（　　）
2. Android 是非开源的，要想得到 Android 的源代码需要向 Google 公司购买。
　　　　　　　　　　　　　　　　　　　　　　　　　　　　　　　（　　）
3. Android 手机可以安装后缀名是 .ipa 的手机 App。　　　　　　　　（　　）
4. Android 平台的 Linux 内核层自带了一个核心 App 集合。　　　　　（　　）
5. Android 平台与传统的 Linux 发行版本不一样。　　　　　　　　　（　　）
6. 两个模拟器之间能模拟电话呼叫。　　　　　　　　　　　　　　　（　　）
7. 模拟器不支持模拟 USB 连接。　　　　　　　　　　　　　　　　　（　　）
8. 模拟器不支持模拟手机 SD 卡。　　　　　　　　　　　　　　　　（　　）

【第 1 章参考答案】

第 2 章 Android App 结构

在开发 App 时，编译流程大多会被认为是直接通过集成开发环境（IDE）的按钮或命令行生成 apk 文件的。事实上，生成 apk 文件是一个非常复杂的过程。Android 资源打包工具将 App 的资源文件进行编译，生成一个 R.java 文件；AIDL（Android Interface Definition Language）将 App 中所有的.aidl 文件转换为 Java 接口（有的工程没有用到 AIDL，这个过程可以省略）；Java 编译器将应用中所有 Java 代码（包括 R.java 和 aidl 接口）编译输出为 class 类文件；dex 工具（Dalvik VM Executes）将 App 编译输出的 class 类文件转换为 Dalvik VM 支持的 dex 文件（该文件为字节码）；apkbuild 工具将所有非编译资源、编译资源和 dex 文件打包成一个 apk；最后对生成的 apk 文件进行签名后就可以在 Android 设备上运行了。

掌握使用 Android Studio 创建 Android App 的方法。

掌握 Android App 的目录结构、开发过程中各类资源的使用方法和 App 配置文件的结构。

掌握开发 Android App 中的四大组件及各个组件的功能。

知识要点	能力要求	相关知识
如何创建 App	掌握使用 Android Studio 创建 App 的方法	集成开发环境
剖析 Android App	（1）掌握 Android App 的目录结构及各目录文件的功能 （2）掌握 Android App 开发过程中各类资源的使用方法 （3）掌握 AndroidManifest.xml 配置文件的结构及修改方法	目录和文件
Android App 的四大组件	掌握 Activity、BroadcastReceiver、Service、ContentProvider 四大组件的功能和特点	组件

2.1 创建 App

在 Android Studio 环境下，可以在创建工程时一起创建 App，这个在第 1 章中已经进行了详细介绍。在原有工程项目里也可以单独创建 App，下面介绍其创建步骤。

（1）打开 File 菜单下的 New→New Module... 命令，弹出如图 2.1 所示窗口。

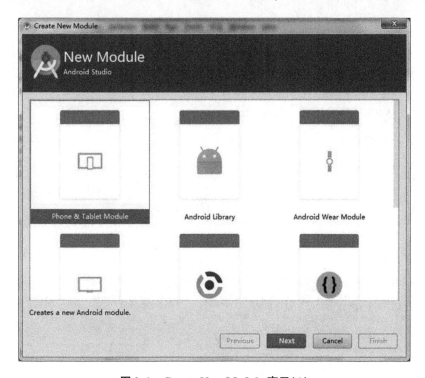

图 2.1　Create New Module 窗口（1）

（2）在图 2.1 中选中 Phone & Tablet Module 选项，然后单击 Next 按钮，弹出如图 2.2 所示窗口，该窗口中有如下选项。

① Application/Library name：应用程序/库名称。
② Module name：模块名称。
③ Minimum SDK：最低兼容到的 Android 版本。

（3）在图 2.2 窗口中进行相应设置后，单击 Next 按钮，出现如图 2.3 所示窗口，选择 Empty Activity 选项后，单击 Next 按钮，出现如图 2.4 所示窗口，该窗口中有如下选项。

① Activity Name：Activity 名称。
② Layout Name：布局文件名称。

如果在图 2.4 中没有选择 Backwards Compatibility（AppCompat），那么创建的 MainActivity 直接继承自 Activity，需要使用下面的语句导入包：

import android. app. Activity；

如果选择了"Backwards Compatibility（AppCompat）"，那么创建的 MainActivity 会继承

自 AppCompatActivity，需要使用下面的语句导入包：

import android. support. v7. app. AppCompatActivity；

图 2.2　Create New Module 窗口(2)

图 2.3　Create New Module 窗口(3)

第2章 Android App结构

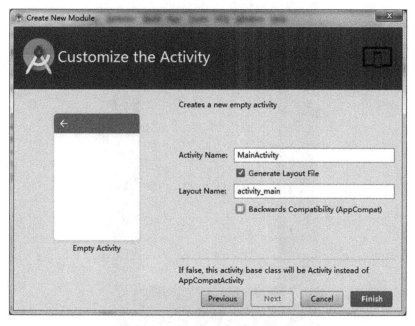

图2.4 Create New Module 窗口(4)

（4）在图2.4所示窗口中进行相应设置后，单击Finish按钮，系统就会自动在当前的工程中创建sampleapp，创建完后的开发环境界面如图2.5所示。

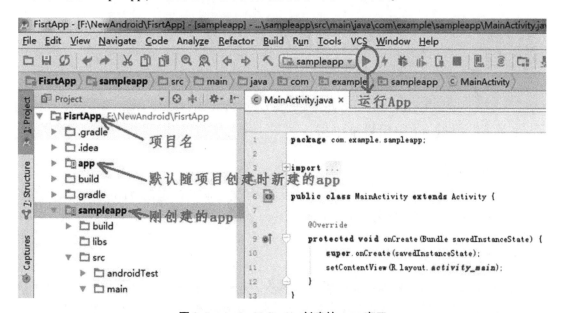

图2.5 Android Studio 创建的App窗口

（5）在图2.5所示窗口单击"运行App"图标，就可以将刚刚建立的sampleapp安装到模拟器或真机上运行。运行后的效果如图2.6所示。

23

图 2.6　sampleapp 的运行效果

2.2　剖析 Android App

【Android App 结构组成】

本节将通过 2.1 节创建的 sampleapp 来介绍 Android App 的目录结构。Android App 的目录结构如图 2.7 所示。

2.2.1　Android 应用目录剖析

1. build 目录

build 目录存放编译时自动生成的文件,编译后的 apk 文件也以 "app 名-debug.apk" 默认文件名(本例中的文件名为 sampleapp-debug.apk)保存在 build 目录下的 outputs/apk 目录里。

2. libs 目录

libs 目录是第三方依赖库存放目录,如果某个项目中使用了第三方 jar 包,就需要把这些 jar 包都放在 libs 目录下,放在这个目录下的 jar 包都会被自动添加到构建路径里去。

3. src/androidTest 目录

androidTest 目录中存放 Android Studio 生成的 androidTest 测试用例,以便对项目进行一些自动化测试。此目录通常可以删除。

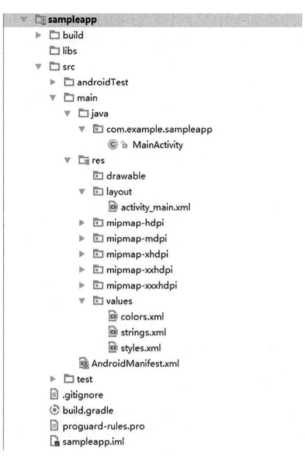

图 2.7 Android App 的目录结构

4. src/test 目录

test 目录存放开发者编写的 Unit Test（单元测试）用例，用于对项目进行自动化测试，如果不需要可以删除此目录。

5. src/main/java 目录

java 目录存放 Android App 的所有源代码，即所有允许用户修改的 Java 文件和用户自己添加的 Java 文件都保存在这个目录中。MainActivity.java 的源程序代码如下。

```java
package com.example.sampleapp;
import android.app.Activity;
import android.os.Bundle;
public class MainActivity extends Activity{
    @Override
    public void onCreate(Bundle savedInstanceState){
        super.onCreate(savedInstanceState);
        setContentView(R.layout.main);
    }
}
```

在 MainActivity.java 文件中导入了两个类（android.app.Activity 和 android.os.Bundle），该类继承自 Activity 且重写了 onCreate 方法。在重写父类的 onCreate()时，在方法前面加上@Override，可以使系统帮助检查方法的正确性。

例如，public void onCreate（Bundle savedInstanceState）{…} 这种写法是正确的，如果写成 public void oncreate（Bundle savedInstanceState）{…} 编译器会报错误：The method oncreate（Bundle）of type HelloWorld must override or implement a supertype method。而如果不加@Override，则编译器将不会检测出错误，而是会认为新定义了一个方法 oncreate()。

6. src/main/res 目录

res 目录中存放有项目中使用的图片、布局和字符串等资源。向此目录添加资源时，会被 R.java（在项目目录的 build\generated\source\r\debug\项目包名下）自动记录。新建一个 Android App 后，会自动在 res 目录下生成以下几类资源文件目录。

（1）图片资源文件目录。

drawable：存放各种位图文件（如.png，.jpg，.png，.gif 等），也可以存放 drawable 类型的 XML 文件。

mipmap-hdpi：存放高分辨率图片。

mipmap-mdpi：在放中等分辨率图片。

mipmap-xhdpi：存放超高分辨率图片。

mipmap-xxhdpi：存放超超高分辨率图片。

系统通常会根据移动终端的分辨率分别到这几个文件夹里面去找对应的图片。为了让开发的 App 能够兼容不同分辨率的屏幕，建议将不同分辨率的图片资源文件放入对应的文件夹中。

（2）layout 目录。

layout 目录用来存放布局文件。

（3）values 目录。

values 目录中默认存放了 strings.xml、colors.xml、styles.xml 文件，这些文件的内容是基于 XML 格式的 key-value 键值对。

strings.xml：定义字符串资源。

colors.xml：定义颜色资源。

styles.xml：定义样式资源。

开发者也可以根据项目设计的需要在此目录下添加一些额外的资源，如数组资源文件 arrays.xml、尺寸资源文件 dimens.xml 等。

2.2.2 资源的使用

1. Android 中的常用资源文件

资源文件（即 resource 文件）在 values 目录下，它们都是在 values 目录下的 XML 文件，且都是以 resource 作为根节点。

(1) strings.xml（定义字符串）。

代码格式如下。

```xml
<resources>
    <string name="hello">Hello World!</string>
    <string name="app_name">我的应用程序</string>
</resources>
```

(2) colors.xml（定义颜色）。

颜色通过红（red）、绿（green）、蓝（blue）三种颜色及透明度（alpha）来表示。颜色值总是以"#"开头，如果没有alpha值，默认完全不透明。其颜色定义形式如下。

#RGB：红绿蓝三原色值，最小为0，最大为f。
#ARGB：透明度、红绿蓝值，最小为0，最大为f。
#RRGGBB：红绿蓝三原色值，最小为0，最大为f。
#AARRGGBB：透明度、红绿蓝值，最小为0，最大为f。

代码格式如下。

```xml
<resources>
    <color name="solid_red">#f00</color>
    <color name="solid_blue">#f0f0</color>
    <color name="solid_green">#0000ff</color>
    <color name="solid_yellow">#ffffff00</color>
</resources>
```

(3) arrays.xml（定义数组）。

数组资源文件中使用<array>子元素标签定义普通类型数组，使用<string-array>子元素标签定义字符串数组，使用<integer-array>子元素标签定义整数数组。代码格式如下。

```xml
<resources>
    <string-array name="planets">
        <item>Mercury</item>
        <item>Venus</item>
        <item>Earth</item>
        <item>Mars</item>
        <item>Jupiter</item>
        <item>Saturn</item>
        <item>Uranus</item>
        <item>Neptune</item>
        <item>Pluto</item>
    </string-array>
    <integer-array name="numbers">
        <item>100</item>
        <item>500</item>
        <item>800</item>
    </integer-array>
</resources>
```

(4) dimen.xml（定义尺寸）。

代码格式如下。

```
<resources>
    <dimen name="cwidth">320px</dimen>
    <dimen name="double_density">2dp</dimen>
    <dimen name="fontsize">16sp</dimen>
</resources>
```

Android 的常用度量单位见表 2-1。

表 2-1 Android 的常用度量单位

单 位	说 明
px（像素）	相当于实际屏幕的像素，由于不同屏幕的像素数差异比较大，因此该单位不推荐用于尺寸单位
dp/dip（密度无关像素）	一种与屏幕密度无关的尺寸单位，当屏幕密度是 160dpi 时，1dp＝1px。当运行在高 dpi 的屏幕上时，dp 就会按比例放大，当运行在低 dpi 的屏幕上时，dp 就会被按比例缩小。因此 dp 是一种简单地解决 view 在不同大小屏幕上显示问题的解决方案
sp	与 dp 相似，但是它会随着用户对系统字体大小的设置进行比例缩放，即它能够跟随用户系统字体大小变化，所以它更加适合作为字体大小的单位

(5) styles.xml（定义样式）。

在实际开发时，每个 App 可能有多个用户界面，同时也希望把每个用户界面的标题栏格式设置为统一的风格。例如，将每个用户界面标题栏显示内容的字体、字号、颜色及对齐方式等都设置为一样的格式，这样就可以将字体、字号、颜色及对齐方式保存在样式文件中，然后在每个用户界面的布局文件中引用自定义的样式。样式文件的代码格式如下。

```
<resources>
    <style name="text_font">
        <item name="android:textColor">#05b</item>
        <item name="android:textSize">18sp</item>
        <item name="android:textStyle">bold</item>
    </style>

    <style name="content_font">
        <item name="android:textColor">#0f5</item>
        <item name="android:textSize">18sp</item>
        <item name="android:textStyle">normal</item>
    </style>

    <style name="hint_text_font" parent="text_font">
        <item name="android:textColor">#f00</item>
    </style>
</resources>
```

引用样式文件的代码如下。

```xml
<TextView style="@style/content_font"
    android:id="@+id/textView1"
    android:layout_width="fill_parent"
    android:layout_height="wrap_content"
    android:text="@string/hello_world" />

<Button style="@style/text_font"
    android:id="@+id/button1"
    android:layout_width="fill_parent"
    android:layout_height="wrap_content"
    android:text="@string/test_IntentService" />

<TextView style="@style/hint_text_font"
    android:id="@+id/hint"
    android:text="@string/hint_text"
    android:layout_height="wrap_content"
    android:layout_width="fill_parent" />
```

2. Android 中资源的使用

Android Studio 会为 res 目录下的所有资源在 R.java 文件下生成对应 id，开发者可以直接通过资源 id 访问对应的资源。一般情况下，开发人员并不需要管理这个类，更不需要修改这个类，只需要直接使用 R 类中的 id 即可。R.java 文件源代码如图 2.8 所示。

```
10  public final class R {
11      public static final class attr {
12          public static final int color=0x7f0100a1;
13          public static final int icon=0x7f010009;
14      }
15      public static final class color {
16          public static final int foreground=0x7f0a001b;
17      }
18      public static final class id {
19          public static final int btnAdd=0x7f0b005d;
20      }
21      public static final class drawable {
22          public static final int notification_bg=0x7f020054;
23      }
24      public static final class string {
25          public static final int app_name=0x7f060021;
26          public static final int search_title=0x7f060013;
27      }
28  }
```

图 2.8　R.java 文件源代码

从 R 类中很容易看出，Android Studio 为 res 目录中的每一个子目录（如 layout）或 XML 格式的资源文件（如 strings.xml）都生成了一个静态的子类，同时为 XML 布局文件中每个指定 id 属性的组件生成了唯一的 id，并封装在 id 子类中，这样就可以在 Android App 中通过 id 引用这些组件。

（1）Java 代码引用资源。

格式：[packageName.] R.resourceType.resourceName

其中：packageName 表示 R.java 类所在的包名，R 类可能来自 App 本身的资源文件，也可能来自 Android 自带的资源文件。resourceType 表示 R 类中资源类型名称；R.string.app_name 表示引用 string 类中的静态变量 app_name。resourceName 表示引用资源的名称；R.layout.main 表示引用 layout 类中的 main.xml 资源文件。例如，在 App 中获得 btnShowDate 按钮对象，可以使用如下代码。

```
Button btnShowDate = (Button)findViewById(R.id.btnShowDate);
```

可以看到，在使用资源时直接引用了 R.id.btnShowDate 的 id 值。当然，直接使用 0x7f050000 也可以，不过为了使程序更容易维护，一般会直接使用在 R 的内嵌类中定义的变量名。

另外，Android SDK 中的很多方法都支持直接使用 id 值来引用资源。例如，android.app.Activity 类的 setTitle 方法除了支持以字符串的方式设置 Activity 的标题外，还支持以字符串资源 id 的方式设置 Activity 的标签，即可以通过如下代码设置 Activity 的标题。

```
setTitle(R.string.hello);
```

（2）XML 文件中引用资源。

格式：@［packageName :］resourceType / resourceName

其中：packageName、resourceType 和 resourceName 使用含义与上述相同。例如，资源文件内容如下。

```
<?xml version ="1.0" encoding ="utf-8"?>
<resources>
    <string name ="app_name">ResourceTest</string>
    <string name ="action_settings">Settings</string>
    <string name ="hello_world">Hello world!</string>
    <color name ="red">#FF4000</color>
</resources>
```

引用资源代码如下。

```
<TextView
    android:layout_width ="wrap_content"
    android:layout_height ="wrap_content"
    android:text ="@string/hello_world"
    android:textColor ="@color/red"/>
```

2.2.3 AndroidManifest.xml 文件的结构

每一个 Android App 必须有一个 AndroidManifest.xml 文件（不能改成其他的文件名），从图 2.7 中可以看到 AndroidManifest.xml 文件存放在 src/main 目录中。在这个文件中配置了 App 运行需要的组件、权限及一些相关信息。AndroidManifest.xml 是 Android App 的入口文件，其中定义了 App 包含的 Activity、Service、ContentProvider 和 BroadcastReceiver 组件实现及各种能被处理的数据和启动位置等信息。AndroidManifest.xml 文件部分源代码如下。

```
<?xml version ="1.0" encoding ="utf-8"?>
<manifest xmlns:android ="http://schemas.android.com/apk/res/android"
    package ="cn.edu.nnutc"
```

```
            android:versionCode ="1"
            android:versionName ="1.0" >
        <uses-sdk android:minSdkVersion ="9" />
        <application android:icon ="@drawable/icon" android:label =
"@string/app_name" >
            <activity android:name =".HelloWorld"
                    android:label ="@string/app_name" >
                <intent-filter >
                    <action android:name ="android.intent.action.MAIN" />
                    <category android:name ="android.intent.category.LAUNCHER" />
                <data/ >
                </intent-filter >
            </activity >
        </application >
    </manifest >
```

1. manifest

xmlns：android 定义 android 命名空间，一般为 http://schemas.android.com/apk/res/android；package 指定 App Java 程序的包名；versionCode 是给设备程序识别版本（升级）用的，必须有一个 integer 值代表 App 更新过多少次，如第 1 版一般为 1，之后若要更新版本就设置为 2、3 等；versionName 是给用户看到的版本名，通常用 1.2、2.0 等格式表示版本名。

2. application

一个 AndroidManifest.xml 中必须含有一个 application 标签，这个标签声明了每一个 App 的组件及其属性。icon、lable 分别指明 App 安装后在设备终端上显示的图标和文字信息；theme 指明 App 的主题风格，也就是定义一个默认的主题风格给所有的 activity。

3. activity

App 显示的每一个 Activity 都要求有一个 activity 标签，name 指明 activity 对应的类名称；label 指明该 name 指定的 Activity 运行后的标签名。

4. intent-filter

intent filter 内通常包含 action、category 与 data 三种标签。action 标签中只有 android：name 属性，常见的 android：name 值为 android.intent.action.MAIN，表明此 activity 是作为 App 的入口；category 标签中也只有 android：name 属性，常见的 android：name 值为 android.intent.category.LAUNCHER，从而决定 App 是否显示在程序列表里；data 标签用于指定一个 URI 和数据类型（MIME 类型）。

2.3　Android 的四大组件

Android 没有使用常见的 App 入口点的方法（如 Java App 中的 main()方法），它的 App 就是由组件组成的，组件包括活动（Activity）、广播接收器（BroadcastReceiver）、服务（Service）、内容提供者（ContentProvider）。一个 Android App 必定至少包含一个 Activity，其他的三个组件为可选部分。组件是可以通过 Intent 调用的相互独立的基本功能模块。

2.3.1 Activity

Activity 是 Android App 的表现层，显示可视化的用户界面，并接收与用户交互所产生的界面事件。一个 Activity 表示一个可视化的用户界面，关注一个用户从事的事件。例如，一个 Activity 可能表示一个用户可选择的菜单项列表。一个文本短信 App 可能有多个 Activity，一个 Activity 用于显示联系人的名单；另一个 Activity 用于写信息给选定的联系人。虽然 App 工作时形成一个整体的用户界面，但是每个 Activity 都是独立于其他 Activity 的。每一个 Activity 都是作为 Activity 基类的一个子类的实现。以下两个方法是几乎在所有的 Activity 子类都需要实现的。

（1）onCreate（Bundle）：初始化 Activity。在这个方法里通常使用 setContentView（int）方法将布局资源（layout resource）定义到用户界面上，然后使用 findViewById（int）在用户界面中检索需要编程交互的小部件（widgets）。即 setContentView（）方法用于指定由哪个文件指定布局（main. xml），把界面显示出来，然后通过界面上的组件或触发事件进行相关操作。

（2）onPause（）：处理当离开 Activity 时要做的事情。在使用组件构建 Android App 时，只要用到组件就必须在 AndroidManifest. xml 文件中声明及指定它们的特性和要求。

2.3.2 BroadcastReceiver

BroadcastReceiver 是用来接收并响应广播消息的组件，它不包含任何用户界面，可以通过启动 Activity 或者 Notification 通知用户接收到重要信息。即 BroadcastReceiver 不做任何事情，仅是接收广播公告并做出相应的反应。许多广播源自系统代码，如公告时区的改变、电池电量低、检测到无线信号等。App 也可以自己发起广播（即自定义广播事件）。例如，让其他程序知道某些数据已经下载到设备且可以使用这些数据。一个 App 可以有任意数量的 BroadcastReceiver 去反应任何它认为重要的公告。所有的 BroadcastReceiver 继承自 BroadcastReceiver 基类。使用时可以用 Context. registerReceiver（）方法在程序代码中动态地注册这个类的实例，也可以通过 AndroidManifest. xml 中 < receiver > 标签静态声明。具体应用在后面章节中将有详细介绍。

2.3.3 Service

Service 用于没有可视化用户界面，但需要长时间在后台运行的应用。例如，用户在浏览网页时可以听到音乐声，此时就是将播放音乐作为一个 Service，即在后台播放音乐的同时，不影响用户浏览网页内容；或者也可能是边浏览网页，边从网络下载文件或进行软件升级。每个 Service 都继承自 Service 基类。使用时每个 Service 类在 AndroidManifest. xml 中有相应的 < service > 声明。Service 可以通过 Context. startService（）和 Context. bindService（）启动。

2.3.4 ContentProvider

ContentProvider 是 Android 提供的一种标准的共享数据机制，App 可以通过它访问其他 App 的私有数据。Android 提供了一些内置的 ContentProvider，能够为 App 提供重要的数据信息。ContentProvider 继承自 ContentProvider 基类并实现了一个标准的方法集，使得其他 App 可以检索和存储数据。然而，App 并不直接调用这些方法，而是使用一个 ContentResolver 对象并调用它的方法。ContentResolver 能与任何 ContentProvider 通信，它与 ContentProvider 合作来管理参与进来的进程间的通信。

本 章 小 结

本章介绍了 Android App 使用的四大组件，读者可以了解四大组件在 Android App 中的作用及使用方法；详细列出了 App 的目录结构并介绍了每个目录的功能；初步讲解了 AndroidManifest.xml 的配置方法和各类 XML 文件的创建方法。通过本章的学习，读者也可以创建出第 1 个 Android 项目。

习　　题

一、选择题

1. Android 具有四个重要的组件，下面选项中不属于四大组件的是（　　）。
   A. Activity   B. Service
   C. BroadcastReceiver   D. Intent

2. 在 AndroidManifest.xml 中描述一个 Activity 时，该 Activity 的 label 属性是指（　　）。
   A. 指定该 Activity 的图标
   B. 指定该 Activity 的显示标签
   C. 指定该 Activity 和类相关联的类型
   D. 指定该 Activity 的唯一标识

3. Activity 生命周期中，第 1 个需要执行的方法是（　　）。
   A. onStart( )   B. onCreate( )
   C. onReStart( )   D. onRresume( )

4. 可以运行于后台的、可以无界面的程序，在 Android 中，可以使用以下（　　）技术来实现。
   A. 程序必须有界面，没有无界面的程序   B. Activity
   C. Service   D. Intent

5. Android 工程项目下面的图片资源文件目录的作用是（　　）。
   A. 放置应用到的图片资源
   B. 主要放置一些文件资源，这些文件会被原封不动打包到 apk 里面
   C. 放置字符串、颜色、数组等常量数据
   D. 放置一些与用户界面相应的布局文件，都是 XML 文件

6. Android 会为 App 中的每一个资源生成一个唯一的 id，所有资源生成的 id 在（　　）文件中。

　　A. R. java　　　　　　　　　　　B. AndroidManifest. xml
　　C. main. xml　　　　　　　　　　D. string. xml

二、填空题

1. Android 项目中有许多文件夹，其中用于存放布局文件的文件夹是_____。

2. 在创建 Service 时，需要在_____文件中注册。

3. Android App 由组件构成，其中的四大组件分别为 Activity、Service、_____和_____。

4. Android 图形用户界面框架基于_____设计模式。

5. 如果需要将 Activity 设置为横屏，需要设置 Activity 中的属性 android：screenOrientation =_____。

三、判断题

1. 在创建 Android 程序时，JDK 会自动创建一些目录和文件，如 src、gen、res 等。
　　　　　　　　　　　　　　　　　　　　　　　　　　　　　　（　　）

2. 布局文件的命名只能是小写字母，不能是大写字母。　　　　　　（　　）

3. "@ +id/tvMsg"中的"+"表示该控件是一个新的控件，需要建立资源名称，并添加到 R. java 中。　　　　　　　　　　　　　　　　　　　　　　　（　　）

4. 在创建 Android App 时，填写的 Application Name 表示应用名称。（　　）

5. values 目录中默认存放了 strings. xml、colors. xml、styles. xml 文件，这些文件的内容是基于 XML 格式的 key-value 键值对。　　　　　　　　　　　（　　）

【第2章参考答案】

第3章
基本界面组件与布局

一个功能再强大的App,如果没有美观、易用的用户界面,往往也很难吸引用户,移动应用终端的App的用户界面更是如此。因此在Android App开发中的用户界面设计显得尤为重要。为了方便开发者开发Android App,Android SDK提供了许多组件(如文本框、命令按钮、复选框等)和标准的用户界面布局(如线性布局、相对布局等),方便开发者构建用户界面。Android App的绝大部分界面组件都放在android.widget包及其子包、android.view包及其子包中,Android App的所有用户界面组件都继承于View类。Android使用布局的概念来管理组件在容器视图中的摆放方式及摆放位置。本章将结合具体的案例介绍常用组件和基本布局的使用方法。

教学目标

了解Android中图形界面的MVC设计模式和六种布局管理器的继承关系、展现效果;熟悉线性布局、帧布局、相对布局、表格布局等基本布局;掌握应用开发中使用率较高的基本组件的常用属性和使用方法;掌握Toast、Handler和CountDownTimer等类的使用方法。

教学要求

知识要点	能力要求	相关知识
用户界面基础	了解Android App开发的设计模式和布局管理器	MVC
计算器的设计与实现	(1) 掌握TextView、Button的常用属性和使用方法 (2) 掌握LinearLayout的使用方法和应用场景	监听事件、嵌套布局
高仿QQ登录界面的设计与实现	(1) 掌握EditText、ImageView的常用属性和使用方法 (2) 掌握RelativeLayout的使用方法和应用场景 (3) 掌握Toast的使用方法	Toast

（续）

知识要点	能力要求	相关知识
注册界面的设计与实现	（1）掌握 RadioButton、RadioGroup、CheckBox、Spinner 及 RatingBar 等组件的常用属性和使用方法 （2）理解 RadioButton 和 RadioGroup 的关系	Adapter
考试系统界面的设计与实现	（1）掌握 TabHost/TabWidget 的常用属性和使用方法 （2）掌握 FrameLayout 的使用方法和应用场景	LayoutInflater
打老鼠游戏的设计与实现	（1）掌握 TableLayout 的使用方法和应用场景 （2）掌握 CountDownTimer 的使用方法 （3）掌握横屏与竖屏的实现方法	Handler、Runnable 和 Message
猜扑克游戏的设计与实现	（1）掌握 ScrollView 实现水平滚屏和垂直滚屏的方法 （2）掌握 ToggleButton、Switch 和 ProgressBar 组件的常用属性和使用方法	Handler、Runnable 和 Message

3.1 用户界面基础

3.1.1 组件和布局管理器

Android 用户界面框架（Android UI Framework）采用的是比较流行的 MVC（Model-View-Controller）框架模型。MVC 框架模型提供了处理用户输入的控制器（Controller），显示用户界面的视图（View），以及保存数据和代码的模型（Model）。为了方便人们进行 Android App 的用户界面设计，Android SDK 自带了一些系统组件和布局管理器。

系统组件是 Android 提供给用户的已经封装的界面组件，可以帮助用户快速开发 App，同时也能够使 Android App 的用户界面保持一致性。布局管理器本身就是一个用户界面组件，方便开发者调整组件的大小、位置。目前，Android 提供了六种布局管理器，其继承关系如图 3.1 所示。

（1）线性布局（LinearLayout）：它是一种最常用的布局方式，可以使用垂直和水平两种方式放置组件，假如组件的宽度和高度超过了屏幕的宽度和高度，那么超出的组件不会显示在屏幕上。

（2）相对布局（RelativeLayout）：它是可以让 App 在屏幕大小不同、分辨率不同的 Android 终端屏幕上友好显示的一种布局方式。放置在该布局管理器中的组件的位置都是相对位置。

（3）表格布局（TableLayout）：它是和 TableRow 配合使用的一种常用的布局管理方式，类似于 HTML 里面的 Table。

（4）帧布局（FrameLayout）：它是一种在 Android 终端屏幕上开辟一块空白区域的布局方式，放置在空白区域的组件必须对齐到屏幕的左上角。

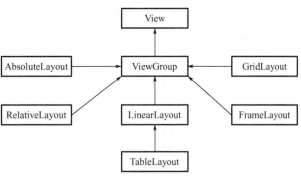

图 3.1　布局管理器继承关系

（5）网格布局（GridLayout）：它是 Android 在 4.0 之后的版本中新加的布局。该布局使用虚细线将布局划分为行、列和单元格。

（6）绝对布局（AbsoluteLayout）：它是一种需要开发者通过设定组件坐标决定组件摆放位置的布局方式。由于布局内所有元素位置都是固定的，所以不能保证所有 Android 终端上显示的效果都一样，目前已经过时不用，本书不再介绍。

所有的布局管理器都可以作为容器类对象使用，开发 App 时将布局管理器管理的组件以布局资源文件（XML 文件）的方式保存在 App 项目的 src/res/layout 目录下。

3.1.2　View 类和 ViewGroup 类

Android App 的用户界面都是通过 View 类和 ViewGroup 类及其派生子类对象构建的，View 类是所有可视化组件的基类。例如，本章即将介绍的 TextView、ImageView、ProgressBar 等组件是它的直接子类，Button、EditText、CheckBox 是它的间接子类。

ViewGroup 类也是 View 类的直接子类，但它可以作为其他组件的容器。Android 中六种布局管理器和一些高级组件都是它的直接或间接子类，如将要在第 4 章介绍的 GridView、Spinner、Gallery、ImageSwitcher 等。

一般来说，开发 Android App 的用户界面都不会直接使用 View 类和 ViewGroup 类，而是使用它们的派生类。View 类的派生子类见表 3-1，ViewGroup 类的派生子类见表 3-2。

表 3-1　View 类的派生子类

子类类型	类　名
直接子类	AnalogClock、ImageView、KeyboardView、ProgressBar、SurfaceView、TextView、ViewGroup、ViewStub
间接子类	AbsListView、AbsSeekBar、AbsSpinner、AbsoluteLayout、AdapterView < T extends Adapter >、AdapterViewAnimator、AdapterViewFlipper、AppWidgetHostView、AutoCompleteTextView、Button、CalendarView、CheckBox、CheckedTextView、Chronometer、CompoundButton

表 3-2 ViewGroup 类的派生子类

子类类型	类 名
直接子类	AbsoluteLayout，AdapterView ＜ T extends Adapter ＞，FrameLayout，LinearLayout，RelativeLayout，SlidingDrawer
间接子类	AbsListView，AbsSpinner，AppWidgetHostView，DatePicker，DialerFilter，ExpandableListView，Gallery，GridView，HorizontalScrollView，ImageSwitcher，ListView，MediaController，RadioGroup，ScrollView，Spinner，TabHost，TabWidget，TableLayout，TableRow，TextSwitcher，TimePicker，TwoLineListItem，ViewAnimator，ViewFlipper，ViewSwitcher，WebView，ZoomControls

表 3-1 和表 3-2 中的大部分子类是 App 开发过程中用得比较多的类，其使用方法将在后面章节中详细介绍。

3.2 计算器的设计与实现

计算器在各种设备中的应用非常普遍，本节使用 LinearLayout 布局管理器及 TextView（文本框）和 Button（命令按钮）组件设计一款如图 3.2 所示的计算器。

图 3.2 计算器的显示效果

3.2.1 预备知识

1. TextView

TextView 是一个文本显示组件，具备基本的显示文本功能。因为大多数用户界面系统组件都需要展示信息，所以它也是大多数用户界面系统组件的父类。要想在 Activity 中显示 TextView，就需要在相应的布局文件（即开发环境默认创建的 Activity 类——MainActivity.java 对应的 activity_main.xml 布局文件）添加相应的组件标签，这些 XML 格式的标签可以确定组件的大小、位置及颜色等属性。例如，下列代码运行后的效果如图 3.3 所示。

```
<TextView
        android:id ="@ +id/txtname"
        android:layout_width ="match_parent"
        android:layout_height ="wrap_content"
        android:gravity ="center"
        android:textSize ="30sp"
        android:textColor ="#be0e0a"
        android:text ="这是第1个TextView!"/>
</TextView >
```

【TextView、Button、LinearLayout】

图 3.3　TextView 显示效果

其中，android：id 属性代表着 TextView 的 id 为 txtname，它是 TextView 的唯一标识，在 java 代码中可以通过 findViewById（R. id. txtname）方法来获取绑定这个 TextView 对象；android：layout_ width 和 android：layout_ height 分别用来设置组件在屏幕上显示的宽度和高度，设置的值可以使用度量单位，也可以使用 fill_ parent（填充整个屏幕）或 wrap_ content（随着文字内容的不同改变视图的宽度/高度）。TextView 的常用属性和方法见表 3－3。

表 3－3　TextView 的常用属性和方法

属 性 名	对 应 方 法	说　明
android：autoLink	setAutoLinkMask(int)	设置是否当文本为 URL 链接、E－mail、电话号码、Map 时，文本显示为可单击的链接
android：gravity	setGravity(int)	设置文本框内文本的对齐方式
android：layout_gravity		设置组件本身相对于父组件的显示位置

(续)

属 性 名	对 应 方 法	说　　明
android:lines	setLines(int)	设置文本的行数，设置两行就显示两行，即使第 2 行没有数据
android:text	setText(CharSequence)	设置文本框显示的文本内容
android:textColor	setTextColor(int)	设置文本显示的颜色
android:textSize	setTextSize(float)	设置文字大小，推荐度量单位"sp"
android:textStyle	setTypeface(Typeface)	设置字形：粗体、斜体等
android:layout_width		设置文本框在屏幕上的宽度
android:layout_height		设置文本框在屏幕上的高度

gravity 和 layout_gravity 的常用属性值及说明见表 3-4。

表 3-4　gravity 和 layout_gravity 的常用属性值及说明

属 性 值	说　　明
top	顶部对齐
bottom	底部对齐
left	左边对齐
right	右边对齐
center_horizontal	横向中央位置对齐

在 TextView 中可以实现跑马灯效果，其实现代码如下。

```
1    <TextView
2        android:layout_width ="wrap_content"
3        android:layout_height ="wrap_content"
4        android:focusable ="true"
5        android:focusableInTouchMode ="true"
6        android:singleLine ="true"
7        android:ellipsize ="marquee"
8        android:marqueeRepeatLimit ="marquee_forever"
9        android:text ="跑马灯效果跑马灯效果跑马灯效果跑马灯效果跑马灯效果"
10       />
```

要实现跑马灯效果，必须设置 TextView 可以获得焦点（focusable 的属性值为 true）、只能为单行显示文本（singleline 的属性值为 true）、实现跑马灯效果和循环无限次（ellipsize 的属性值为 marquee、marqueeRepeatLimit 的属性值为 marquee_forever）及 Text 属性显示的文本内容必须超过屏幕宽度。

2. Button

Button 是 TextView 的子类，所以 TextView 上很多属性也可以直接应用到 Button 上。实际开发中对于 Button 操作主要是按下后执行何种操作。例如，本节案例项目中按下数字按钮能在 TextView 上显示该数字、按下"="按钮可以执行计算操作。表 3-5 列出了 Button 的常用属性和方法。

表 3-5 Button 的常用属性和方法

属 性 名	对 应 方 法	说 明
android:clickable	setClickable（boolean）	设置是否允许单击按钮（true/false）
android:background	setBackgroundResource（int）	通过资源文件设置背景色
android:onClick	setOnClickListener（OnClickListener）	设置单击事件

例如，下列代码运行后的效果如图 3.4 所示。

```
1   <Button
2       android:id="@+id/btn_click_one"
3       android:layout_width="wrap_content"
4       android:layout_height="wrap_content"
5       android:text="Button 单击事件写法 1" />
6   <Button
7       android:id="@+id/btn_click_two"
8       android:layout_width="wrap_content"
9       android:layout_height="wrap_content"
10      android:onClick="myClick"
11      android:text="Button 单击事件写法 2" />
12  <Button
13      android:id="@+id/btn_click_three"
14      android:layout_width="wrap_content"
15      android:layout_height="wrap_content"
16      android:background="@mipmap/ic_launcher"
17      android:text="Button 设置背景图片" />
18  <Button
19      android:id="@+id/btn_click_four"
20      android:layout_width="wrap_content"
21      android:layout_height="wrap_content"
22      android:background="@android:color/holo_red_dark"
23      android:text="Button 设置背景颜色" />
```

上述第 10 行代码表示给 btn_click_two 的单击事件绑定 myClick（）方法；第 16 行表示给 btn_click_three 设置背景图片 ic_launcher，该图片需要放置在 src/main/res/ mipmap-*文件夹下；第 22 行表示给 btn_click_ four 设置背景色为系统资源自带的颜色资源 holo_red_dark。

每一个添加到用户界面上的 Button 一般都需要给它绑定监听事件，给 Button 绑定监听事件通常有以下两类方法。

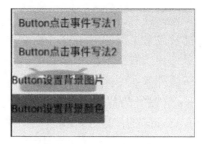

图 3.4 Button 显示效果

(1) 内部类。

① 匿名内部类。

```
1    bt = (Button) findViewById(R.id.btn_click_one);
2    bt.setOnClickListener(new OnClickListener() {
3        @Override
4        public void onClick(View v) {
5            Log.i("info","匿名内部类单击事件写法");
6        }
7    });
```

② 定义内部类，实现 OnClickListener 接口。

```
1    bt = (Button) findViewById(R.id.btn_click_three);
2    bt.setOnClickListener(new MyListener());
3    class MyListener implements OnClickListener{
4        @Override
5        public void onClick(View v) {
6            Log.i("info ","定义内部类单击事件写法");
7        }
8    }
```

【实例化对象和监听事件】

(2) 在布局文件中给 Button 增加了 android：onClick ="myclick" 属性（如上述布局文件的第 10 行代码），然后在该布局文件对应的 Acitivty 中实现该方法。需要注意的是该方法必须符合以下三个条件。

① 方法的修饰符是 public。

② 返回值是 void 类型。

③ 只有一个参数 View，这个 View 就是被单击的这个控件。

其实现代码如下。

```
1    public void myClick(View v){
2        switch (v.getId()){
3            case R.id.btn_click_two:
4                Log.i("info"," 自定义方法实现单击事件写法");
5                break;
6        }
7    }
```

3. LinearLayout

LinearLayout 是 Android App 开发时较常见的一种布局类型，它可以将用户界面上的组件摆放成水平（horizontal）或垂直（vertical）的形式。如果 LinearLayout 方向设置为水平，它里面的所有子组件被摆放在同一行中；如果 LinearLayout 方向设置为垂直，它里面的所有子组件被摆放在同一列中。Android 提供了 android.widget.LinearLayout 类实现 LinearLayout，该类的常用属性如 layout_gravity、gravity、layout_width、layout_height 等使用方法与 TextView 的使用方法相同，其他属性与对应方法见表 3-6。

第3章 基本界面组件与布局

表 3-6 LinearLayout 的常用属性和对应方法

属 性 名	对 应 方 法	说　　明
android:orientation	setOrientation(int)	设置线性布局的方向,有 horizontal 和 vertical 两个值,必须设置
android:layout_gravity		设置组件本身相对于父组件的显示位置
android:divider		设置分割线的图片
android:layout_weight		设置权重,即让一行或一列的组件按比例显示

(1) 垂直布局方式。

在新建 Android 项目时,开发环境在项目文件夹 res/layout 下自动生成 activity_main.xml 布局文件,文件代码修改和说明如下。

```
1    <?xml version="1.0" encoding="utf-8"?>
2    <!--LinearLayout 布局方式为垂直布局,宽度和高度与父容器匹配-->
3    <LinearLayout
xmlns:android="http://schemas.android.com/apk/res/android"
4        android:layout_width="match_parent"
5        android:layout_height="match_parent"
6        android:orientation="vertical">
7        <!--在该布局容器中增加四个 Button 组件,该组件的高度按组件实际内容显示、宽度与父容器匹配-->
8        <Button
9            android:layout_width="match_parent"
10           android:layout_height="wrap_content"
11           android:text="按钮1"/>
12       <Button
13           android:layout_width="match_parent"
14           android:layout_height="wrap_content"
15           android:text="按钮2"/>
16       <Button
17           android:layout_width="match_parent"
18           android:layout_height="wrap_content"
19           android:text="按钮3"/>
20       <Button
21           android:layout_width="match_parent"
22           android:layout_height="wrap_content"
23           android:text="按钮4"/>
24   </LinearLayout>
```

该代码的运行效果如图 3.5 所示。

(2) 水平布局方式。

如果要设计图 3.6 所示的用户界面,即将四个 Button 组件水平布局在一行上,那么就必须分别将垂直布局代码文件中的第 6 行代码和第 9、13、17、21 行代码修改为如下代码。

```
android:orientation="horizontal"
android:layout_width="wrap_content"
```

图3.5 垂直布局

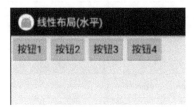
图3.6 水平布局

第6行代码的修改就是直接将线性布局管理器的布局方式设置为水平布局方式；而原来第9、13、17、21行中的android：layout_width 属性值是 match_parent，该属性值表示填充布局上该行的剩余空间，如果不修改其属性值，那么运行的效果图中就不会出现按钮2及后面的组件，读者可以自行编写代码测试。

（3）混合（嵌套）布局方式。

在实际开发中，单独使用水平布局和垂直布局两种方式通常不能满足用户界面设计的需要。例如，图3.2所示的计算器界面就需要在一个布局文件中既使用垂直布局又使用水平布局进行设计，这种方式称为混合（嵌套）布局方式。

图3.2所示用户界面布局文件的部分源代码如下。

```
1    <?xml version ="1.0" encoding ="utf-8"? >
2    <LinearLayout
xmlns:android ="http://schemas.android.com/apk/res/android"
3        android:layout_margin ="10dp"
4        android:layout_width ="match_parent"
5        android:layout_height ="match_parent"
6        android:orientation ="vertical" >
7    <!--设置计算器第1行用于显示结果的 TextView -->
8        <TextView
9        android:background ="#e9f4e9"
10       android:layout_margin ="4dp"
11       android:id ="@ +id/txtresult"
12       android:layout_width ="match_parent"
13       android:layout_height ="wrap_content"
14       android:lines ="2"
15       android:gravity ="center_vertical|right"
16       android:hint ="34" />
17   <!--设置计算器第2行所有组件 -->
18       <LinearLayout
19       android:layout_width ="match_parent"
20       android:layout_height ="wrap_content"
21       android:orientation ="horizontal" >
22       <Button
23           android:id ="@ +id/btn_clear"
24           android:layout_width ="0dp"
25           android:layout_height ="wrap_content"
26           android:layout_weight ="1"
27           android:text ="C" />
28       <Button
```

```
29          android:id = "@ + id/btn_del"
30          android:layout_width = "0dp"
31          android:layout_height = "wrap_content"
32          android:layout_weight = "2"
33          android:text = "DEL" />
34       <Button
35          android:id = "@ + id/btn_div"
36          android:layout_width = "0dp"
37          android:layout_height = "wrap_content"
38          android:layout_weight = "1"
39          android:text = "÷" />
40    </LinearLayout>
41    <!—设置计算器第 3 行所有组件 -->
42    <LinearLayout
43       android:layout_width = "match_parent"
44       android:layout_height = "wrap_content"
45       android:orientation = "horizontal" >
46       <Button
47          android:id = "@ + id/btn_1"
48          android:onClick = "calClick"
49          android:layout_width = "wrap_content"
50          android:layout_height = "wrap_content"
51          android:layout_weight = "1"
52          android:text = "1" />
53       <Button
54          android:id = "@ + id/btn_2"
55          android:onClick = "calClick"
56          android:layout_width = "wrap_content"
57          android:layout_height = "wrap_content"
58          android:layout_weight = "1"
59          android:text = "2" />
60       <Button
61          android:id = "@ + id/btn_3"
62          android:onClick = "calClick"
63          android:layout_width = "wrap_content"
64          android:layout_height = "wrap_content"
65          android:layout_weight = "1"
66          android:text = "3" />
67       <Button
68          android:id = "@ + id/btn_add"
69          android:layout_width = "wrap_content"
70          android:layout_height = "wrap_content"
71          android:layout_weight = "1"
72          android:text = "+" />
73    </LinearLayout>
74    <!—其他三行的按钮布局方式与上述代码类似,读者可以参见代码包中 Calculator
文件夹里的内容 -->
75 </LinearLayout>
```

以上布局文件中,最外层的 LinearLayout 在第 6 行代码中设置了整个界面布局方式为"vertical",该行代码让整体布局的所有组件垂直布置在用户界面上;而第 18 行开始内嵌的 LinearLayout 在第 21 行代码中设置为"horizontal",表示该内嵌布局为水平布局方式,即里面的 btn_clear、btn_del 和 btn_div 三个 Button 按钮水平放置在用户界面上，所以图 3.2 中的"C"按钮、"DEL"按钮和"÷"按钮水平放置在用户界面上。混合布局就是将水平布局和垂直布局方式作为一个用户界面。从代码中可以看出，它们的整体结构是外层 LinearLayout 里面放置了另外的 LinearLayout，这种形式就构成了布局的嵌套结构。开发者在进行用户界面设计时，既可以在水平布局方式里嵌套垂直布局，也可以在垂直布局方式里嵌套水平布局，但是，一定要注意嵌套结构的层次性，这样才能设计出让用户满意的用户界面。

TextView 中显示的文本内容默认对齐方式为右对齐，为了达到图 3.2 所示的效果，上述代码的第 15 行代码中的 gravity 属性设置为"center_vertical | right"，表示 TextView 中显示内容的对齐方式为垂直居中并且靠右对齐。

进行组件的 layout_width 和 layout_height 属性设置时，可以使用 weight（权重）属性来控制组件大小在行或列上的占比。如果设置时按比例划分组件在水平方向的占比，就需要将涉及组件的 layout_width 属性设置为 0dp，并将相应组件的 weight 属性值设置为对应的比例值。例如，图 3.2 中的"C"按钮、"DEL"按钮和"÷"按钮的占比为 1∶2∶1，所以对应 Button 的 weight 属性值分别设置为 1、2、1。

3.2.2 计算器的实现

1. 主界面的设计

【计算器的实现】

从图 3.2 可以看出，计算器主要实现了加、减、乘、除的计算功能，主界面上用 11 个 Button 组件显示数字 0~9 和"."，用于输入数字 0~9 和小数点；用四个 Button 组件显示"+""-""×""÷"运算符，用于实现加、减、乘、除运算；用三个 Button 组件显示"=""C""DEL"，"="用于计算结果、"C"用于将显示结果的 TextView 清零、"DEL"用于删除输入的一个字符。读者可以参见代码包中 Calculator 文件夹里的布局文件中的详细代码。

2. 功能实现

为了在代码中使用布局文件定义的组件，必须使用 findViewById() 方法。例如，要引用显示结果的 TextView 组件 txtResult，需要使用下面的语句实现。

```
TextView txtResult = (TextView)findViewById(R.id.txtResult)
```

（1）定义变量和初始化组件。

由于本案例中使用的 Button 按钮较多，并且绑定在按钮上的监听事件类别相同，为了减少代码冗余，可以通过定义 Button 类型的数组来实现。定义变量代码如下。

```
    private Button[] btnDigit = new Button[11];//用于存放0~9和小数点按钮
    private Button btnAdd,btnDec,btnMul,btnDiv,btnEqu,btnClear,btnDel;//用于存放其他功能按钮
    private int[] btnDigitId = {R.id.btn_0, R.id.btn_1, R.id.btn_3, R.id.btn_4, R.id.btn_5, R.id.btn_6, R.id.btn_7, R.id.btn_8, R.id.btn_9, R.id.btn_dot};
    private TextView txtResult;
     private StringBuffer digitA = new StringBuffer(), digitB = new StringBuffer();
```

```
private boolean isChar = false;//标记是否按下运算符
private int operator = 0;//默认运算符为 +
private boolean isDigitA = true;//标记第1个操作数,用于退格删除 txtResult 中内容
```

本案例中定义了一个 init( )方法实现初始化组件,其详细代码如下。

```
void init() {
    for (int i = 0; i <btnDigitId.length; i ++) {
        btnDigit[i] = (Button) this.findViewById(btnDigitId[i]);
    }
    btnAdd = (Button) this.findViewById(R.id.btn_add);// +
    btnDec = (Button) this.findViewById(R.id.btn_dec);// -
    btnMul = (Button) this.findViewById(R.id.btn_mul);// ×
    btnDiv = (Button) this.findViewById(R.id.btn_div);// ÷
    btnEqu = (Button) this.findViewById(R.id.btn_equ);// =
    btnClear = (Button) this.findViewById(R.id.btn_clear);//C
    btnDel= (Button) this.findViewById(R.id.btn_del);//DEL
    txtResult = (TextView) this.findViewById(R.id.txtresult);//显示结果
}
```

(2) 数字 0~9 及小数点的监听事件定义了 calClick( )方法来实现,其主要代码如下。

```
1    public void calClick(View view) {
2        switch (view.getId()) {
3            case R.id.btn_0://按下 0 键
4                if (isChar) {
5                    txtResult.setText("");
6                    digitB.append("0");
7                    txtResult.setText(digitB.toString());
8                } else {
9                    digitA.append("0");
10                   txtResult.setText(digitA.toString());
11               }
12               break;
13           case R.id.btn_1://按下 1 键
14               ……
15               ……
16           case R.id.btn_dot:
17               ……
18        }
19    }
```

上述第 4~11 行代码表示如果没有按下运算符(isChar 是布尔型,初始值为 false 表示没有按下运算符),那么就将"0"添加到 StringBuffer 类型的 digitA 中(即参与运算的第 1 个数),并将 digitA 中的内容输出到显示区;否则将显示区域 txtResult 清空,并将"0"添加到 digitB 中(即参与运算的第 2 个数)。数字 1~9 及小数点字符的输入方法与 0 字符输入相同,限于篇幅不再赘述。

(3) 运算符"+"的监听事件,代码如下。

```
1    btnAdd.setOnClickListener(new View.OnClickListener() {
2        @Override
```

```
3           public void onClick(View v) {
4               operator = 0;//运算符 +
5               isChar = true;//按下运算符
6               isDigitA = false;//标记已不是第 1 个操作数
7           }
8       });
```

为了方便实现"+""-""×""÷"运算,本案例采取了间接实现方法,当按下"+",operator 的值为 0;当按下"-",operator 的值为 1;当按下"×",operator 的值为 2;当按下"÷",operator 的值为 3。第 4 行代码表示如果按下"+",isChar 为 true 表示已经按下了运算符,isDigitA 为 false 表示 DEL 键对应的删除的内容为第 2 个操作数的内容。

(4) 清空按钮"C"的监听事件,代码如下。

```
1   btnClear.setOnClickListener(new View.OnClickListener() {
2       @Override
3       public void onClick(View v) {
4           digitA = new StringBuffer();
5           digitB = new StringBuffer();
6           isChar = false;
7           txtResult.setText("");
8       }
9   });
```

(5) 等于号"="的监听事件,代码如下。

```
1   btnEqu.setOnClickListener(new View.OnClickListener() {
2       @Override
3       public void onClick(View v) {
4           float da = Float.parseFloat(digitA.toString());
5           float db = Float.parseFloat(digitB.toString());
6           switch (operator) {
7               case 0:
8                   txtResult.setText(da + db + "");
9                   break;
10              case 1:
11                  txtResult.setText(da - db + "");
12                  break;
13              case 2:
14                  txtResult.setText(da * db + "");
15                  break;
16              case 3:
17                  txtResult.setText(da / db + "");
18                  break;
19          }
20          isChar = true;
21          isDigitA = true;
22      }
23  });
```

第 4~5 行代码将 digitA、digitB 中的字符串转换成 float 类型,第 6~19 行代码表示根据 operator 的值实现"+""-""×""÷"运算。

(6)退格删除"DEL"的监听事件,代码如下。

```
1     btnDel.setOnClickListener(new View.OnClickListener() {
2         @Override
3         public void onClick(View v) {
4             String temp = txtResult.getText().toString();
5             if (isDigitA) {
6                 digitA = new StringBuffer();
7                 if (temp.length() > 0)
8                     if (temp.length() == 1) {
9                         temp = "0";
10                    } else {
11                        temp = temp.substring(0, temp.length() - 1);
12                    }
13                txtResult.setText(temp);
14                digitA.append(txtResult.getText().toString());
15            } else {
16                ……
17            }
18        }
19    });
```

第 5～14 行代码表示如果 isDigitA 为 true(此时 DEL 按钮退格删除的是第 1 个操作数),那么按一次 DEL 按钮就删除字符串末尾的一个字符,删除后将结果显示在 txtResult 显示区域;如果 isDigitA 为 false 表示按一次 DEL 按钮,删除第 2 个操作数末尾的一个字符,其代码与第 5～14 行类似,限于篇幅不再详述。全部功能实现代码请读者参见代码包中Calculator文件夹里的内容。

3.3 高仿 QQ 登录界面的设计与实现

移动终端 App 大多需要用户登录后才能使用,如 QQ、微信、酷狗音乐等。本节将通过高仿 QQ 登录界面(图 3.7)的实现过程介绍 RelativeLayout、EditText 和 ImageView 的使用方法。

图 3.7 高仿 QQ 登录界面

3.3.1 预备知识

1. EditText

【EditText、ImageView、RelativeLayout】

在 App 开发中 EditText 是经常用到的组件,也是一个比较必要的组件。它是用户与 Android App 进行数据传输的窗口。例如,用户想要登录一个界面,需要输入账号、密码,然后获取输入的内容,提交后台进行判断。EditText 是 TextView 的子类,所以其大部分属性和方法与 TextView 相同。表 3-7 列出了 EditText 的常用属性和对应方法。

表 3-7 EditText 的常用属性和对应方法

属 性 名	对 应 方 法	说　　明
android:hint	setHint(int)	设置编辑框内容为空时显示的文本
android:inputType	setInputType(int)	设置编辑框中限制输入的类型:number—整数类型、numberDecimal—小数点类型、date—日期类型、text—文本类型(默认值)、phone—拨号键盘、textPassword—密码、textVisiblePassword—可见密码、textUri—网址
android:maxLength		限制显示的文本长度,超出部分不显示
android:drawableLeft android:drawableRight android:drawableTop android:drawableBottom		设置编辑框文本的左边、右边、顶部、底部显示的 drawable
android:drawablePadding		设置编辑框文本与 drawable 的间隔,与 drawableLeft、drawableRight、drawableTop 和 drawableBottom 一起使用,可设置为负数,单独使用没有效果
android:digits		设置允许输入哪些字符,如 "1234567890"

例如,要显示图 3.8 所示手机号登录界面的效果,可以使用下列代码实现。

```
1    <?xml version = "1.0" encoding = "utf-8"? >
2    <LinearLayout xmlns:android = "http://schemas.android.com/apk/res/android"
3        android:layout_width = "match_parent"
4        android:layout_height = "match_parent"
5        android:orientation = "vertical" >
6        <EditText
7            android:layout_width = "match_parent"
8            android:layout_height = "60dp"
9            android:layout_marginLeft = "10dp"
10           android:layout_marginRight = "10dp"
```

```
11          android:layout_marginTop ="15dp"
12          android:drawableLeft ="@mipmap/mobile"
13          android:drawablePadding ="5dp"
14          android:hint ="请输入手机号" />
15      <EditText
16          android:layout_width ="match_parent"
17          android:layout_height ="60dp"
18          android:layout_marginLeft ="10dp"
19          android:layout_marginRight ="10dp"
20          android:drawableLeft ="@mipmap/lock"
21          android:drawablePadding ="5dp"
22          android:hint ="请输入密码"
23          android:inputType ="numberPassword" />
24      <Button
25          android:layout_width ="match_parent"
26          android:layout_height ="60dp"
27          android:layout_marginLeft ="10dp"
28          android:layout_marginRight ="10dp"
29          android:background ="#f7e635ae"
30          android:text ="登录"/ >
31  </LinearLayout >
```

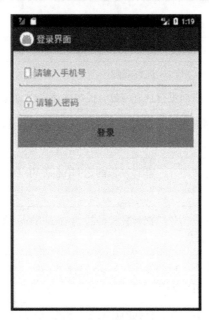

图3.8 手机号登录界面

以上第9行、第10行代码分别表示 EditText 离左边缘、右边缘的间距为10dp；第12行、第20行表示引用 src/main/mipmap-* 文件夹下的 mobile 和 lock 图片在 EditText 显示文本的左侧显示，这些图片需要在引用之前复制到 src/main/mipmap-* 文件夹下；第29行表示设置 Button 的背景色。

2. ImageView

ImageView 直接继承自 View 类,它主要用于显示图片,其图片可以来自于资源文件、Drawable 对象和 ContentProvider。ImageView 的常用属性和对应方法见表 3-8。

表 3-8 ImageView 常用属性和对应方法

属 性 名	对 应 方 法	说 明
android:adjustViewBounds	setAdjustViewBounds(boolean)	设置 ImageView 是否自动调整边界来适应所显示图片的长宽比
android:maxHeight	setMaxHeight(int)	设置 ImageView 的最大高度
android:maxWidth	setMaxWidth(int)	设置 ImageView 的最大宽度
android:scaleType	setScaleType(ImageViwe.ScaleType)	设置 ImageView 所显示的图片如何缩放或移动以适应 ImageView 的大小,其属性值及说明见表 3-9
android:background		设置 ImageView 的背景颜色或图片
android:src	setImageResource(int)	设置 ImageView 所显示的 Drawable 对象

表 3-9 scaleType 的属性值及说明

属 性 值	说 明
matrix	使用 matrix 方式对图片进行缩放
fitXY	对图片横向及纵向进行独立缩放,使得图片完全适应 ImageView(横纵比可能改变)
fitStart	保持纵横比缩放图片,直到较长的边与 ImageView 的边长相等,并且让图片显示在 ImageView 的左上角
fitCenter	保持纵横比缩放图片,直到较长的边与 ImageView 的边长相等,并且让图片显示在 ImageView 的中间
fitEnd	保持纵横比缩放图片,直到较长的边与 ImageView 的边长相等,并且让图片显示在 ImageView 的右下角
center	保持原图的大小,把图片放在 ImageView 的中间,当原图尺寸大于 ImageView 的尺寸时,超过部分裁剪处理
centerCrop	保持纵横比缩放图片,直到图片完全覆盖 ImageView,可能出现图片显示不完整
centerInside	保持纵横比缩放图片,直到 ImageView 能完整显示图片

ImageView 有两个可以设置图片的属性:src 和 background。通常,src 指的是 ImageView 上显示的内容,background 指的是 ImageView 的背景。使用 src 填入图片时,按照图片大小直接填充,并不进行拉伸;而使用 background 填入图片时,按照 ImageView 给定的宽度来进行拉伸。例如,下列代码运行后的效果如图 3.9 所示。

```
 1    <LinearLayout xmlns:android="http://schemas.android.com/apk/res/android"
 2        android:layout_width="match_parent"
 3        android:layout_height="match_parent"
 4        android:orientation="vertical">
 5        <ImageView
 6            android:layout_width="wrap_content"
 7            android:layout_height="wrap_content"
 8            android:background="@mipmap/img"/>
 9        <ImageView
10            android:layout_width="200dp"
11            android:layout_height="wrap_content"
12            android:background="@mipmap/img"/>
13        <ImageView
14            android:layout_width="wrap_content"
15            android:layout_height="wrap_content"
16            android:src="@mipmap/img"/>
17        <ImageView
18            android:layout_width="200dp"
19            android:layout_height="wrap_content"
20            android:background="#ff00ff"
21            android:src="@mipmap/img"/>
22    </LinearLayout>
```

上述代码中加载的图片分辨率为 246×132，从显示效果可以看出，宽度和高度属性都设置为 wrap_content 时，ImageView 中使用 src 和 background 加载的图片显示效果一样（即为原图的大小）；但是如果指定了宽度或高度的值，src 和 background 加载的图片显示效果就不一样了，background 加载的图片完全填充了整个 ImageView，而 src 加载的图片的尺寸只有指定的值那么大，并且显示在 ImageView 中间。

3. RelativeLayout

在 3.2 节计算器的界面布局中已经对线性布局方式的使用方法进行了介绍，实际开发中可以使用 weight 属性进行等比例划分以达到屏幕适配

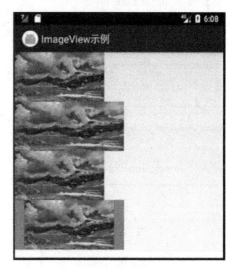

图 3.9 ImageView 显示效果

的更好效果，但是当用户界面比较复杂时需要嵌套多层的 LinearLayout，这样不仅会降低界面渲染的效率，也会占用更多的系统资源。如果使用 RelativeLayout 的话，可能仅仅需要一层就可以完成比较复杂的用户界面设计。

相对布局就是相对于父容器或其他兄弟组件控制组件对象在用户界面上的位置，它可以更细致地布局用户界面。例如，ImageView、TextView 和 Button 这三个组件，可以分别将 ImageView、Button 放在 TextView 的左下角、右下角；也可以将 ImageView、Button 放在

TextView 的右下角、左下角。即通过这种布局方式，可以将一个组件放在另外一个组件的顶部、底部、左边或右边；也可以相对于父容器放置，包括放置在父容器的顶部、底部、左边或右边。Android 提供了 android.widget.RelativeLayout 类实现相对布局，该类的常用属性见表 3-10～表 3-13。

表 3-10　设置组件与组件之间关系和位置的相关属性

属 性 名	说　　明	备　　注
android:layout_above	将该组件的底部置于给定 ID 组件的上面	属性值为某个组件的 ID，如 android:layout_above ="@id/inputname"，其中 inputname 为 EditText 组件的 ID
android:layout_below	将该组件的底部置于给定 ID 组件的下面	
android:layout_toLeftOf	将该组件的右边缘与给定 ID 组件的左边缘对齐	
android:layout_toRightOf	将该组件的左边缘与给定 ID 组件的右边缘对齐	

表 3-11　设置组件与组件之间对齐方式的相关属性

属 性 名	说　　明	备　　注
android:layout_alignBaselineabove	将该组件的基线与给定 ID 组件的基线对齐	属性值为某个组件的 ID，如：android:layout_alignTop ="@id/inputname"，其中 inputname 为 EditText 组件的 ID
android:layout_alignTop	将该组件的顶部与给定 ID 组件的顶部对齐	
android:layout_alignBottom	将该组件的底部与给定 ID 组件的底部对齐	
android:layout_alignLeft	将该组件的左边边缘与给定 ID 组件的左边边缘对齐	
android:layout_alignRight	将该组件的右边边缘与给定 ID 组件的右边边缘对齐	

表 3-12　设置组件与父组件之间对齐方式的相关属性

属 性 名	说　　明	备　　注
android:layout_alignParentTop	将该组件的顶部与父组件的顶部对齐	属性值可选为"true"或"false"
android:layout_alignParentBottom	将该组件的底部与父组件的底部对齐	
android:layout_alignParentLeft	将该组件的左边缘与父组件的左边缘对齐	
android:layout_alignParentRight	将该组件的右边缘与父组件的右边缘对齐	

表3-13 设置组件方向的相关属性

属 性 名	说　明	备　注
android:layout_centerHoriaontal	将该组件置于水平方向的中央	属性值可选为"true"或"false"
android:layout_centerVertical	将该组件置于垂直方向的中央	
android:layout_centerInParent	将该组件置于父组件水平方向和垂直方向的中央	

设计用户界面时，开发者可以通过组合这些属性来实现各种各样的布局。例如，下面代码的显示效果如图3.10所示。

```
 1    <?xml version="1.0" encoding="utf-8"?>
 2    <RelativeLayout xmlns:android="http://schemas.android.com/apk/res/android"
 3        android:layout_width="match_parent"
 4        android:layout_height="match_parent">
 5     <ImageView
 6        android:id="@+id/imgMain"
 7        android:layout_width="wrap_content"
 8        android:layout_height="wrap_content"
 9        android:layout_centerInParent="true"
10        android:src="@mipmap/people" />
11     <ImageView
12        android:id="@+id/imgtop"
13        android:layout_width="wrap_content"
14        android:layout_height="wrap_content"
15        android:layout_above="@id/imgMain"
16        android:layout_alignLeft="@id/imgMain"
17        android:src="@mipmap/circle" />
18     <ImageView
19        android:layout_width="wrap_content"
20        android:layout_height="wrap_content"
21        android:layout_alignTop="@id/imgMain"
22        android:layout_toLeftOf="@id/imgMain"
23        android:src="@mipmap/circle" />
24     <ImageView
25        android:layout_width="wrap_content"
26        android:layout_height="wrap_content"
27        android:layout_alignTop="@id/imgMain"
28        android:layout_toRightOf="@id/imgMain"
29        android:src="@mipmap/circle" />
30     <ImageView
31        android:layout_width="wrap_content"
32        android:layout_height="wrap_content"
33        android:layout_alignLeft="@id/imgMain"
34        android:layout_below="@id/imgMain"
35        android:src="@mipmap/circle" />
36    </RelativeLayout>
```

在使用 RelativeLayout 进行布局设置时选择第 1 个参照对象非常重要，第 1 个参照对象一旦选定，其他对象就可以相对于该参照对象进行摆布。例如，上述代码第 5～9 行就是首先指定 imgMain 对象在布局的中央，即 layout_ centerInParent 属性值设置为 true，然后在它的上、左、右、下分别摆放四个 ImageView 以显示 circle.jpg 图片。

图 3.10 相对布局显示效果

3.3.2 高仿 QQ 登录界面的实现

1. 主界面的设计

根据图 3.7 可以将用户界面的设计分解成如图 3.11 所示。整个界面最外层使用 RelativeLayout，然后里面嵌套了 RelativeLayout-rlayouttop 用于显示最上部的背景及放置头像图片，嵌套了 RelativeLayout-rlayoutmiddle 用于放置 EditText（输入 QQ 号码）和向下的箭头 ImageView，嵌套了水平线性布局 llmiddle 用于放置 EditText（输入密码）放置了 Button（登录按钮），最底部嵌套 RelativeLayout 用于放置显示"无法登录？"和"新用户"的 TextView。

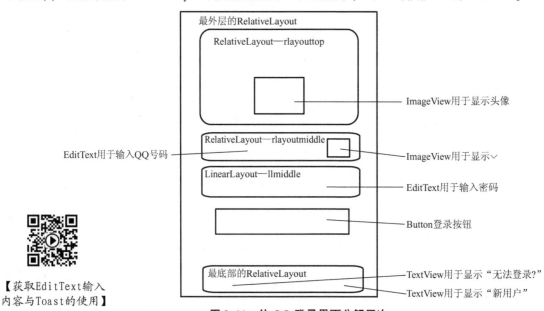

【获取 EditText 输入内容与 Toast 的使用】

图 3.11 仿 QQ 登录界面分解层次

RelativeLayout – rlayouttop 的代码如下。

```
1       <RelativeLayout
2           android:id ="@ +id/rlayouttop"
3           android:layout_width ="match_parent"
4           android:layout_height ="240dp"
5           android:background ="#b6dbe6" >
6           <ImageView
7               android:layout_width ="100dp"
8               android:layout_height ="100dp"
9               android:layout_alignParentBottom ="true"
10              android:layout_centerHorizontal ="true"
11              android:layout_marginBottom ="10dp"
12              android:src ="@mipmap/meinv" />
```

RelativeLayout – rlayoutmiddle 的代码如下，第 5 行表示此布局放置在 rlayouttop 的正下方，第 18 行表示图片调整为与 ImageView 大小相等。

```
1       <RelativeLayout
2           android:id ="@ +id/rlayoutmiddle"
3           android:layout_width ="match_parent"
4           android:layout_height ="50dp"
5           android:layout_below ="@ id/rlayouttop"
6           android:background ="#ffffff" >
7           <EditText
8               android:id ="@ +id/edtUserName"
9               android:layout_width ="match_parent"
10              android:layout_height ="50dp"
11              android:layout_marginLeft ="10dp"
12              android:background ="@ null"
13              android:hint ="请输入 QQ 号" />
14          <ImageView
15              android:layout_width ="50dp"
16              android:layout_height ="50dp"
17              android:layout_alignParentRight ="true"
18              android:scaleType ="center"
19              android:src ="@mipmap/row_down" />
20      </RelativeLayout>
```

LinearLayout – llmiddle 的代码如下，其中第 13 行表示 EditText 自带的背景设为空，第 15 行表示接收输入的是数字密码。

```
1       <LinearLayout
2           android:id ="@ +id/llmiddle"
3           android:layout_width ="match_parent"
4           android:layout_height ="50dp"
5           android:layout_below ="@ id/rlayoutmiddle"
6           android:layout_marginTop ="1dp"
7           android:background ="#ffffff" >
8           <EditText
9               android:id ="@ +id/edtPWD"
```

```
10        android:layout_width ="match_parent"
11        android:layout_height ="50dp"
12        android:layout_marginLeft ="10dp"
13        android:background ="@null"
14        android:hint ="请输入密码"
15        android:inputType ="numberPassword" />
16    </LinearLayout>
```

登录按钮 Button 组件放置在 LinearLayout – llmiddle 的正下方，离父容器左边、右边、顶部的距离都为20dp，其详细代码如下。

```
1    <Button
2        android:id ="@ +id/btnLogin"
3        android:layout_width ="match_parent"
4        android:layout_height ="50dp"
5        android:layout_below ="@ id/llmiddle"
6        android:layout_marginLeft ="20dp"
7        android:layout_marginRight ="20dp"
8        android:layout_marginTop ="20dp"
9        android:background ="#e6aeed"
10       android:text ="登 录"
11       android:textColor ="#ffffff"
12       android:textSize ="24sp" />
```

将 RelativeLayout 放置在父容器的最底部，由 layout_alignParentBottom 属性值"true"设置，在这个布局中最左侧放置了一个 TextView、最右侧放置了一个 TextView，其详细代码如下。

```
1    <RelativeLayout
2        android:layout_width ="match_parent"
3        android:layout_height ="wrap_content"
4        android:layout_alignParentBottom ="true" >
5    <TextView
6        android:layout_width ="wrap_content"
7        android:layout_height ="wrap_content"
8        android:paddingLeft ="10dp"
9        android:paddingBottom ="10dp"
10       android:text ="无法登录?"
11       android:textSize ="20sp"/ >
12   <TextView
13       android:layout_width ="wrap_content"
14       android:layout_height ="wrap_content"
15       android:layout_alignParentRight ="true"
16       android:paddingRight ="10dp"
17       android:paddingBottom ="10dp"
18       android:text ="新用户"
19       android:textSize ="20sp"/ >
20   </RelativeLayout >
```

从图 3.7 上可以看出登录界面没有标题栏，这种效果需要修改配置文件 AndroidManifest.xml（位于 qqui/src/main 目录下），即将本项目配置文件中的 activity 标签默认的 theme

修改为 android：theme =" @android：style/Theme. Light. NoTitleBar"。

2. 高仿 QQ 登录界面的功能实现

本案例只是模仿 QQ 登录界面的设计及功能的实现，所以仅实现了当输入的 QQ 号是 "50501212"、密码是 "832777" 时，就显示 "登录成功"，否则显示 "QQ 号或密码出错，请重输！"，其关键代码如下。

```
1    btnLogin.setOnClickListener(new View.OnClickListener() {
2        @Override
3        public void onClick(View v) {
4            String sUserName = edtUserName.getText().toString();
5            String sPWD = edtPWD.getText().toString();
6            if(sUserName.equals("50501212")&&sPWD.equals("832777")){
7                Toast.makeText(MainActivity.this,"登录成功",Toast.LENGTH_LONG).show();
8            }else {
9                Toast.makeText(MainActivity.this,"QQ号或密码出错,请重输!",Toast.LENGTH_LONG).show();
10           }
11       }
12   });
```

上述代码中使用了 Toast 组件来告诉用户登录是否成功。Toast 组件是 Android 提供的一种简单消息提示框机制，可以在用户做了某种操作后，给用户一些提示信息。该提示信息不能被用户操作，根据用户设置的显示时间显示后会自动消失，其运行后的效果如图 3.12 所示。创建并显示 Toast 通常有两种方式。

图 3.12　Toast 显示效果

（1）默认 Toast。

Toast.makeText(Context context, int resId, int duration)

Toast.makeText(Context context, CharSequencetext, int duration)

以上方法中 context 为上下文，通常为当前 activity，即可用 getApplicationContext() 或 this；resId 为要显示字符串的 id（如 R.string.info）；text 为要显示的字符串；duration 为内容显示的时间，该时间值通常使用 Toast.LENGTH_SHORT 或 Toast.LENGTH_LONG。例如：

Toast.makeText(this, "this is string", Toast.LENGTH_SHORT).show();

【QQ登录界面的实现】

（2）自定义 Toast。

如果要在 Toast 显示信息中包含一个指定的图片，可以使用下列代码实现。

```
1    Toast toast = new Toast(this);
2    //定义一个 ImageView
3    ImageView imageView = new ImageView(this);
4    imageView.setImageResource(R.drawable.ic_launcher);
5    //定义一个 LinearLayout，也可以是其他布局
6    LinearLayout layout = new LinearLayout (this);
7    layout.setOrientation (LinearLayout.HORIZONTAL);
8    //将 ImageView 放到 Layout 中
9    layout.addView (imageView);
10   //设置 View
11   toast.setView (layout);
12   //设置显示时间
13   toast.setDuration (20);
14   toast.show();
```

如果要指定 Toast 的显示位置，需要使用方法 setGravity（int gravity, int xOffset, int yOffset），该方法有以下三个参数。

① int gravity：指定 Toast 的初始位置，其值见表 3-14。

表 3-14 Toast 常用方法

值	说　　明
Gravity.TOP	初始位置在屏幕垂直中轴线的最上面，但不会遮住通知栏
Gravity.BOTTOM	初始位置在屏幕垂直中轴线最下面
Gravity.LEFT	初始位置在屏幕水平中轴线最左边
Gravity.RIGHT	初始位置在屏幕水平中轴线最右边

② int xOffset：决定了离初始显示位置的水平偏移量，单位是 px，左负右正。

③ int yOffset：决定了离初始显示位置的垂直偏移量，单位是 px，上负下正。

限于篇幅，完整的布局代码和功能代码读者可以参阅 qqui 文件夹中的内容。

3.4 注册界面的设计与实现

3.3 节中介绍了移动端 App 的登录界面的设计与实现功能。实际应用中，大多数 App

都需要用户必须先注册,然后才能登录或者才能使用其中的一些功能。这就需要为 App 设计一个注册界面,并可以实现用户名、性别、爱好、出生地、自我评价等功能,如图 3.13 所示。要实现图 3.13 所示注册界面的功能,除了需要使用前面介绍的布局及组件外,还需要使用 RadioButton(单选按钮)、RadioGroup(单选组合框)、CheckBox(复选框)、Spinner(下拉列表框)及 RatingBar(星级评价)等组件。

图 3.13　注册界面效果

3.4.1　预备知识

1. RadioButton 与 RadioGroup

RadioButton 在开发中提供了一种"多选一"的操作模式,是 Android 开发中常用的一种组件。例如,在用户注册时选择的性别只能从"男"或"女"中选择一个。在 Android 应用开发中实现 RadioButton 有以下两种方式。

【RadioButton、RadioGroup、RatingBar】

(1) 在布局文件中直接定义 RadioButton 组件,使用这种方式定义 RadioButton 时,如果界面上有多个 RadioButton,则表示可以多个、甚至全部选中。

(2) 与 RadioGroup 配合使用,使用这种方式定义的 RadioButton 只可以选中一个,并可以使用 setOnCheckedChangeListener 来对 RadioButton 进行事件监听。

RadioGroup 类继承于 LinearLayout 类,它只是提供 RadioButton 的容器,实际开发时可以在该容器中添加多个 RadioButton。RadioButton 类是 Button 类的子类,因此该组件可以直接使用 Button 支持的各种属性和方法,使用时RadioButton组件必须放在 RadioGroup 的组件中才能达到预期效果。RadioGroup 的常用属性和方法见表 3-15。

表 3-15　RadioGroup 的常用属性和方法

属性/方法名	说　　明
android:orientation	设置里面的 RadioButton 摆布方式：horizontal、vertical
void addView( )	使用指定的布局参数添加一个子视图，它有三个参数： View child——添加的子视图 int index——添加的位置 ViewGroup.LayoutParams params——添加的子视图的布局参数
voidcheck( )	用于指定该组中要勾选的单选按钮的唯一标识符（id），它有一个参数： int id——如果为-1，则清除单选按钮组的勾选状态，相当于调用 clearCheck( )
voidclearCheck( )	清除单选按钮组中所有单选按钮的选中状态
intgetCheckedRadioButtonId ( )	返回该单选按钮组中所选单选按钮的标识 id，如果没有勾选则返回-1
void setOnCheckedChangeListener	注册一个当该单选按钮组中的单选按钮勾选状态发生改变时所要调用的回调函数

例如，要实现文化程度选择界面的设计，可以使用下面的代码。

```
1    <RadioGroup
2        android:id = "@ + id/rgWhcd"
3        android:layout_width = "match_parent"
4        android:layout_height = "wrap_content"
5        android:orientation = "horizontal" >
6        <RadioButton
7            android:id = "@ + id/radZx"
8            android:text = "中学"
9            android:layout_width = "wrap_content"
10           android:layout_height = "wrap_content" />
11       <RadioButton
12           android:id = "@ + id/radZk"
13           android:text = "专科"
14           android:layout_width = "wrap_content"
15           android:layout_height = "wrap_content" />
16       <RadioButton
17           android:id = "@ + id/radBk"
18           android:text = "本科"
19           android:layout_width = "wrap_content"
20           android:layout_height = "wrap_content" />
21       <RadioButton
22           android:id = "@ + id/radSs"
23           android:text = "研究生"
```

第3章 基本界面组件与布局

```
24              android:layout_width ="wrap_content"
25              android:layout_height ="wrap_content" />
26      </RadioGroup >
```

上述第 5 行代码指明了按钮组中的每个按钮是水平放置的,显示效果如图 3.14 所示。当选中按钮组中的某个按钮时,其监听事件实现代码如下。

```
1   radioGroup = (RadioGroup) this.findViewById(R. id. rgWhcd);
2   radZx = (RadioButton) this.findViewById(R. id. radZx);
3   ……
4   radioGroup.setOnCheckedChangeListener(new
RadioGroup.OnCheckedChangeListener() {
5       @ Override
6       public void onCheckedChanged(RadioGroup group, int checkedId) {
7           switch (checkedId){
8             case R. id. radZx://当选中了中学及以下的按钮
9               Toast.makeText(MainActivity.this,"你选择的是" + radZx.getText
().toString(),Toast.LENGTH_LONG).show();
10              break;
11              ……
12          }
13      }
14  });
```

图 3.14 单选按钮组界面

2. CheckBox

CheckBox 可以用来实现多个选项同时选中的功能,它也是 Button 的子类,也支持使用 Button 的所有属性。

```
1   <TextView
2           android:id ="@ +id/textView"
3           android:layout_width ="match_parent"
4           android:layout_height ="wrap_content"
5           android:text ="请选择你的爱好:" />
```

【CheckBox、Spinner】

```
6       <CheckBox
7           android:id = "@ + id/chkTour"
8           android:layout_width = "match_parent"
9           android:layout_height = "wrap_content"
10          android:text = "旅游" / >
11      <CheckBox
12          android:id = "@ + id/chkRead"
13          android:layout_width = "match_parent"
14          android:layout_height = "wrap_content"
15          android:checked = "true"
16          android:text = "阅读" / >
17      <CheckBox
18          android:id = "@ + id/chkFootBall"
19          android:layout_width = "match_parent"
20          android:layout_height = "wrap_content"
21          android:text = "足球" / >
22      <Button
23          android:id = "@ + id/btnOK"
24          android:layout_width = "match_parent"
25          android:layout_height = "wrap_content"
26          android:text = "提交" / >
```

上述第 15 行代码指定阅读复选框默认为选中状态，第 22～26 行定义一个 Button 用于提交在复选框中选中的选项，并用 Toast 显示。其关键代码如下。

```
1   btnOk.setOnClickListener(new View.OnClickListener() {
2       @Override
3       public void onClick(View v) {
4           String str = "";   //存放选中的选项的值
5           if (chkRead.isChecked())
6               str + = chkRead.getText().toString() + " ";
7           if (chkTour.isChecked())
8               str + = chkTour.getText().toString() + " ";
9           if (chkFootBall.isChecked())
10              str + = chkFootBall.getText().toString() + " ";
11          Toast.makeText(CheckActivity.this, "您选择爱好为:" + str, Toast.LENGTH_LONG).show();
12      }
13  });
```

从上述第 5 行代码可以看出，要判断复选框是否选中，只需要使用 isChecked( )，如果复选框选中返回 true，否则返回 false。复选框界面显示效果如图 3.15 所示。

3. Spinner

Spinner 提供了从一个数据集合中快速选择一项值的办法。默认情况下，Spinner 显示的是当前选择的值，单击 Spinner 会弹出一个包含所有可选值的 dropdow 菜单，从该菜单中可以为 Spinner 选择一个新值。它是 ViewGroup 的间接子类，因此可以作为容器使用。Spinner 的常用属性和方法见表 3 - 16。

图 3.15 复选框界面

表 3-16　Spinner 的常用属性和方法

属性/方法名	说　　明
android:dropDownHorizontalOffset	设置列表框的水平距离
android:dropDownVerticalOffset	设置列表框的竖直距离
android:dropDownSelector	列表框被选中时的背景
android:dropDownWidth	设置下拉列表框的宽度
android:popupBackground	设置列表框的背景
android:prompt	设置对话框模式列表框的提示信息（标题），只能够引用 string.xml 中的资源 id，不能直接写字符串
android:spinnerMode	列表框的模式，有两个可选值 dialog：对话框风格的窗口； dropdown：下拉菜单风格的窗口（默认）
android:entries	使用数组资源设置该下拉列表框的列表项目
voidsetSelection(int n)	默认下拉列表框的第 1 项为选中条目，用这个方法指定第 n 项条目为选中条目
Object getSelectedItem()	返回列表框中选中项

（1）下拉列表（Spinner）的使用一般按如下步骤进行。

① 在布局文件中定义控件，代码如下。

```
<Spinner
    android:id="@+id/spinProvince"
    android:layout_height="wrap_content"
    android:layout_width="match_parent"/>
```

② 在 Activity 中引用，代码如下。

```
Spinner spinner = (Spinner)this.findViewById(R.id.spinProvince);
```

③ 创建一个适配器（ArrayAdapter）为 Spinner 提供数据，ArrayAdapter 中的数据来源有字符串数组和 XML 两种方式。

方式一：使用字符串数组作为数据来源，代码如下。

```
String[] provinces = new String[]{"江苏省","浙江省","上海市","北京市","湖南省","黑龙江省","辽宁省"};
```

设置 ArrayAdapter，代码如下。

```
ArrayAdapter<String> weekArray = new ArrayAdapter<String>(this,
android.R.layout.simple_dropdown_item_1line, provinces);
```

参数说明：ArrayAdapter 有三个参数：第 1 个参数为 Context，第 2 个参数为布局文件，第 3 个参数为数组（即显示在下拉列表中的内容，不能用 int 型数组）。此例第 2 个参数应用了 Android 自带的布局文件，也可以使用用户自定义的布局文件来控制列表每个条目的样式。

方式二：使用 XML 作为数据来源，需要使用 XML 文件将下拉菜单列出的所有内容放到 values 目录下的 strings.xml 资源文件中，strings.xml 的代码如下。

```xml
<?xml version="1.0" encoding="utf-8"?>
<resources>
    <string-array name="provinces_array">
        <item>江苏省</item>
        <item>浙江省</item>
        <item>上海市</item>
        <item>北京市</item>
        <item>湖南省</item>
        <item>黑龙江省</item>
        <item>辽宁省</item>
    </string-array>
</resources>
```

设置 ArrayAdapter，代码如下。

```
ArrayAdapter provinceArray = ArrayAdapter.createFromResource(this,
        R.array.provinces_array, android.R.layout.simple_spinner_item);
provinceArray.setDropDownViewResource(android.R.layout.simple_spinner_dropdown_item);
```

④ 将适配器与 Spinner 相关联，代码如下。

```
spinner.setAdapter(provinceArray);
```

⑤ 创建一个监听器，代码如下。

```
1    class SpinnerListener implements OnItemSelectedListener{
2        @Override
3        public void onItemSelected(AdapterView<?> arg0, View arg1, int arg2, long arg3){
4            String selected = arg0.getItemAtPosition(arg2).toString();
5            Toast.makeText(MainActivity.this,"你所在的省份是:"+selected,1).show();
6        }
7        @Override
8        public void onNothingSelected(AdapterView<?> arg0){
9        }
10    }
```

当用户选定了一个条目时，就会调用该方法。第 1 个参数指整个列表的 View 对象，第 2 个参数指被选中条目的 View 对象，第 3 个参数指被选中条目的位置，第 4 个参数指被选中条目的 id。上述第 4 行代码根据 AdapterView 中选中条目的位置返回值，这条语句也可以修改为 String selected = spinner.getSelectedItem()。

⑥ 绑定监听器，代码如下。

```
spinner.setOnItemSelectedListener(new SpinnerListener());
```

说明：在使用 Spinner 时如果已经可以确定下拉列表框里的列表项，则完全不需要编写代码,也就不需要设置 ArrayAdapter 适配器,而直接使用 android:entries 属性来设置数组资源作为下拉列表框的列表项目,即可以省略上述步骤(3)和(4),而将布局文件的代码更改如下。

```
<Spinner
    android:id="@+id/spinner"
    android:layout_height="wrap_content"
    android:layout_width="match_parent"
    android:entries="@array/provinces_array"/>
```

其中,provinces_array 数组是在 strings.xml 资源文件中定义的。使用以上步骤实现的省份选择效果如图 3.16 所示。

(2)如果要在该布局实现每个条目左侧用 ImageView 显示一个图片,右侧用 TextView 显示省份名,如图 3.17 所示,其实现步骤如下。

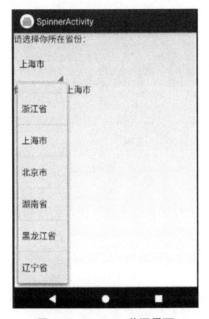

图 3.16　Spinner 普通界面

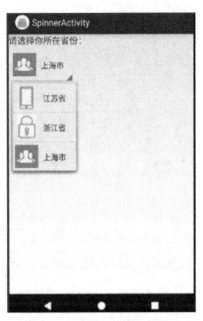

图 3.17　带图的 Spinner 界面

① 在 layout 目录下创建 province.xml 布局文件,代码如下。

```
1    <?xml version="1.0" encoding="utf-8"?>
2    <LinearLayout
xmlns:android="http://schemas.android.com/apk/res/android"
3        android:layout_width="match_parent"
4        android:layout_height="match_parent"
5        android:orientation="horizontal"
6        android:padding="5dp">
7        <ImageView
8            android:id="@+id/img_icon"
9            android:layout_width="48dp"
```

```
10          android:layout_height="48dp"
11          android:src="@mipmap/circle"/>
12      <TextView
13          android:id="@+id/txt_name"
14          android:layout_width="wrap_content"
15          android:layout_height="wrap_content"
16          android:layout_marginLeft="10dp"
17          android:layout_marginTop="15dp"
18          android:text="江苏省"
19          android:textSize="16sp"/>
20  </LinearLayout>
```

② 在java/包名目录下创建 Province.java，用于封装每个条目对象，即 ImageView 和 TextView，主要代码如下。

```
1   public class Province {
2       private int hIcon;
3       private String hName;
4       public Province() { }
5       public Province(int hIcon, String hName) {
6           this.hIcon = hIcon;
7           this.hName = hName;
8       }
9       public int gethIcon() {         return hIcon;       }
10      public String gethName() {      return hName;       }
11      public void sethIcon(int hIcon) {       this.hIcon = hIcon;     }
12      public void sethName(String hName) {        this.hName = hName;     }
13  }
```

③ 在java/包名目录下创建 MyAdapter.java，用于自定义适配器以便封装数据源，代码如下。

```
1   public class MyAdapter extends BaseAdapter {
2       private List<Province> mList;
3       private Context mContext;
4       public MyAdapter(Context context, List<Province> pList) {
5           this.mContext = context;
6           this.mList = pList;
7       }
8       @Override
9       public int getCount() {         return mList.size();        }
10      @Override
11      public Object getItem(int position) {return mList.get(position); }
12      @Override
13      public long getItemId(int position) {       return position;        }
14      @Override
15      public View getView(int position, View convertView, ViewGroup parent) {
16          LayoutInflater _LayoutInflater = LayoutInflater.from(mContext);
17          //将 province.xml 转换为 View
18          convertView = _LayoutInflater.inflate(R.layout.province, null);
19          if (convertView != = null) {
```

```
20            ImageView imageView = (ImageView) convertView.findViewById
(R.id.img_icon);
21            //从List中取出对应的图片作为ImageView中加载的图片
22            imageView.setImageResource(mList.get(position).gethIcon());
23            TextView  textView= (TextView) convertView.findViewById(R.id.txt_pro);
24            //从List中取出对应的省份作为TextView中加载的省份信息
25            textView.setText(mList.get(position).gethName());
26         }
27         return convertView;
28      }
29   }
```

④ 在 java/包名目录下 SpinnerActivity.java 的主调用模块用如下代码实现。

```
1        ArrayList<Province> mData = new ArrayList<Province>();
2        mData.add(new Province(R.mipmap.mobile, "江苏省"));
3        mData.add(new Province(R.mipmap.lock, "浙江省"));
4        mData.add(new Province(R.mipmap.people, "上海市"));
5        MyAdapter myAdapter = new MyAdapter(this,mData);
6        spinner.setAdapter(myAdapter);
```

4. RatingBar

RatingBar 即评分条，用星形来表示评分等级，设置时默认可以选择三种不同的风格，每种风格都可以通过 isIndicator 属性设置是否可以与用户交互。RatingBar 的常用属性和方法见表 3-17。

表 3-17 RatingBar 的常用属性和方法

属性/方法名	说　明
android:isIndicator	设置是否可以与用户交互（默认 true 表示不可与用户交互）
android:rating	默认的评分数
android:stepSize	单击一次增长的长度
android:numStars	表示星星的数量，超出显示范围时会以最大数量显示，然后把星星分成 numStars/stepSize 份
android:style	设置风格，可以为 "?android:ratingBarStyleSmall" 及 "?android:ratingBarStyleIndicator" 或 "?android:ratingBarStyle"
void setIsIndicator(boolean)	设置是否为指示器模式及不可交互
void setMax（int）	设置评分条的最大范围
void setNumStars（int）	设置星星数量（只有当布局的宽被设置为 wrap_content 时，设置的星星数量才能生效）
void setRating（float）	设置当前等级
void setStepSize（float）	设置步长
intget NumStars()	获取星星数量
floatget StepSize()	获取步长
floatget Rating()	获取当前评分，与参数 rating 一致

Android 自带的默认风格效果如图 3.18 所示，布局关键代码如下。

```
1      <TextView
2          android:layout_width ="wrap_content"
3          android:layout_height ="wrap_content"
4          android:text ="请给本软件打分:"
5          android:textSize ="20dp" />
6      <RatingBar
7          android:id ="@ +id/ratingBar1"
8          style ="? android:ratingBarStyle"
9          android:layout_width ="wrap_content"
10         android:layout_height ="wrap_content"
11         android:isIndicator ="false"
12         android:numStars ="5"
13         android:rating ="1.0"
14         android:stepSize ="1.0" />
15     <RatingBar
16         android:id ="@ +id/ratingBar4"
17         style ="? android:ratingBarStyleSmall"
18         android:layout_width ="wrap_content"
19         android:layout_height ="wrap_content"
20         android:isIndicator ="false"
21         android:numStars ="5"
22         android:rating ="1.0"
23         android:stepSize ="1.0" />
24     <RatingBar
25         android:id ="@ +id/ratingBar3"
26         style ="? android:ratingBarStyleIndicator"
27         android:layout_width ="wrap_content"
28         android:layout_height ="wrap_content"
29         android:numStars ="5"
30         android:rating ="1.0" />
31     <TextView
32         android:id ="@ +id/tvinfo"
33         android:layout_width ="wrap_content"
34         android:layout_height ="wrap_content"
35         android:text ="你的评分结果为:"
36         android:textSize ="20dp" />
37  </LinearLayout >
```

上述第 11 行代码将 RatingBar 的 isIndicator 属性设置为 false 后，就可以通过单击星形来改变评分值。如果没有这一行代码，其默认值为 true，就没有办法实现此功能。RatingBar 还有一个在星级值改变时的回调方法——RatingBar.OnRatingBarChangeListener( )，在星级进度改变时触发事件。其实现代码如下。

```
1  ratingBar.setOnRatingBarChangeListener(new RatingBar.OnRatingBarChangeListener() {
2      @Override
3      public void onRatingChanged(RatingBar ratingBar, float rating, boolean fromUser) {
4          tvInfo.setText("你的评分结果为:" +ratingBar.getRating() +"");//获取当前评分
5      }
6  });
```

上述第 4 行代码用 getRating()方法获得 RatingBar 上的星级评分值,并显示在 TexView 上。

除了系统自带的星形评分条图标外,还可以使用用户自定义图标。例如图 3.19 所示带图的 Sprinner 界面,其实现步骤如下。

图 3.18 Spinner 普通界面

图 3.19 带图的 Spinner 界面

(1) 在 res/drawable 目录中导入两张分别代表选中和未选中的图片 star1.png、star2.png。

(2) 在 drawable 里创建 myratingbar.xml 文件。

```
1    <?xml version ="1.0" encoding ="utf-8"? >
2    <layer-list xmlns:android ="http://schemas.android.com/apk/res/android" >
3        <!-- 定义未选中图片作为背景 -->
4        <item
5            android:id ="@android:id/background"
6            android:drawable ="@drawable/star2" >
7        </item>
8        <item
9            android:id ="@android:id/secondaryProgress"
10           android:drawable ="@drawable/star2" >
11       </item>
12       <item
13           android:id ="@android:id/progress"
14           android:drawable ="@drawable/star1" >
15       </item>
16   </layer-list>
```

(3) 打开 res/values 目录下的 styles.xml 文件,在文件中添加以下代码用于定义 RatingBar 的 style。

```
1    <style name ="RadingStyle" parent ="@android:style/Widget.RatingBar" >
2        <!-- 定义星星图片 -->
3        <item name ="android:progressDrawable" >@drawable/myratingbar </item>
4        <!-- 根据自定义星星图片的大小,设置相应的值,否则可能显示不全 -->
5        <item name ="android:numColumns" >5 </item>
6    </style>
```

(4) 在布局文件中应用自定义 RatingBar 的 style。

```
1    <RatingBar
2        style ="@ style/RadingStyle"
3        android:layout_width ="wrap_content"
4        android:layout_height ="wrap_content"
5        android:clickable ="true"
6        android:isIndicator ="false"
7        android:maxHeight ="45dp"
8        android:rating ="3"
9        android:stepSize ="0.5" />
```

3.4.2 注册界面的实现

1. 主界面的设计

【注册界面的实现】

（1）由于图 3.13 所示的注册界面上需要输入的元素较多，在移动终端屏幕上显示时需要上下滚动才能将界面上的内容显示完整，所以在界面设计时需要使用 ScrollView 方式进行布局。关于 ScrollView 的使用方法后面章节会详细介绍。从图 3.13 中可以看出，在实际实现注册时还需要以下详细功能。

① 需要使用跑马灯效果显示"欢迎注册南京师范大学泰州学院智慧校园平台，注册成功后，你可以方便获得校园内资讯！"。

② 用户名长度不能低于四个字符，若低于四个字符，在用户名编辑框失去焦点时给出提示。

③ 登录密码长度为 6～12 位，如果不满足，在密码编辑框失去焦点时给出提示，并且清空密码编辑框中内容；登录密码与确认密码不一样不能注册。

④ 出生地使用 string-array 实现，省份和城市不要求级联。

⑤ 爱好至少选一项才能注册。

⑥ 在满足上面要求的情况下，注册才能成功。

（2）根据注册界面显示效果分析，整个界面要实现滚动显示就需要使用 ScrollView 布局方式。显示的界面可以分解成顶层、跑马灯层、用户名层、密码层、确认密码层、性别选择层、出生地层、兴趣爱好层、自我评价层和立即注册层等部分。它们以垂直线性布局方式排布在界面上，其代码如下。

```
1    <ScrollView xmlns:android ="http://schemas.android.com/apk/res/android"
2        android:layout_width ="match_parent"
3        android:layout_height ="match_parent"
4        android:fillViewport ="true"
5        android:scrollbars ="none" >
6        <LinearLayout
7            android:layout_width ="match_parent"
8            android:layout_height ="wrap_content"
9            android:orientation ="vertical" >
10           <!--顶层 -->
11           <!--跑马灯 -->
12           <!--用户名层 -->
```

```
13              <!--密码层 -->
14              <!--确认密码层 -->
15              <!--性别选择层 -->
16              <!--出生地层 -->
17              <!--兴趣爱好层 -->
18              <!--自我评价层 -->
19              <!--立即注册层 -->
20         </LinearLayout >
21    </ ScrollView >
```

① 顶层布局。

```
1     <RelativeLayout
2         android:layout_width ="match_parent"
3         android:layout_height ="wrap_content"
4         android:background ="#33B5E4" >
5         <Button
6             android:id ="@ +id/btn_back"
7             android:layout_width ="wrap_content"
8             android:layout_height ="wrap_content"
9             android:layout_centerVertical ="true"
10            android:layout_margin ="10dp"
11            android:background ="@ null"
12            android:padding ="10dp"
13            android:text ="返回"
14            android:textColor ="#ffffff" />
15        <TextView
16            android:layout_width ="wrap_content"
17            android:layout_height ="wrap_content"
18            android:layout_centerInParent ="true"
19            android:text ="用户登录"
20            android:textColor ="#ffffff"
21            android:textSize ="20sp" />
22        <Button
23            android:id ="@ +id/btn_register"
24            android:layout_width ="wrap_content"
25            android:layout_height ="wrap_content"
26            android:layout_alignParentRight ="true"
27            android:layout_centerVertical ="true"
28            android:layout_margin ="10dp"
29            android:background ="@ null"
30            android:padding ="10dp"
31            android:text ="注册"
32            android:textColor ="#ffffff" />
33    </RelativeLayout >
```

③ 跑马灯层布局。

```
1     <TextView
2         android:layout_width ="fill_parent"
3         android:layout_height ="wrap_content"
```

```
4         android:background ="#55333333"
5         android:ellipsize ="marquee"
6         android:focusable ="true"
7         android:focusableInTouchMode ="true"
8         android:marqueeRepeatLimit ="marquee_forever"
9         android:paddingLeft ="10dp"
10        android:paddingRight ="10dp"
11        android:singleLine ="true"
12         android:text ="欢迎注册南京师范大学泰州学院智慧校园平台,注册成功后,你可以方便获得校园内资讯!"
13        android:textColor ="#FFFF00"
14        android:textSize ="20dp" />
```

③ 用户名层布局。

```
1     <LinearLayout
2         android:layout_width ="match_parent"
3         android:layout_height ="match_parent"
4         android:layout_marginTop ="10dp"
5         android:orientation ="horizontal" >
6         <TextView
7             android:id ="@ +id/txt_username"
8             android:layout_width ="wrap_content"
9             android:layout_height ="wrap_content"
10            android:layout_marginLeft ="10dp"
11            android:text ="用户名:"
12            android:textColor ="#000000"
13            android:textSize ="20dp" />
14        <EditText
15            android:id ="@ +id/edt_name"
16            android:layout_width ="fill_parent"
17            android:layout_height ="wrap_content"
18            android:layout_marginLeft ="20dp"
19            android:layout_marginRight ="10dp"
20            android: maxLines ="1"
21            android:hint ="请输入用户名" >
22        </EditText>
23    </LinearLayout>
```

④ 密码层布局。

```
1     <LinearLayout
2         android:layout_width ="fill_parent"
3         android:layout_height ="wrap_content"
4         android:orientation ="horizontal" >
5         <TextView
6             android:layout_width ="wrap_content"
7             android:layout_height ="wrap_content"
8             android:layout_marginLeft ="10dp"
9             android:text ="输入密码:"
10            android:textColor ="#000000"
```

```
11              android:textSize="20dp" />
12          <EditText
13              android:id="@+id/edt_pass"
14              android:layout_width="fill_parent"
15              android:layout_height="wrap_content"
16              android:layout_marginRight="10dp"
17              android:hint="请输入密码6-12位"
18              android:inputType="textPassword"
19              android:maxLines="1"
20              android:text="" >
21          </EditText>
22      </LinearLayout>
```

确认密码层布局代码与密码层布局代码几乎一样,限于篇幅不再赘述,读者可以参见代码包中 RegisterUI 文件夹里的 activity_register.xml 文件。

⑤ 性别选择层布局。

```
1       <LinearLayout
2           android:layout_width="fill_parent"
3           android:layout_height="wrap_content"
4           android:gravity="center_vertical"
5           android:orientation="horizontal" >
6           <TextView
7               android:layout_width="wrap_content"
8               android:layout_height="wrap_content"
9               android:layout_marginLeft="10dp"
10              android:text="性别:"
11              android:textColor="#000000"
12              android:textSize="20dp" />
13          <RadioGroup
14              android:id="@+id/rg_sex"
15              android:layout_width="fill_parent"
16              android:layout_height="wrap_content"
17              android:checkedButton="@+id/aur_rb_boy"
18              android:orientation="horizontal" >
19              <RadioButton
20                  android:id="@+id/rb_boy"
21                  android:layout_width="wrap_content"
22                  android:layout_height="wrap_content"
23                  android:layout_marginLeft="25dp"
24                  android:text="男"
25                  android:textColor="#000000" />
26              <RadioButton
27                  android:id="@+id/rb_gril"
28                  android:layout_width="wrap_content"
29                  android:layout_height="wrap_content"
30                  android:layout_marginLeft="50dp"
31                  android:text="女"
```

```
32              android:textColor ="#000000" />
33          </RadioGroup>
34      </LinearLayout>
```

⑥ 出生地层布局。

由于出生地包含了省份和地区，所以需要使用两个Spinner组件，本案例中为Spinner装载的数据源用string-array方式实现，所以需要在strings.xml文件中添加如下数组定义代码。

```
1   <string-array name ="provinces_array">
2       <item>江苏省</item>
3       ……
4   </string-array>
5   <string-array name ="citys_array">
6       <item>泰州市</item>
7       ……
8   </string-array>
```

出生地层布局代码如下。

```
1   <LinearLayout
2       android:layout_width ="fill_parent"
3       android:layout_height ="wrap_content"
4       android:gravity ="center_vertical"
5       android:orientation ="horizontal">
6       <TextView
7           android:id ="@ +id/tv_address"
8           android:layout_width ="wrap_content"
9           android:layout_height ="wrap_content"
10          android:layout_marginLeft ="10dp"
11          android:text ="出生地:"
12          android:textColor ="#000000"
13          android:textSize ="20dp" />
14      <Spinner
15          android:id ="@ +id/s_provinces"
16          android:layout_width ="wrap_content"
17          android:layout_height ="40dp"
18          android:layout_weight ="1"
19          android:entries ="@array/provinces_array" />
20      <Spinner
21          android:id ="@ +id/s_citys"
22          android:layout_width ="wrap_content"
23          android:layout_height ="40dp"
24          android:layout_weight ="1"
25          android:entries ="@array/citys_array" />
26      </LinearLayout>
```

⑦ 兴趣爱好层布局。

```
1   <TextView
2       android:layout_width ="wrap_content"
```

3	android:layout_height ="wrap_content"
4	android:layout_marginLeft ="10dp"
5	android:text ="我的爱好:"
6	android:textColor ="#000000"
7	android:textSize ="20dp" />
8	<LinearLayout
9	android:layout_width ="fill_parent"
10	android:layout_height ="wrap_content"
11	android:layout_marginLeft ="15dp"
12	android:orientation ="horizontal" >
13	<CheckBox
14	android:id ="@ +id/cb_reading"
15	android:layout_width ="wrap_content"
16	android:layout_height ="wrap_content"
17	android:text ="读书"
18	android:textColor ="#000000" />
19	<!--CheckBox 定义类似,此处省略 -->
20	</LinearLayout >
21	<!--下一行代码类似,此处省略 -->

⑧ 自我评价层布局。

1	<LinearLayout
2	android:layout_width ="match_parent"
3	android:layout_height ="wrap_content"
4	android:layout_margin ="10dp"
5	android:orientation ="vertical"
6	android:padding ="8dp" >
7	<TextView
8	android:layout_width ="wrap_content"
9	android:layout_height ="wrap_content"
10	android:layout_marginLeft ="2dp"
11	android:text ="自我评价:"
12	android:textColor ="#000000"
13	android:textSize ="20dp" />
14	<LinearLayout
15	android:layout_width ="wrap_content"
16	android:layout_height ="wrap_content"
17	android:gravity ="center_vertical"
18	android:orientation ="horizontal" >
19	<TextView
20	android:layout_width ="wrap_content"
21	android:layout_height ="wrap_content"
22	android:layout_marginLeft ="2dp"
23	android:text ="勤劳"
24	android:textColor ="#000000"
25	android:textSize ="18sp" />
26	<RatingBar
27	android:id ="@ +id/rb_qinlao"
28	android:layout_width ="wrap_content"

```
29                  android:layout_height ="wrap_content"
30                  android:layout_marginLeft ="2dp"
31                  android:numStars ="5"
32                  android:rating ="2"
33                  android:stepSize ="0.5" />
34              </LinearLayout>
35              <!--其他评价定义类似,此处省略 -->
36          </LinearLayout>
```

最后的立即注册层只要单独使用 Button 组件就可以了,限于篇幅不再赘述,读者可以参见代码包中 RegisterUI 文件夹里的 activity_register.xml 文件。

2. 注册界面的功能实现

(1) 用户名输入框失去焦点的监听事件。

```
1   loginNameET.setOnFocusChangeListener(new View.OnFocusChangeListener() {
2       @Override
3       public void onFocusChange(View v, boolean hasFocus) {
4           //用户名编辑框失去焦点时,检测是否大于4位
5           if (! hasFocus) {
6               if (loginNameET.getText().length() <4) {
7                   Toast.makeText(RegisterActivity.this, "用户名不能低于4位", Toast.LENGTH_SHORT).show();
8               }
9           }
10
11      }
12  });
```

(2) 确认密码输入框失去焦点的监听事件。

```
1   qpassET.setOnFocusChangeListener(new View.OnFocusChangeListener() {
2       @Override
3       public void onFocusChange(View v, boolean hasFocus) {
4           if (! hasFocus) {
5               String pass = qpassET.getText().toString().trim();
6
7               if (pass.length() <6) {
8                   Toast.makeText(RegisterActivity.this, "确认密码不能低于6位", Toast.LENGTH_SHORT).show();
9                   return;
10              }
11
12              if (pass.length() > 12) {
13                  Toast.makeText(RegisterActivity.this, "确认密码不能大于12位", Toast.LENGTH_SHORT).show();
14                  return;
15              }
16
17              if (! pass.equals(passET.getText().toString().trim())) {
```

```
18                    Toast.makeText(RegisterActivity.this,"两次密码不一样",
Toast.LENGTH_SHORT).show();
19                    return;
20                }
21
22            }
23        }
24    });
```

密码输入框中的判断逻辑与确认密码输入框一样，限于篇幅不再赘述，读者可以参见代码包中 RegisterUI 文件夹里的 Register Activity.java 文件。

（3）性别选择的监听事件。

```
1    sexRG.setOnCheckedChangeListener(new RadioGroup.OnCheckedChangeListener() {
2        @Override
3        public void onCheckedChanged(RadioGroup group, int checkedId) {
4            if (checkedId == R.id.aur_rb_boy) {
5                sex = "男";
6                Toast.makeText(RegisterActivity.this,"选择了男",Toast.LENGTH_SHORT).show();
7            } else if (checkedId == R.id.aur_rb_gril) {
8                sex = "女";
9                Toast.makeText(RegisterActivity.this,"选择了女",Toast.LENGTH_SHORT).show();
10           }
11       }
12   });
```

（4）爱好复选框的监听事件。

```
1    //游泳复选框监听事件
2    swimCB.setOnCheckedChangeListener(new CompoundButton.OnCheckedChangeListener() {
3        @Override
4        public void onCheckedChanged(CompoundButton button, boolean isChecked) {
5            if (swimCB.isChecked()) {
6                hobby = hobby + swimCB.getText().toString() + "、";
7            } else {
8                hobby = hobby.replaceAll(swimCB.getText().toString() + "、", "");
9            }
10       }
11   });
```

其他复选框监听事件与游泳复选框类似，限于篇幅不再赘述，读者可以参见代码包中 RegisterUI 文件夹里的 Register Activity.java 文件。

（5）自我评价监听事件。

```
1    //勤劳监听事件
2    qinlaoRB.setOnRatingBarChangeListener(new RatingBar.OnRatingBarChangeListener() {
3        @Override
```

```
    4            public void onRatingChanged(RatingBar ratingBar, float rating,
boolean fromUser) {
    5                qinlao = rating + "颗星";
    6            }
    7        });
```

其他评价监听事件与勤劳监听事件类似，限于篇幅不再赘述，读者可以参见代码包中 RegisterUI 文件夹里的 Register Activity.java 文件。

（6）立即注册监听事件。

```
 1    registerBN.setOnClickListener(new View.OnClickListener() {
 2        @Override
 3        public void onClick(View v) {
 4            String name = loginNameET.getText().toString();
 5            if (name.length() <4) {
 6                Toast.makeText(RegisterActivity.this, "用户名不能低于4位", Toast.LENGTH_SHORT).show();
 7                return;
 8            }
 9            //密码及密码确认、密码相等、省份、城市与上述代码类似,此处省略
10            //性别过滤
11            if (TextUtils.isEmpty(sex)) {
12                Toast.makeText(RegisterActivity.this, "请选择性别", Toast.LENGTH_SHORT).show();
13                return;
14            }
15            //出生地过滤
16            if (TextUtils.isEmpty(province)) {
17                Toast.makeText(RegisterActivity.this, "请选择出生地", Toast.LENGTH_SHORT).show();
18                return;
19            }
20            //爱好过滤
21            if (hobby.length() <= 0) {
22                Toast.makeText(RegisterActivity.this, "至少勾选一个爱好", Toast.LENGTH_SHORT).show();
23                return;
24            }
25            //自我评价过滤
26            if ("0.0 颗星".equals(qinlao) && "0.0 颗星".equals(shangjing) && "0.0 颗星".equals(yonggan)) {
27                Toast.makeText(RegisterActivity.this, "至少有一个自我评价", Toast.LENGTH_SHORT).show();
28                return;
29            }
30        }
31    });
32
```

限于篇幅，完整的布局代码和功能代码读者可以参阅 RegisterUI 文件夹中的内容。

3.5 考试系统界面的设计与实现

一个 App 可能有多个选项卡页面。例如,Android 自带的通讯录 App,打开后会出现"收藏"和"全部"两个选项卡标签。"收藏"用于显示保存收藏的联系人,"全部"用于显示通讯录中的全部联系人。用户可以单击"收藏"和"全部"切换定义的标签页面内容。如果要实现多个标签页显示内容,并能通过单击标签切换标签页内容,可以使用 Android 提供的 FrameLayout 帧布局管理器和 TabHost/TabWidget 组件。本节介绍用这两个组件设计一个考试系统界面的方法,以实现考试系统界面上的单选题、多选题、填空题和判断题等不同的题型页面之间的切换,如图 3.20 所示。

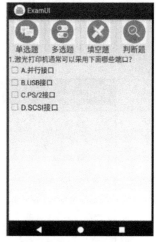

图 3.20 考试系统界面

3.5.1 预备知识

1. FrameLayout

FrameLayout 继承自 ViewGroup,使用 FrameLayout 布局管理器时,整个界面被当成一块空白备用区域,所有的子元素都不能被指定放置的位置,它们全部放在这块区域的左上角,并且后面的子元素直接覆盖在前面的子元素之上,将前面的子元素部分或全部遮挡,即帧布局的大小由子元素中尺寸最大的那个子元素来决定。如果子元素一样大,则同一时刻只能看到最上面的子元素。下面的代码产生的效果如图 3.21 和图 3.22 所示。

【FrameLayout、TabHost】

```
1   <?xml version ="1.0" encoding ="utf-8"?>
2   <FrameLayout xmlns:android ="http://schemas.android.com/apk/res/android"
3       android:orientation ="vertical"
4       android:layout_width ="match_parent"
5       android:layout_height ="match_parent" >
6       <TextView
7           android:layout_width ="match_parent"
8           android:layout_height ="match_parent"
9           android:background ="#ff000000"
10          android:gravity ="center"
11          android:text ="1" />
12      <TextView
13          android:layout_width ="match_parent"
14          android:layout_height ="match_parent"
15          android:background ="#d386d7"
16          android:gravity ="center"
17          android:text ="2" />
```

```
18      <TextView
19          android:layout_width ="match_parent "
20          android:layout_height ="match_parent "
21          android:background ="#fffedcba"
22          android:gravity ="center"
23          android:text ="3" />
24  </FrameLayout >
```

图 3.21　帧布局显示效果(1)

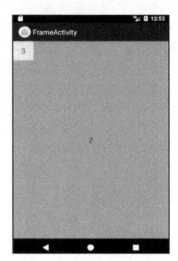
图 3.22　帧布局显示效果(2)

上述代码的 6～11 行、12～17 行、18～23 行分别定义了 TextView 组件。三个 TextView 组件的背景色设置为不同的颜色，组件的 text 属性值分别为 1、2、3，其他属性值相同。观察图 3.21，可能认为只有 18～23 行设置的代码显示在效果图上。其实，前两个 TextView 也显示在了效果图上，只是设置 TextView 的先后顺序造成 6～11 行设置的 TextView 组件被 12～17 行设置的 TextView 组件遮挡住，12～17 行设置的 TextView 组件被 18～23 行设置的 TextView 组件遮挡住。此时将 19～20 行的代码改修改如下。

```
android:layout_width ="50dp"
android:layout_height ="50dp"
```

修改后的代码产生的效果如图 3.22 所示。因为此时第 3 个 TextView 组件的宽度和高度分别设置为 50dp，它的宽度和高度明显比第 2 个 TextView 组件的宽度和高度小，而且组件在帧布局显示时，一定从左上角开始，所以即使第 3 个 TextView 对第 2 个 TextView 有遮挡，但由于宽度和高度较小，所以不能全部遮挡，最终出现了如图 3.22 所示的效果。

使用 FrameLayout 布局管理器时，可以根据用户的需要设计一些特殊效果的用户界面。例如，图 3.23 所示的霓虹灯效果，可以使用以下步骤实现。

（1）修改 values 目录下的 colors.xml 颜色配置文件。

霓虹灯需要有七种颜色的配置文件，其代码如下。

```
<?xml version ="1.0" encoding ="utf-8"? >
<resources >
```

```xml
    <color name="color1">#ffff00</color>
    <color name="color2">#ff00ff</color>
    <color name="color3">#00ffff</color>
    <color name="color4">#0ffff0</color>
    <color name="color5">#326864</color>
    <color name="color6">#00ff00</color>
    <color name="color7">#ff0000</color>
</resources>
```

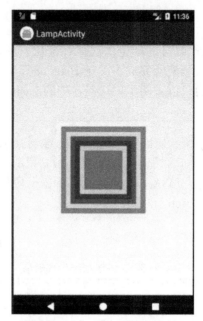

图 3.23 霓虹灯效果图

(2) 用户界面布局文件的设计。

为了达到霓虹灯的效果,本案例项目中用七个 TextView 分别表示七个不同的霓虹灯管,七个 TextView 的宽度和高度依次递减 20dp 并设置为不同的背景色。关键代码如下。

```xml
<?xml version="1.0" encoding="utf-8"?>
<FrameLayout xmlns:android="http://schemas.android.com/apk/res/android"
    android:layout_width="match_parent"
    android:layout_height="match_parent">
    <TextView
        android:id="@+id/view1"
        android:layout_width="200dp"
        android:layout_height="200dp"
        android:layout_gravity="center"
        android:background="@color/color1" />
    <TextView
        android:id="@+id/view2"
        android:layout_width="180dp"
        android:layout_height="180dp"
```

```
            android:layout_gravity="center"
            android:background="@color/color2"/>
        <!--其他TextView组件的设置与此处代码类似-->
        ...
    <TextView
        android:id="@+id/view7"
        android:layout_width="80dp"
        android:layout_height="80dp"
        android:layout_gravity="center"
        android:background="#ff0000"/>
</FrameLayout>
```

如果要让图 3.23 所示的霓虹灯的颜色依次循环变化，从而产生一种动态的效果，则需要使用消息处理机制和子线程配合完成。关于 Handler 和 Thread 的内容在 3.6 节中详细介绍，读者可以参照实现，也可以查阅示例代码包中的 LampActivity.java 文件。

2. TabHost

TabHost 继承自 FrameLayout，是带 Tab 选项卡的容器，包含 TabWidget 和 FrameLayout 两个部分。TabWidget 是每个 Tab 选项卡标签按钮，FrameLayout 是每个 Tab 选项卡的内容。Android App 开发中提供了继承 TabActivity 和继承 Activity 两种实现方法，其中 TabActivity 已经过时，本节主要讲述继承 Activity 的使用步骤。

（1）定义布局文件。

Tab 选项卡的用户界面布局文件的设计需要遵循表 3-18 的要求进行定义。

表 3-18 Tab 选项卡定义规范

属性/方法名	说　　明
TabHost	可自定义 id
TabWidget	必须设置 android:id 为 @android:id/tabs
FrameLayout	必须设置 android:id 为 @android:id/tabcontent

详细代码如下。

```
1   <?xml version="1.0" encoding="utf-8"?>
2   <TabHost xmlns:android="http://schemas.android.com/apk/res/android"
3       android:id="@+id/myTabHost"
4       android:layout_width="match_parent"
5       android:layout_height="match_parent">
6       <LinearLayout
7           android:layout_width="match_parent"
8           android:layout_height="match_parent"
9           android:orientation="vertical">
10          <TabWidget
11              android:id="@android:id/tabs"
12              android:layout_width="match_parent"
13              android:layout_height="wrap_content"/>
14          </TabWidget>
15          <FrameLayout
```

```
16        android:id ="@ android:id/tabcontent"
17        android:layout_width ="match_parent"
18        android:layout_height ="match_parent" >
19     <!--红色背景 Tab 内容 -->
20      <LinearLayout
21          android:id ="@ +id/layout_red"
22          android:layout_width ="match_parent"
23          android:layout_height ="match_parent"
24          android:background ="#ff0000"
25          android:orientation ="vertical" >
26          <TextView
27              android:layout_width ="wrap_content"
28              android:layout_height ="fill_parent"
29              android:textSize ="20dp"
30              android:text ="红色"/ >
31      </LinearLayout >
32      <!--黄色背景 Tab 内容 -->
33      <LinearLayout
34        android:id ="@ +id/layout_yellow"
35        android:layout_width ="match_parent"
36        android:layout_height ="match_parent"
37        android:background ="#FCD209"
38        android:orientation ="vertical" >
39        <TextView
40            android:layout_width ="wrap_content"
41            android:layout_height ="fill_parent"
42            android:textSize ="20dp"
43            android:text ="黄色"/ >
44        </LinearLayout >
45     </FrameLayout >
46   </LinearLayout >
47 </TabHost >
```

上述第 20 ~ 31 行代码定义"红色"选项卡显示的内容,第 33 ~ 44 行代码定义"黄色"选项卡显示的内容。TabWidget 组件定义的选项标签按钮可以放在用户界面的上部或底部,本例中是放在用户界面的上部(图 3.24),也可以使用相对布局方式将它定义在用户界面的底部(图 3.25)。如果要将选项卡标签放在用户界面的底部,只需要修改上述第 6 行代码,将其布局方式修改为 RelativeLayout,然后将 TabWidget 组件代码修改如下。

```
1   <TabWidget
2       android:layout_alignParentBottom ="true"
3       android:id ="@ android:id/tabs"
4       android:layout_width ="match_parent"
5       android:layout_height ="wrap_content" >
6   </TabWidget >
```

(2) 功能代码。

当用户单击"红色"选项卡时,显示 layout_red 布局中定义的内容;当用户单击"黄色"选项卡时,显示 layout_yellow 布局中定义的内容。其实现代码如下。

```
1    tabHost = (TabHost) this.findViewById(R.id.myTabHost);
2    tabHost.setup();
3    //设置选项卡标签显示内容及选项卡上显示的内容
4    TabHost.TabSpec tabSpecRed = tabHost.newTabSpec("tabred").setIndicator
("红色").setContent(R.id.layout_red);
5    tabHost.addTab(tabSpecRed);
6    TabHost.TabSpec tabSpecYellow = tabHost.newTabSpec("tabyellow").
setIndicator("黄色").setContent(R.id.layout_yellow);
7    tabHost.addTab(tabSpecYellow);
```

图 3.24　标签在上部的显示效果　　　　图 3.25　标签在底部的显示效果

上述第 4 行代码定义了 TabSpec，用来显示选项卡标签及单击选项卡后显示的内容。TabSpec 的常用方法和功能说明见表 3-19。

表 3-19　TabSpec 的常用方法和功能说明

属性/方法名	功 能 说 明
setIndicator(CharSequence label)	用于指定选项卡显示的标签名
setIndicator(CharSequence label, Drawable icon)	用于指定选项卡显示的标签名及图片
setIndicator(View view)	用于指定选项卡显示的标签 View
setContent(int viewId)	用于指定选项卡显示的内容，通常在布局文件中定义
setContent(Intent intent)	用于指定选项卡显示的内容，通常 Intent 用来封装启动另一个 Activity
setCurrentTab(int index)	用于指定默认选项卡

3.5.2　考试系统界面的实现

1. 主界面的设计

图 3.20 所示考试系统界面的选项卡上既显示了图片，又显示了标签文字，虽然 TabSpec 提供了 setIndicator( ) 重载方法用于既能设置图

【考试界面的实现】

片,又能设置标签文字,但是如果使用".setIndicator("多选题",this.getResources().getDrawable(R.mipmap.multi))"语句进行设置的话,根本没有办法实现如图3.20所示的效果。为了解决这个问题,需要使用下列步骤实现。

(1) 创建选项卡布局文件。

在src/main/res/layout目录下创建选项卡布局文件examtab.xml,其详细代码如下。

```
1    <?xml version="1.0" encoding="utf-8"?>
2    <LinearLayout xmlns:android="http://schemas.android.com/apk/res/android"
3        android:layout_width="match_parent"
4        android:layout_height="match_parent"
5        android:gravity="center"
6        android:orientation="vertical">
7        <ImageView
8            android:id="@+id/tabimg"
9            android:layout_width="wrap_content"
10           android:layout_height="wrap_content"  />
11       <TextView
12           android:id="@+id/tabtxtinfo"
13           android:layout_width="wrap_content"
14           android:layout_height="wrap_content"
15           android:textSize="20dp" />
16   </LinearLayout>
```

上述第7~10行代码用于显示选项卡上的图片,第11~15行代码用于显示选项卡上的标签。

(2) 创建选项卡设置需要的View对象。

在考试系统主界面对应的MainActivity.java文件中定义一个方法,用于实现TabSpec.setIndicator(View view)方法需要的View对象,其详细代码如下。

```
1    public View getMenuItem(int imgID, String textID) {
2        LinearLayout ll = (LinearLayout) LayoutInflater.from(this).inflate(R.layout.examtab, null);
3        ImageView imgView = (ImageView) ll.findViewById(R.id.tabimg);
4        imgView.setBackgroundResource(imgID);
5        TextView textView = (TextView) ll.findViewById(R.id.tabtxtinfo);
6        textView.setText(textID);
7        return ll;
8    }
```

(3) 创建考试系统界面的布局文件。

在src/main/res/layout目录下创建考试系统界面的布局文件activity_main.xml,其详细代码如下。

```
1    <?xml version="1.0" encoding="utf-8"?>
2    <TabHost xmlns:android="http://schemas.android.com/apk/res/android"
3        android:layout_width="match_parent"
4        android:layout_height="match_parent"
5        android:id="@+id/examTabHost">
6        <LinearLayout
7            android:layout_width="fill_parent"
```

```
 8        android:layout_height="wrap_content"
 9        android:orientation="vertical">
10        <TabWidget
11            android:id="@android:id/tabs"
12            android:layout_width="fill_parent"
13            android:layout_height="wrap_content"
14            android:orientation="horizontal"/>
15        <FrameLayout
16            android:id="@android:id/tabcontent"
17            android:layout_width="fill_parent"
18            android:layout_height="fill_parent">
19            <!--单选题布局-->
20            <LinearLayout
21                android:id="@+id/singleChoice"
22                android:layout_width="fill_parent"
23                android:layout_height="fill_parent"
24                android:orientation="vertical">
25                <TextView
26                    android:layout_width="fill_parent"
27                    android:layout_height="wrap_content"
28                    android:text="1. 负责管理计算机的硬件资源和软件资源,为应用程序
29 开发和运行提供高效平台的软件是?"
30                    android:textSize="18sp"/>
31                <RadioGroup
32                    android:id="@+id/singleRG"
33                    android:layout_width="wrap_content"
34                    android:layout_height="wrap_content">
35                    <RadioButton
36                        android:id="@+id/optionA"
37                        android:layout_width="wrap_content"
38                        android:layout_height="wrap_content"
39                        android:text="A. 操作系统"
40                        android:textSize="18sp"/>
41                    <!--其他选项定义与A类似-->
42                </RadioGroup>
43            </LinearLayout>
44            <!--多选题布局-->
45            <LinearLayout
46                android:id="@+id/multiChoice"
47                android:layout_width="fill_parent"
48                android:layout_height="wrap_content"
49                android:orientation="vertical">
50                <TextView
51                    android:layout_width="fill_parent"
52                    android:layout_height="wrap_content"
53                    android:text="1. 激光打印机通常可以采用下面哪些端口?"
54                    android:textSize="18sp"/>
55                <CheckBox
56                    android:id="@+id/checkBoxA"
57                    android:layout_width="wrap_content"
```

```xml
58          android:layout_height="wrap_content"
59          android:text="A. 并行接口"
60          android:textSize="18sp" />
61       <!--其他选项定义与A类似 -->
62     </LinearLayout>
63     <!--填空题布局 -->
64     <LinearLayout
65       android:id="@+id/fill"
66       android:layout_width="fill_parent"
67       android:layout_height="wrap_content"
68       android:orientation="vertical" >
69       <TextView
70          android:id="@+id/fanswerValue"
71          android:layout_width="fill_parent"
72          android:layout_height="wrap_content"
73          android:text="1. 一幅分辨率为512×512的彩色图像,其R、G、
74  B 3个分量分别用8个二进位表示,则未进行压缩时该图像的数据量是多少 kB?"
75          android:textSize="18sp" />
76       <EditText
77          android:id="@+id/fillValue"
78          android:layout_width="fill_parent"
79          android:layout_height="wrap_content"
80          android:hint="请输入答案"
81          android:textSize="18sp" />
82     </LinearLayout>
83     <!--判断题布局 -->
84     <LinearLayout
85       android:id="@+id/judge"
86       android:layout_width="fill_parent"
87       android:layout_height="wrap_content"
88       android:orientation="vertical" >
89       <TextView
90          android:layout_width="fill_parent"
91          android:layout_height="wrap_content"
92          android:text="1. 程序就是算法,算法就是程序。"
93          android:textSize="18sp" />
94       <RadioGroup
95          android:id="@+id/judgeRG"
96          android:layout_width="wrap_content"
97          android:layout_height="wrap_content" >
98          <RadioButton
99             android:id="@+id/judgeoptionA"
100            android:layout_width="wrap_content"
101            android:layout_height="wrap_content"
102            android:text="对"
103            android:textSize="18sp" />
104         <!--其他选项定义与"对"类似 -->
105      </RadioGroup>
106    </LinearLayout>
```

```
107            </FrameLayout>
108        </LinearLayout>
109  </TabHost>
```

2. 考试系统界面的功能实现

为了实现选项卡上既显示图片又显示文字的功能,可以在设置 TabSpec 的属性时使用 .setIndicator（getMenuItem（R. mipmap. single,"单选题"））方法,其详细代码如下。

```
1    protected void onCreate(Bundle savedInstanceState) {
2        super.onCreate(savedInstanceState);
3        setContentView(R.layout.activity_main);
4        tabHost = (TabHost) findViewById(R.id.examTabHost);// 获得 TabHost 对象
5        tabHost.setup();//通过 setup()方法加载启动 TabHost
6        tabHost.addTab(tabHost
7               .newTabSpec("tab01")
8               .setIndicator(getMenuItem(R.mipmap.single,"单选题"))
9               .setContent(R.id.singleChoice));
10       tabHost.addTab(tabHost
11              .newTabSpec("tab02")
12              .setIndicator(getMenuItem(R.mipmap.multi,"多选题"))
13              .setContent(R.id.multiChoice));
14       tabHost.addTab(tabHost
15              .newTabSpec("tab03")
16              .setIndicator(getMenuItem(R.mipmap.blank,"填空题"))
17              .setContent(R.id.fill));
18       tabHost.addTab(tabHost
19              .newTabSpec("tab04")
20              .setIndicator(getMenuItem(R.mipmap.judge,"判断题"))
21              .setContent(R.id.judge));
22   }
```

以上代码运行后默认显示的第 1 个选项卡页面是"单选题"页面,如果需要将其他页面作为默认页,需要在上述第 21 行代码后增加"tabHost.setCurrentTab（int i）"语句,其中 i 表示每个选项卡的序号(从 0 开始)。例如,需要将"填空题"页面作为默认页,需要使用 tabHost.setCurrentTab（2）语句。

如果在选项卡页面切换时需要触发其他事件,可以给 TabHost 设置监听事件,其代码如下。

```
1    tabHost.setOnTabChangedListener(new TabHost.OnTabChangeListener() {
2        @Override
3        public void onTabChanged(String tabId) {
4            if (tabId.equals("tab01")){
5                Toast.makeText(MainActivity.this,"这是切换到单项选择题",
Toast.LENGTH_LONG).show();
6            }
7            if (tabId.equals("tab02")){
8                Toast.makeText(MainActivity.this,"这是切换到多项选择题",
Toast.LENGTH_LONG).show();
```

```
 9                    }
10                    if (tabId.equals("tab03")){
11                        Toast.makeText(MainActivity.this,"这是切换到填空题",
Toast.LENGTH_LONG).show();
12                    }
13                    if (tabId.equals("tab04")){
14                        Toast.makeText(MainActivity.this,"这是切换到判断题",
Toast.LENGTH_LONG).show();
15                    }
16                }
17            });
```

限于篇幅，完整的布局代码和功能代码读者可以参阅 examui 文件夹中的内容。

3.6　打老鼠游戏的设计与实现

"打老鼠"游戏：在游戏机上投入一元硬币后，游戏柜上面有多个洞口，随时会有"老鼠"从洞口探出头来，玩家只需要抢起锤子击打"老鼠"，击中了就会加分；在规定时间内，如果分值达到预期目标，那么游戏继续，否则游戏结束。本节将模拟"打老鼠"游戏，设计一款在 Android 终端设备上运行的打老鼠游戏。本游戏采用 TableLayout 布局管理器设计一个 5×5 的表格，在表格中随机出现"老鼠"，要求玩家立即触摸，触摸到表示击中，如果玩家在限定的时间里击中七只"老鼠"，那么游戏继续，否则游戏结束。

3.6.1　预备知识

1. TableLayout

在实际应用开发中，经常会出现数据的显示以表格布局方式展现的情况，如图 3.26 所示。要达到类似的界面显示效果，可以使用 TableLayout 方式。

【TabLayout、CountDown Timer、横竖屏控制】

TableLayout 是由一系列行和列组成的网格，在这些网格的单元格中可以显示 View 组件。从用户界面设计的角度看，一个 TableLayout 由一系列 TableRow 组成，每个 TableRow 对应表格里的一行。TableRow 的内容由单元格中的 View 组件组成。Android 中使用 android.widget.TableLayout 类实现表格布局。

TableLayout 以行和列的形式对单元格中的 View 组件进行管理，每一行可以是一个 TableRow 对象，也可

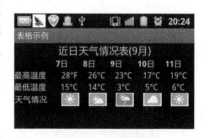

图 3.26　表格布局

以是一个 View 组件。每一行为 TableRow 对象时，可以在 TableRow 下添加子组件，默认情况下，每个子组件占据一列；每一行为 View 组件时，该 View 组件独占一行。在应用开发时，TableLayout 的行数由开发者直接指定，即有多少个 TableRow（或 View 组件）就有多少行。TableLayout 的列数由包含最多子组件的 TableRow 的列数决定。例如，有一个 TableLayout 布局管理器，它的第 1 个 TableRow 包含两个子组件，第 2 个 TableRow 包含三个子

组件，第 3 个 TableRow 包含四个子组件，那么该 TableLayout 的列数就是 4。

android.widget.TableLayout 类可设置的属性包括全局属性（即列属性）和单元格属性，详细说明见表 3-20 和表 3-21。

表 3-20　全局属性

属性名	说明	备注
android：stretchColumns	设置允许被拉伸的列的序号	这三个属性的列号都是从 0 开始算，如 shrinkColunmns = "2"，对应的是第 3 列；
android：shrinkColumns	设置允许被收缩的列的序号	可以设置多个列，但必须用逗号隔开，如"0,2"；如果是所有列都生效，则用"*"号即可
android：collapseColumns	设置要隐藏的列的序号	

表 3-21　单元格属性

属性名	说明	备注
android：layout_column	指定该单元格在第几列显示	android：layout_column = "1" 该组件显示在第 1 列
android：layout_span	指定该单元格占据的列数，默认为 1	android：layout_span = "2"，合并两个单元格，即该组件占据两个单元格

（1）在使用 TableLayout 布局管理器设计 App 界面布局时，需要注意以下要点。

① 如果直接在 TabLayout 中添加组件，那么这个组件占满一行。

② 如果一行上要放置多个组件，就必须添加一个 TabRow 容器，并将组件放置在 TabRow 容器中。

③ TableRow 中组件的个数决定了该行有多少列，而且列的宽度由该列中最宽的单元格决定。

④ TableRow 的 layout_width 属性默认是 match_parent，如果设置成其他值不会生效；而 layout_height 默认是 wrap_content，开发者可以根据需要设置其大小。

⑤ 整个表格布局的宽度取决于父容器的宽度（占满父容器本身）。

⑥ 整个表格的行数由 TableRow 的数目及单独组件数目决定，列数由 TableRow 中最多个组件数决定。

表格的某列可以同时具有 stretchColumns 和 shrinkColumns 属性，在这种情况下，当该列的内容很多，导致不能用一行全部显示时，它们将"多行"显示。但是这种多行，只是表示在该单元格列的多行，整个表格的 TableRow 还是原来的值，只是此时系统会根据需要自动调节该行的 layout_height 的值。

（2）对于图 3.26 所示的用户界面布局文件代码组成，下面分四个部分介绍。

① TableLayout 的定义，代码如下。

```
1    <TableLayout
2      xmlns:android = "http://schemas.android.com/apk/res/android"
3      android:id = "@ + id/tableLayout1"
4      android:layout_width = "match_parent"
5      android:layout_height = "match_parent"
6      android:shrinkColumns = "*"
7      android:stretchColumns = "*" >
8    </TableLayout >
```

第 6 行和第 7 行代码表示该表格的任意行可以伸展，任意列可以收缩，也就是当列内容不能在一行中全部显示时，可以在该列对应的单元格中分行显示，同时整个表格会自动调整该单元格所在行的 layout_height 值。

② 表格第 1 行 TableRow 的定义，代码如下。

```
1    <TableRow
2        android:id = "@ + id/tableRow1"
3        android:layout_height = "wrap_content"
4        android:layout_width = "match_parent"
5        android:gravity = "center_horizontal" >
6        <TextView
7          android:id = "@ + id/textView9"
8          android:layout_width = "match_parent"
9          android:layout_height = "wrap_content"
10         android:textSize = "18dp"
11         android:text = "近日天气情况表(9 月)"
12         android:gravity = "center"
13         android:layout_span = "6"/ >
14   </TableRow >
```

表格的第 1 行只包含一个 TextView 组件，也就是第 1 行只有一个单元格，而要显示如图 3.26 所示的效果，必须使用 layout_span 属性进行设置，第 13 行代码表示该单元格占据 6 列。

③ 表格第 2 行 TableRow 的定义，代码如下。

```
1    <TableRow
2        android:id = "@ + id/tableRow2"
3        android:layout_height = "wrap_content"
4        android:layout_width = "match_parent" >
5        <TextView
6          android:id = "@ + id/TextView05"
7          android:text = ""/ >
8        <TextView
9          android:id = "@ + id/TextView04"
10         android:text = "7 日"
11         android:textStyle = "bold" / >
12       <TextView
13         android:id = "@ + id/TextView03"
```

```
14            android:text ="8 日"
15            android:textStyle ="bold" />
16        <TextView
17            android:id ="@ +id/TextView02"
18            android:text ="9 日"
19            android:textStyle ="bold" />
20        <TextView
21            android:id ="@ +id/TextView01"
22            android:text ="10 日"
23            android:textStyle ="bold"/>
24        <TextView
25            android:text ="11 日"
26            android:id ="@ +id/textView00"
27            android:textStyle ="bold" />
28    </TableRow>
```

该行包含 6 列，每一列包含一个 TextView 组件，每个 TextView 组件正好对应该行上的每一个单元格。表格的第 3、4 行 TableRow 的定义中只需要将 TextView 组件的 text 属性值做相应修改，其他代码与第 2 行类似，不再详述。

④ 表格第 5 行 TableRow 的定义，代码如下。

```
1    <TableRow
2        android:id ="@ +id/tableRow5"
3        android:layout_height ="wrap_content"
4        android:layout_width ="match_parent"
5        android:gravity ="center" >
6        <TextView
7            android:id ="@ +id/textView8"
8            android:text ="天气情况"
9            android:textStyle ="bold" > </TextView >
10       <ImageView
11           android:id ="@ +id/imageView1"
12           android:src ="@drawable/sun"/ >
13       <ImageView
14           android:id ="@ +id/imageView2"
15           android:src ="@drawable/cloud"/ >
16       <ImageView
17           android:id ="@ +id/imageView3"
18           android:src ="@drawable/rain"/ >
19       <ImageView
20           android:id ="@ +id/imageView4"
21           android:src ="@drawable/shade"/ >
22       <ImageView
23           android:id ="@ +id/imageView5"
24           android:src ="@drawable/sun"/ >
25    </TableRow >
```

该行中显示的内容除了第 1 列使用 TextView 组件显示"天气情况"文本信息外，其余都是用天气状况图来表示。为了显示图片，本案例中使用了 ImageView 组件，所以需要将天气状况图保存在 res/drawable 文件夹下,然后修改 ImageView 组件的 src 属性值,如代码中的第 12、15、18、21、24 行所示。

如果要在 TableRow 之间增加间隔线,可以在两个 TableRow 之间添加 View,并设置该 View 的 layout_height 和 background 属性。要实现如图 3.27 所示效果,只需要在第 1 行 TableRow 和第 2 行的 TableRow 之间加入如下代码。

图 3.27 表格布局(线)

```
<View
    android:layout_height="2sp"
    android:background="#f00f00"/>
```

在实际应用开发时,读者只需要对本案例布局代码进行部分修改,一般都可以设计出布局合理的用户界面。

2. CountDownTimer

CountDownTimer 是 Android 提供的一个具有倒计时功能的抽象类 CountDownTimer 的常用方法和功能说明见表 3-22。

表 3-22 CountDownTimer 的常用方法和功能说明

方 法 名	功能说明
CountDownTimer(long time, long interval)	构造方法,time 表示从开始调用 start( )到倒计时完成并调用 onFinish( )方法的总时间; interval 表示调用 onTick( )方法的间隔时间,单位为毫秒
cancel( )	取消倒计时
onFinsih( )	抽象方法,倒计时完成被调用
start( )	倒计时开始
onTick(longtime)	抽象方法,每个间隔时间 interval 一到就会调用一次,time 表示剩余时间

下面通过一个简单的倒计时器(图 3.28)来介绍它的用法。倒计时器可以在 "设定时间" 后输入倒计时的时间,然后单击 "开始倒计时" 按钮,倒计时器开始倒计时,布局设计比较简单,读者可以参阅 sampleapp 文件夹下的 src/main/res/layout/timer_layout.xml 文件。

(1) 创建了一个继承 CountDownTimer 的内部类 MyCount,并重写相关方法,具体代码如下。

图 3.28 倒计时器

```
//倒计时功能
private class MyCount extends CountDownTimer {
    public MyCount(long time, long interval) {
        super(time, interval);
    }
    @Override
    public void onFinish() {
        txtTime.setText("时间到!");
    }
    @Override
    public void onTick(long millisUntilFinished) {
        int hour = (int) millisUntilFinished / 1000 / 3600;//时
        int minute = (int) millisUntilFinished / 1000 % 3600 / 60;//分
        int second = (int) millisUntilFinished / 1000 % 3600 % 60;//秒
        txtTime.setText("倒计时:" + hour + ":" + minute + ":" + second);
    }
}
```

（2）给"开始倒计时"按钮设置监听事件，代码如下。

```
btnStart.setOnClickListener(new View.OnClickListener() {
    @Override
    public void onClick(View v) {
        long time = Integer.parseInt(edtTime.getText().toString())* 1000;
        new MyCount(time,1000).start();
    }
});
```

3. 消息处理机制

当一个 App 的组件启动时，Android 会为该 App 组件创建一个新的线程来执行。默认情况下，同一个 App 的所有组件运行在同一个线程中，该线程即为主线程（Main Thread），它主要用来加载用户界面，完成系统与用户之间的交互，并将交互后的结果展现给用户，所以主线程也称 UI Thread。

通常一个 Android App 的所有组件默认都在主线程中运行，但是某些时候 App 可能需要处理一个耗时的操作（如访问网络，Android 4.0 以上版本中已经不允许在主线程中访问网络），一旦处理耗时操作，UI Thread 就会被阻塞，如果 UI Thread 阻塞时间超过 5s，就会出现 ANR（Application Not Responding）问题，即应用程序会弹出一个提示框，让用户选择是否退出程序。另外，由于 Android UI 组件不是线程安全的，所以不能在 UI Thread 之外的线程中对 UI 组件进行更新、删除等操作。

综上所述，在 Android 的多线程编程中，必须遵循以下两个原则。①不能在 UI Thread 中进行耗时操作，即耗时操作只能在子线程中实现。②不能在 UI Thread 之外的线程中操纵 UI 组件，即操纵 UI 组件只能在主线程中实现。

为了解决以上问题，就需要提供一个子线程中的耗时操作完成后能通知主线程操纵 UI 组件的机制，而 Android 中的异步消息处理机制恰恰能解决这一问题，该机制有 AsyncTask 和 Handler 两种实现方式。

(1) AsyncTask。

AsyncTask（异步任务）是一个抽象类，用于在 UI 主线程运行的时候异步完成一些操作。也就是将耗时的操作放在异步任务当中执行，并随时将任务执行的结果返回给 UI Thread 来更新 UI 组件。AsyncTask 包含三个泛型参数和多个重写方法，见表 3-23 和表 3-24。

表 3-23 AsyncTask 的泛型参数

参数名	说明	备注
Params	启动任务执行的参数	例如，Http 请求的 URL 一般为 Sring 类型
Progress	后台任务执行的百分比	一般用 Interger 类型
Result	后台执行任务最终返回结果	后台任务执行的最终返回结果，一般用 byte[] 或 String

表 3-24 AsyncTask 的重写方法

方法名	说明	备注
onPreExecute()	在执行实际的后台操作前被 UI Thread 调用	通常可以在该方法中做一些准备工作，如在界面上显示进度
doInBackground(Params...)	onPreExecute() 执行后马上在后台线程中执行	主要负责耗时的后台工作，可以调用 publishProgress() 来更新实时的任务进度
onProgressUpdate(Progress...)	在 publishProgress() 方法被调用后，UI Thread 将调用该方法	该方法在主线程中执行，用于在用户界面上展示任务执行的进度情况
onPostExecute(Result)	在 doInBackground() 方法执行完成后，UI Thread 将调用该方法	主要用于更新用户界面操作

下面用一个动态改变 TextView 组件的 text 属性值的实例来讲述 AsyncTask 的使用方法。

首先创建一个布局文件，该布局文件上放置一个 TextView 和一个 Button 组件，读者可以参阅 sampleapp 目录下 src/main/res/layout/asynctask_layout.xml 文件。

接着在 Activity（src/main/java/AsyncActivity.java）中创建一个继承自 AsyncTask 的子类，其详细代码如下：

```
1    public class MyTask extends AsyncTask<String, Integer, String> {
2        @Override
3        protected void onPreExecute() {
4            super.onPreExecute();
5            Toast.makeText(AsyncActivity.this, "开始执行异步线程", Toast.LENGTH_SHORT).show();
6        }
7        @Override
8        protected String doInBackground(String...params) {
```

```
 9            int i = 0;
10            for (i = 10; i <100; i ++) {
11                try {
12                    Thread.sleep(1000);
13                } catch (InterruptedException e) {
14                    e.printStackTrace();
15                }
16                publishProgress(i);//实时更新任务进度
17            }
18            return i + "";
19        }
20        @Override
21        protected void onProgressUpdate(Integer... values) {
22            super.onProgressUpdate(values);
23            tvinfo.setText(values[0] + "");//在TextView上更新任务进度
24        }
25        @Override
26        protected void onPostExecute(String s) {
27            super.onPostExecute(s);
28            Toast.makeText(AsyncActivity.this, "执行结束", Toast.LENGTH_SHORT).show();
29        }
30    }
```

最后实现当用户单击"开始"按钮后,开始更新 TextView 组件上的内容,直到该组件上显示到"99"为止,用 Toast 显示"执行结束",代码如下。

```
1    btn.setOnClickListener(new View.OnClickListener() {
2        @Override
3        public void onClick(View v) {
4            new MyTask().execute();//开始执行异步任务
5        }
6    });
```

(2) Handler。

Handler 是 Android 提供的用于接收、传递和处理 Message(消息)或 Runnable 对象的处理类,它结合 Message、MessageQueue 和 Looper 类及当前线程实现了一个消息循环机制,用于实现任务的异步加载和处理。Handler 异步消息处理流程如图 3.29 所示。

从图 3.29 可以看出,要使用 Handler 进行异步消息处理需要按如下步骤实现。

① 在主线程中创建一个 Handler 对象,并重写 handleMessage()方法,用于处理 Handler 发送来的消息。

② 如果需要子线程来操纵 UI 组件,就需要创建一个子线程,并在子线程中创建一个 Message 对象,然后通过 Handler 将消息发送出去。

在实际应用开发中,为了有利于消息资源的利用,通常不建议直接用构造方法实例化 Message 对象,而是使用静态方法 Message.obtain()或者 Handler.obtainMessage 获得该对象。Handler 进行消息处理时与消息相关的方法及功能说明见表 3-25。

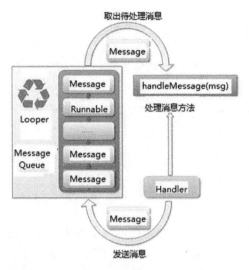

图 3.29　Handler 异步消息处理流程

表 3-25　Handler 进行消息处理时与消息相关的方法及功能说明

属 性 名	功能说明
obtainMessage()	获取一个 Message 对象
sendMessage()	发送一个 Message 对象到消息队列中，并在 UI 线程获得消息后立即执行
sendMessageDelayed()	发送一个 Message 对象到消息队列中，并在 UI 线程获得消息后延迟执行
sendEmptyMessage()	发送一个空 Message 对象到队列中，并在 UI 线程获得消息后立即执行
sendEmptyMessageDelayed()	发送一个空 Message 对象到队列中，并在 UI 线程获得消息后延迟执行
removeMessage()	从消息队列中移除一个未响应的消息

下面使用 Handler 与 Message 实现上面动态改变 TextView 组件的 text 属性值的实例来讲述 Handler 的使用方法。

首先创建一个 Handler 对象用于处理消息，代码如下。

```
1   Handler handler = new Handler(){
2       @Override
3       public void handleMessage(Message msg) {
4           super.handleMessage(msg);
5           tvInfo.setText(msg.arg1 + "");
6       }
7   };
```

然后创建一个 Runnable 对象用于在子线程中实现功能需求，代码如下。

```
1   Runnable runnable = new Runnable() {
2       @Override
3       public void run() {
```

```
4       while ( i < =100){
5           Message msg = handler.obtainMessage();
6           msg.arg1 = i;
7           handler.sendMessage(msg);
8           try {
9               Thread.sleep(1000);
10          } catch (InterruptedException e) {
11              e.printStackTrace();
12          }
13          i ++;
14      }
15    }
16  };
```

最后给"开始"按钮绑定监听事件,在监听事件中创建子线程并启动,代码如下。

```
1   btn.setOnClickListener(new View.OnClickListener() {
2       @Override
3       public void onClick(View v) {
4           Thread thread = new Thread(runnable);
5           thread.start();
6       }
7   });
```

3.6.2 打老鼠游戏的实现

1. 主界面的设计

【游戏的实现】

目前,大多数的 Android App 都是在竖屏的基础上进行用户界面设计的。如果这些 App 允许横屏与竖屏之间的切换操作,并且 Android 设备又设置成允许自动旋转,那么用户在使用这些 App 时会由于 Android 设备摆放状态的改变,导致 App 的用户界面在横屏与竖屏之间来回切换,这样会大大影响用户体验。例如,有些游戏 App 需要在横屏状态下才能让玩家有更好的界面体验,即使 Android 终端设备竖立放置时,也依然要保持游戏画面横屏。所以在设计 App 之初,开发者就应该确定 App 运行时的屏幕状态,即从横屏、竖屏或横屏竖屏都可以的三种状态中选定一种。

如果设计的 App 需要在横屏或竖屏状态下都可以运行,就需要在 App 的 res 目录下建立 layout-land(用于放置横屏布局文件)和 layout-port(用于放置竖屏布局文件)目录,并且 App 对应的横屏布局和竖屏布局文件名相同,横竖屏切换时程序会调用 Activity 的 onCreate() 方法,从而根据当前横竖屏情况自动加载相应的布局文件。

如果设计的 App 需要强制只能在横屏或竖屏的某一种状态下运行,可以使用以下两种方案。

(1) 修改 AndroidManifest.xml 配置文件。在默认设置下,App 的 Activity 对应的布局文件界面是竖屏显示的,如果要强制横屏显示,就必须在配置文件中为 Activity 配置 screenOrientation 属性,screenOrientation 属性值具体配置代码见表 3-26。修改后的配置文件代码如下。

表 3-26 screenOrientation 属性值具体配置代码

属 性 值	说 明
unspecified	默认值，由系统根据设备来判断显示方向
landscape	横屏显示
portrait	竖屏显示
user	用户当前首选的方向
behind	和该 Activity 下面的那个 Activity 的方向一致（在 Activity 堆栈中的）
sensor	由物理感应器来决定，用户旋转设备会导致屏幕的横竖屏切换
nosensor	忽略物理感应器，用户旋转设备不会导致屏幕的横竖屏切换

```
1    <application
2        android:icon ="@drawable/icon"
3        android:label ="@string/app_name" >
4        <activity
5            android:name =".MainActivity"
6            android:screenOrientation ="landscape"
7            android:label ="@string/app_name" >
8            <intent-filter >
9                <action android:name ="android.intent.action.MAIN" />
10               <category android:name ="android.intent.category.LAUNCHER" />
11           </intent-filter >
12       </activity >
13   </application >
```

（2）功能代码实现。在用功能代码实现时，在需要强制横屏显示的 Activity 中的 OnCreate( )方法中增加如下代码。

```
setRequestedOrientation(ActivityInfo.SCREEN_ORIENTATION_LANDSCAPE);
```

从图 3.30 打老鼠游戏界面可以看出，游戏界面使用 TableLayout 布局管理设计器比较

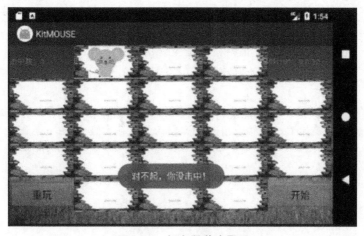

图 3.30 打老鼠游戏界面

容易实现，即将"击中数"和"倒计时"的两个 TextView 组件、"重玩"和"开始"的两个 Button 组件及其他表示鼠洞的多个 ImageView 组件放置在表格布局的单元格中，关键代码如下：

```xml
<?xml version="1.0" encoding="utf-8"?>
<TableLayout
    xmlns:android="http://schemas.android.com/apk/res/android"
    android:id="@+id/tableLayout1"
    android:layout_width="match_parent"
    android:layout_height="match_parent"
    android:background="@mipmap/yewai"
    android:shrinkColumns="*"
    android:stretchColumns="*" >
    <TableRow
        android:id="@+id/tableRow1"
        android:layout_height="wrap_content"
        android:layout_width="match_parent"
        android:gravity="center" >
        <TextView
            android:id="@+id/mcount"
            android:text="击中数:0"
            android:textColor="#ff0000"
            android:textStyle="bold"/>
        <ImageView
            android:id="@+id/im12"
            android:background="@mipmap/dong"/>
        <!---其他ImageView组件的设置与此处代码类似->
        ...
         <TextView
            android:id="@+id/mtime"
            android:text="倒计时:60"
            android:textColor="#ff0000"
            android:textStyle="bold"/>
    </TableRow>
    <TableRow
        android:id="@+id/tableRow2"
        android:layout_height="wrap_content"
        android:layout_width="match_parent"
        android:gravity="center" >
        <ImageView
            android:id="@+id/im21"
            android:background="@mipmap/dong"/>
        <!--其他ImageView组件的设置与此处代码类似-->
        ...
    </TableRow>
        <!--表格布局中此行显示效果代码与上行类似-->
        ...
    <TableRow
        android:id="@+id/tableRow5"
```

```
                android:layout_height ="wrap_content"
                android:layout_width ="match_parent"
                android:gravity ="center" >
                <Button
                    android:id ="@ +id/replay"
                    android:text ="重玩"
                    android:layout_width ="108px"
                    android:layout_height ="wrap_content"/>
                <ImageView
                    android:id ="@ +id/im52"
                    android:background ="@ mipmap /dong"/ >
                <!--其他 ImageView 组件的设置与此处代码类似-->
                ...
                <Button
                    android:id ="@ +id/play"
                    android:text ="开始"
                    android:layout_width ="108px"
                    android:layout_height ="wrap_content"/ >
            </TableRow >
</TableLayout >
```

开发中用到的图片 yewai.jpg、dong.png、mouse.png 需要复制到 kitmouse 文件夹下的 src/main/res/mipmap 目录下。

2. 功能实现

(1) 定义变量。

由于本案例项目涉及变量较多，限于篇幅，这里仅列出部分关键变量定义的代码和说明，其他部分请读者参阅 kitmouse 文件夹中的内容。

```
    Button btnPlay, btnReplay;
    TextView txtCount, txtTime;
    ImageView[] imageViews = new ImageView[21];
    int[] imageID = {R.id.im12, R.id.im13, R.id.im14, R.id.im21, R.id.im22,
R.id.im23, R.id.im24, R.id.im25, R.id.im31, R.id.im32, R.id.im33, R.id.im34,
R.id.im35, R.id.im41, R.id.im42, R.id.im43, R.id.im44, R.id.im45, R.id.im52,
R.id.im53, R.id.im54};
    Drawable dong, mouse;
    int count =0;//记录击中的老鼠数
    int time =60;//倒计时 60 秒
    int oldID =0, newID =0;
    boolean flag = true;//时间到标志
```

(2) 随机在洞中出现"老鼠"。

每隔 1s 发送消息并交给 Handler 处理，首先将前一张老鼠的图片用洞的图片替代，然后产生 0～21 中的任意一个随机数作为 imageViews 数组的下标，并使用 setBackgroundDrawable()方法加载老鼠图片，表示老鼠出洞。

```
Handler handler = new Handler(){
    @Override
    public void handleMessage(Message msg) {
        super.handleMessage(msg);
```

```
            imageViews[oldID].setBackgroundDrawable(dong);
            newID = (int)(0 + Math.random()* 21);//产生随机出现老鼠的数组下标
            imageViews[newID].setBackgroundDrawable(mouse);
            oldID = newID;
        }
    };
    Runnable runnable = new Runnable() {
        @Override
        public void run() {
            while (flag){
                handler.sendEmptyMessage(0);
                try {
                    Thread.sleep(1000);
                } catch (InterruptedException e) {
                    e.printStackTrace();
                }
            }
        }
    };
```

(3) 开始游戏、重玩游戏、打"老鼠"功能代码的实现。

本案例项目中的"老鼠"洞是由多个 ImageView 组件组成的，如果为每个 ImageView 组件单独编写监听事件，则重复代码较多。为了避免出现这类问题，在创建主用户界面 MainActivity时实现OnClickListener 接口，这样就需要重写 onClick()方法，代码如下。

```
    @Override
    public void onClick(View v){
        switch (v.getId()){
        case R.id.play:
            new MyCount(time* 1000,1000).start();//开始倒计时
            Thread thread = new Thread(runnable); //线程开启,"老鼠"出洞
            break;
        case R.id.replay:
            count = 0;//重玩 count 清 0
            flag = true;
            new MyCount(time* 1000,1000).start();//重新开始倒计时
            Thread thread = new Thread(runnable); //重新线程开启,"老鼠"出洞
            break;
        default:
            if (imgView[oldID].getId() == v.getId()){
                count ++ ;
                txtCount.setText("击中数:"+count);
                if(count == 20){
                    Toast.makeText(MainActivity.this,"祝贺你顺利过关!",1).show();
                    flag = false;
                }
            } else {
                Toast.makeText(MainActivity.this,"对不起,你没击中!",1).show();
```

```
            }
            break;
        }
    }
```

代码中的 case R.id.play 分支表示单击"开始"按钮执行的功能代码；case R.id.replay 分支表示单击"重玩"按钮执行的代码；default 分支表示在"老鼠"出洞后，玩家执行的操作，当玩家单击的 ImageView 组件（即当前"老鼠"所在的洞）与随机数中出现的"老鼠"相同，就说明玩家打中了"老鼠"，为玩家计数。

计时功能自定义了一个继承自 CountDownTimer 的 MyCount 类来实现，其详细代码与本书前面介绍的 CountDownTimer 使用方法类似，限于篇幅，这里不再详述，完整的功能代码读者可以参阅 kitmouse 文件夹中的内容。

3.7 猜扑克游戏的设计与实现

基于扑克牌的小游戏很多，本案例实现红桃 A、梅花 A、方块 A 这三张扑克牌按随机顺序摆放在屏幕上，但只能在屏幕中央显示一张扑克牌的背面（图 3.31），由用户在规定的时间内猜测哪张是红桃 A。如果认为是红桃 A 的，就单击该牌，如果认为不是，可以向左或向右滚动并选择另一张牌。本案例使用 ProgressBar（进度条）实现倒计时功能、使用 ScrollView（滚动视图）实现图片的左右滚动功能。另外，为了增强游戏界面的趣味性，本案例中还使用了 ToggleButton（开关按钮）实现背景图片的选择和 RadioGroup（单选按钮组）实现游戏难易度的选择。

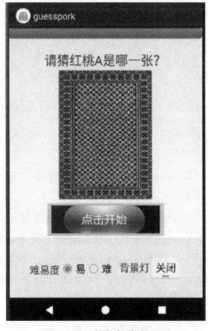

图 3.31 猜扑克游戏界面

3.7.1 预备知识

1. HorizontalScrollView

【HorizontalScrollView、ProgressBar、Switch、ToggleButton】

HorizontalScrollView（水平滚动视图）是用于布局的容器，它里面可以放置多个视图（Views）。当屏幕显示不完时，需要用户使用滚动条来显示完整的视图，也就是使用HorizontalScrollView可以通过滚动屏幕的方式来显示比屏幕区域大的内容。与之对应的还有一个 ScrollView（垂直滚视图）布局容器，使用它可以实现上下滚动屏幕来显示内容。HorizontalScrollView 和 ScrollView 都是 FrameLayout 的子类，它们的使用方法一样。由于篇幅所限，本书只介绍 HorizontalScrollView 的使用方法。如果要在屏幕上实现左右滚动显示四幅图片，需要首先创建一个 HorizontalScrollView 类型的布局文件，然后在布局文件使用如下代码实现。

```
1   <?xml version="1.0" encoding="utf-8"?>
2   <HorizontalScrollView xmlns:android="http://schemas.android.com/apk/res/android"
3       android:layout_width="389px"
4       android:layout_height="match_parent"
5       android:layout_gravity="center">
6       <LinearLayout
7           android:layout_width="wrap_content"
8           android:layout_height="wrap_content">
9           <ImageView
10              android:layout_width="wrap_content"
11              android:layout_height="wrap_content"
12              android:src="@mipmap/weixin_head2" />
13          <ImageView
14              android:layout_width="wrap_content"
15              android:layout_height="wrap_content"
16              android:src="@mipmap/weixin_head3" />
17          <!-- 其他 ImageView 的定义类似 -->
18      </LinearLayout>
19  </HorizontalScrollView>
```

上述第 3 行代码用于指定 HorizontalScrollView 的宽度，用于刚好显示一张图片的宽度大小，该值大小通常与图片的水平分辨率一致。

在使用滚动视图布局时，其子项被滚动查看时是整体移动的，并且子项本身要么是一个单独的组件（如只能是一个 ImageView），要么是一个有复杂层次结构的布局管理器布局的界面（如上面将多个 ImageView 放在 LinearLayout 中）。如果将多个组件直接放在滚动视图布局中，在视图渲染时会提示"ScrollView can host only direct child"的出错信息。

2. ProgressBar

当一个 App 在后台执行时，前台界面没有任何信息，用户不知道该 App 是否在执行、执行进度如何、是否因遇到错误而终止运行等，这时可以使用ProgressBar（进度条）来提示用户后台 App 的执行进度,从而提高用户界面的友好性。本节设计一个模拟文件下载的进度条来讲述 ProgressBar 组件的用法。

为了适应不同的应用环境,Android 内置了多种不同风格的进度条,开发者可以在布局文件中通过 style 属性值来设置 ProgressBar 的风格,style 常用属性值及说明见表 3-27。

表 3-27 ProgressBar 的 style 属性值及说明

属 性 值	说 明
@android:style/Widget.ProgressBar.Horizontal	水平进度条
@android:style/Widget.ProgressBar.Small	旋转动画小进度条
@android:style/Widget.ProgressBar.Large	旋转动画大进度条
@android:style/Widget.ProgressBar.Inverse	旋转动画进度条
@android:style/Widget.ProgressBar.Small.Inverse	旋转动画小进度条
@android:style/Widget.ProgressBar.Large.Inverse	旋转动画大进度条

实际应用开发中用得最多的是 Widget.ProgressBar.Horizontal 风格的进度条,因为只有该进度条设置了进度的递增后才可以直观地看到进度的前进或后退,其常用的属性和方法见表 3-28 和表 3-29。其他风格的进度条只会显示为一个循环的动画。

表 3-28 ProgressBar 的常用属性及说明

属 性	说 明
android:max	设置进度条的最大值
android:progress	设置当前第 1 进度值
android:secondaryProgress	设置当前第 2 进度值是否显示

表 3-29 ProgressBar 的常用方法及说明

方 法	说 明
synchronized int getMax()	获取进度条的最大值
synchronized int getProgress()	获取当前第 1 进度值
synchronized int getSecondaryProgress()	获取当前第 2 进度值
synchronized void setMax(int max)	设置进度条的最大值
synchronized void setProgress(int progress)	设置第 1 进度值
synchronized void setSecondaryProgress(int secondaryProgress)	设置第 2 进度值

所有风格进度条的运行效果如图 3.32 所示。实际应用开发时,如果要使用系统提供风格的进度条,实现起来比较简单,其代码如下。

```
1    <?xml version="1.0" encoding="utf-8"?>
2    <LinearLayout xmlns:android="http://schemas.android.com/apk/res/android"
3      android:layout_width="match_parent"
4      android:layout_height="match_parent"
5      android:orientation="vertical">
6     <ProgressBar
7       style="@android:style/Widget.ProgressBar.Horizontal"
8       android:layout_width="match_parent"
9       android:layout_height="wrap_content"
10      android:progress="50" />
11    <ProgressBar
12      style="?android:attr/progressBarStyleLarge"
13      android:layout_width="match_parent"
14      android:layout_height="wrap_content" />
15    <!--其他 ProgressBar 的定义类似-->
16   </LinearLayout>
```

上述第 7 行代码和第 12 行代码使用了两种不同方法来指定进度条的风格。第 12 行代码使用系统的 attr 来指定进度条的 style。

图 3.32　进度条效果

从表 3-28 可以看出，ProgressBar 有两个进度属性。一个是 android：progress，另一个是 android：secondaryProgress。后者主要为缓存需要设计。例如，在看网络视频的时候都会有一个缓存的进度条和一个播放的进度条，这样 progress 就用来设定播放的进度，而 secondaryProgress 就用来设定缓存的进度。

下面以一个模拟下载的示例来介绍 Widget.ProgressBar.Horizontal 风格进度条的使用步骤。

（1）定义布局文件。在布局文件里添加一个 ProgressBar 组件用于显示进度条、添加一个 TextView 用于显示进度值、一个 Button 用于实现"下载"监听事件，其代码比较简单，读者可以参阅 sampleapp 文件夹下的 src/main/res/layout/pbdownload_layout.xml 文件。

（2）实现"下载"按钮的监听事件。"下载"按钮设置的 Button 的单击监听事件，代码如下。

```
1    btn.setOnClickListener(new View.OnClickListener() {
2       @Override
3       public void onClick(View v) {
4           i=0;
5           new Thread(runnable).start();
6       }
7    });
```

由于本例中要动态更新 TextView 上的显示内容，所以需要使用子线程实现，实现代码如下。

```
1    Handler handler = new Handler(){
2        @Override
3        public void handleMessage(Message msg) {
4            super.handleMessage(msg);
5            pb.setProgress(i);
6            tv.setText(i + "% ");
7        }
8    };
9    Runnable runnable = new Runnable() {
10       @Override
11       public void run() {
12           while (i <= 100){
13               handler.sendEmptyMessage(i);
14               try {
15                   Thread.sleep(500);//每隔0.5s进度条更新一次
16               } catch (InterruptedException e) {
17                   e.printStackTrace();
18               }
19               i++;
20           }
21       }
22   };
```

以上代码使用 Handler 和 Runnable 实现主线程中 TextView 上内容的动态更新和子线程中实现进度值的动态改变。进度条模拟下载效果如图 3.33 所示。

图 3.33 进度条模拟下载效果

3. ToggleButton 和 Switch

ToggleButton（开关按钮）和 Switch（开关）组件都继承自 CompoundButton 类，后者在 Android 4.0 以上的版本才能使用。它们通常用于两种状态改变时要实现的功能，如打开 WiFi 或关闭 WiFi。ToggleButton 的常用属性及说明见表 3-30，Switch 的常用属性及说明见表 3-31。

表 3-30 ToggleButton 的常用属性及说明

属　　性	说　　明
android:disabledAlpha	设置按钮禁用时的透明度
android:textOff	设置按钮没有被选中时显示的文字
android:textOn	设置按钮被选中时显示的文字

表 3−31　Switch 的常用属性及说明

属　　性	说　　明
android:showText	设置 on/off 的时候是否显示文字
android:splitTrack	是否设置一个间隙，让滑块与底部图片分隔
android:switchPadding	设置滑块内文字的间隔
android:switchMinWidth	设置开关的最小宽度
android:track	设置开关背景图片
android:thumb	设置开关滑块图片

使用默认的 ToggleButton 和 Switch 比较简单，只需要在布局文件中定义组件就可以了，定义代码如下。

```
1   <LinearLayout
2       android:layout_gravity ="center"
3       android:layout_width ="wrap_content"
4       android:layout_height ="wrap_content"
5       android:orientation ="horizontal" >
6       <Switch
7           android:id ="@ +id/sw"
8           android:layout_width ="wrap_content"
9           android:layout_height ="wrap_content" />
10      <ToggleButton
11          android:textSize ="20dp"
12          android:id ="@ +id/tb"
13          android:layout_width ="wrap_content"
14          android:layout_height ="wrap_content" />
15  </LinearLayout >
```

由于 ToggleButton 和 Switch 都是继承自 CompoundButton，所以当开关状态改变时，其监听事件可以用如下代码实现。

```
1   sw.setOnCheckedChangeListener (new CompoundButton.OnCheckedChangeListener
() {
2       @Override
3       public void onCheckedChanged(CompoundButton buttonView, boolean isChecked) {
4           if(isChecked){
5               img.setImageResource(R.mipmap.dengl);//设置开灯图片
6           }else{
7               img.setImageResource(R.mipmap.dengh); //设置关灯图片
8           }
9       }
10  });
```

开关灯效果如图 3.34 所示。由图 3.34 所示的运行效果可以看出，使用默认的 ToggleButton 和 Switch 样式不美观，在大多数的 Android 应用开发中，为了获得比较好的用户界面效果，往往需要使用自定义样式来美化开关，其实现步骤如下。

图 3.34 开关灯效果

(1) 复制 switch1.png 和 switch2.png 图片到 drawable 目录下。

(2) 在 drawable 目录下创建两个开关效果文件。

① 创建 thumb_selector.xml 文件用于设置滑块的图标，代码如下。

```
1    <?xml version ="1.0" encoding ="utf-8"?>
2    <selector xmlns:android ="http://schemas.android.com/apk/res/android">
3        <item android:drawable ="@drawable/switch1" android:state_checked ="true" />
4        <item android:drawable ="@drawable/switch2" android:state_checked ="false" />
5    </selector>
```

② 创建 track_selector.xml 文件用于设置开关背景的图标，代码如下。

```
1    <?xml version ="1.0" encoding ="utf-8"?>
2    <selector xmlns:android ="http://schemas.android.com/apk/res/android">
3        <item android:state_pressed ="true" android:drawable ="@drawable/closeswitch"/>
4        <item android:state_pressed ="false" android:drawable ="@drawable/openswitch"/>
5    </selector>
```

(3) 在开关布局代码中应用 (2) 中定义的效果，代码如下。

```
1    <Switch
2        android:id ="@+id/sw"
3        android:thumb ="@drawable/thumb_selector"
4        android:track ="@drawable/track_selector"
5        android:layout_width ="wrap_content"
6        android:layout_height ="wrap_content" />
```

自定义开关图标效果如图 3.35 所示。

图 3.35 自定义开关图标效果

3.7.2 猜扑克游戏的实现

1. 主界面的设计

【游戏的实现】

由图3.30所示的效果可以将用户界面的设计分解成如图3.36所示。从图3.35可以看出，整个用户界面布局垂直线性摆布，接着放置ProgressBar用于显示倒计时功能、用TextView显示"请猜红桃A是哪一张？"；然后放置一个HorizontalScrollView滚动视图组件，在其中嵌套一个水平方向的LinearLayout，并同时放置三个ImageView用于摆放三张扑克牌；最后的命令按钮和游戏的难易设置和背景的实现比较简单，读者可以参阅guesspork文件夹中的内容。

图3.36 用户界面的设计分解图

其余关键代码如下。

```
1   <?xml version ="1.0" encoding ="utf-8"? >
2   <LinearLayout xmlns:android ="http://schemas.android.com/apk/res/android"
3     android:id ="@ + id/llbackground"
4     android:layout_width ="match_parent"
5     android:layout_height ="match_parent"
6     android:orientation ="vertical" >
7     <ProgressBar
8       android:id ="@ + id/pbtime"
9       style ="@android:style/Widget.ProgressBar.Horizontal"
10      android:layout_width ="match_parent"
11      android:layout_height ="wrap_content" />
12    <TextView
13      android:paddingTop ="25dp"
14      android:layout_width ="wrap_content"
```

```
15        android:layout_height = "wrap_content"
16        android:layout_gravity = "center_horizontal"
17        android:text = "请猜红桃 A 是哪一张?"
18        android:textSize = "25dp" />
19    <HorizontalScrollView
20        android:layout_width = "380px"
21        android:layout_height = "wrap_content"
22        android:layout_gravity = "center" >
23        <LinearLayout
24            android:id = "@ + id/linearLayout1"
25            android:layout_width = "wrap_content"
26            android:layout_height = "wrap_content"
27            android:gravity = "center"
28            android:orientation = "horizontal" >
29            <ImageView
30                android:id = "@ + id/mImage01"
31                android:layout_width = "wrap_content"
32                android:layout_height = "wrap_content"
33                android:layout_weight = "0.20"
34                android:padding = "5px"
35                android:src = "@mipmap/pork"/ >
36            <!--其他 ImageView 的定义类似 -->
37        </LinearLayout >
38    </HorizontalScrollView >
39    <ImageButton
40        android:layout_gravity = "center"
41        android:id = "@ + id/btnplay"
42        android:layout_width = "wrap_content"
43        android:layout_height = "wrap_content"
44        android:src = "@mipmap/gamestart" / >
45    <LinearLayout
46        android:gravity = "center"
47        android:layout_width = "match_parent"
48        android:layout_height = "match_parent"
49        android:background = "#e9f2a2"
50        android:orientation = "horizontal" >
51        <LinearLayout
52            android:gravity = "center"
53            android:layout_width = "wrap_content"
54            android:layout_height = "wrap_content"
55            android:orientation = "horizontal"
56            android:paddingTop = "5dp" >
57            <!--游戏难易度设置略 -->
58        </LinearLayout >
59        <LinearLayout
60            android:layout_width = "wrap_content"
61            android:layout_height = "wrap_content"
62            android:orientation = "horizontal"
63            android:paddingLeft = "15dp" >
64            <TextView
```

```
65              android:id ="@ +id/gamelamp"
66              android:layout_width ="wrap_content"
67              android:layout_height ="wrap_content"
68              android:text ="背景灯"
69              android:textSize ="20dp" />
70          <ToggleButton
71              android:id ="@ +id/tblamp"
72              android:layout_width ="wrap_content"
73              android:layout_height ="wrap_content"
74              android:textSize ="20dp" />
75         </LinearLayout >
76     </LinearLayout >
77  </LinearLayout >
```

2. 功能实现

（1）定义变量。

```
1      private int easylever = 15;//默认难易度
2      private int progress = 15;//进度条默认进度
3      private ProgressBar pb;
4      private RadioGroup radioGroup;
5      private RadioButton rbEasy, rbDiff;
6      private ToggleButton tbLamp;
7      private ImageButton btnStart;
8      private LinearLayout llBackground;//用于背景开关选择后设置背景
9      int[] imgporkId = {R.mipmap.blacka, R.mipmap.flowera, R.mipmap.fanga};//红桃
A,梅花A,方块A
10     private ImageView[] img = new ImageView[3];
11     int[] imgViewID = {R.id.mImage01, R.id.mImage02, R.id.mImage03};//3个
ImageView对象的ID
```

（2）随机产生红桃A出现的数组元素下标。

首先用random()产生一个[0，3）的随机数用以确定红桃A所在的数组元素位置，然后将默认的红桃A所在数组元素的位置0与产生的位置交换，实现代码如下。

```
1      private void setImageView() {
2          int s = (int) (Math.random() * 3);
3          int temp = imgporkId[s];
4          imgporkId[s] = imgporkId[0];
5      imgporkId[0] = temp;
6      }
```

（3）使用Handler、Runnable结合，实现进度条的递减效果。

前面已经介绍了进度条增进度值的使用方法，实际开发中也经常会遇到类似倒计时的进度递减的情况，实现时需要一开始就将进度值设置为最大值，然后使用线程实现进度值的递减操作，实现代码如下。

```
1      Handler handler = new Handler() {
2          @Override
3          public void handleMessage(Message msg) {
```

```
4          super.handleMessage(msg);
5          pb.setProgress(msg.arg1);
6          if (msg.arg1 == 0) {
7              //游戏结束开启"开始按钮"和难易度单选
8              btnStart.setEnabled(true);
9              rbEasy.setEnabled(true);
10             rbDiff.setEnabled(true);
11         }
12     }
13 };
14 Runnable runnable = new Runnable() {
15     @Override
16     public void run() {
17         while (progress >= 0) {
18             Message msg = Message.obtain();
19             msg.arg1 = progress;
20             handler.sendMessage(msg);
21             try {
22                 Thread.sleep(1000);
23             } catch (InterruptedException e) {
24                 e.printStackTrace();
25             }
26             progress--;
27         }
28     }
29 };
```

（4）定义实现 View.OnClickListener 接口的类，实现单击 ImageView 后把扑克牌翻开。使用 getId( ) 方法获得该图片的 id，判断单击的是哪一张扑克牌，然后通过 setImage-Resource( ) 方法设置对应的扑克牌图片，实现代码如下。

```
1  class MyimageClistener implements View.OnClickListener {
2      @Override
3      public void onClick(View v) {
4          switch (v.getId()) {
5              case R.id.mImage01:
6                  img[0].setImageResource(imgporkId[0]);
7                  break;
8              case R.id.mImage02:
9                  img[1].setImageResource(imgporkId[1]);
10                 break;
11             case R.id.mImage03:
12                 img[2].setImageResource(imgporkId[2]);
13                 break;
14             default:
15                 break;
16         }
17     }
18 }
```

(5) 启动子线程，开始游戏。

```
1    btnStart.setOnClickListener(new View.OnClickListener() {
2      @Override
3      public void onClick(View v) {
4        pb.setMax(easylever);//通过难易度设置进度条的最大值
5        new Thread(runnable).start();//开启线程计时
6        //游戏开始后不能进行开始按钮、单选按钮的操作
7        btnStart.setEnabled(false);
8        rbDiff.setEnabled(false);
9        rbEasy.setEnabled(false);
10       setImageView();
11       for (int i = 0; i <img.length; i ++) {
12         img[i].setImageResource(R.mipmap.pork);//让每个 ImageView 都显示扑克的背面
13         img[i].setEnabled(true);
14       }
15     }
16   });
```

游戏难易度设置及背景开关设置的实现代码比较简单，限于篇幅不再赘述，读者可以参阅 guesspork 文件夹中的内容。游戏开始运行后的效果如图 3.37 所示。

图 3.37　游戏开始运行的效果

本 章 小 结

本章结合六个典型案例介绍了 Android 中常用基本组件及应用开发中一些常用类的使用方法，如 Toast、Handler、Message 和子线程等。读者通过本章的学习，结合 Java 的编程

知识就可以开发出一些满足用户要求的 Android App，同时也为后续的 Android App 开发打下良好的基础。

习　　题

一、选择题

1. 在设置界面布局的时候，如果将 TextView 的宽度设置为等于文字的内容的长度，则 layout_width 的值为（　　）。

　　A. wrap_content　　　　B. fill_parent　　　　C. match_parent　　　　D. match_content

2. Android 中包含多种界面布局，如果需要根据其他控件的位置设置当前控件的位置需要使用（　　）。

　　A. LinearLayout　　　　B. FrameLayout　　　　C. AbsolutLayout　　　　D. RelativeLayout

3. 下列哪个属性可以设置 EditText 编辑框内容为空时文本框中的提示信息？（　　）

　　A. android：inputType　　　　　　　　B. android：text

　　C. android：digits　　　　　　　　　　D. android：hint

4. 下面定义 style 的方式正确的是（　　）。

　　A. ＜resources＞
　　　　　＜style name = "myStyle"＞
　　　　　　　＜item name = "android：layout_width"＞fill_parent＜/item＞
　　　　　＜/style＞
　　＜/resources＞

　　B. ＜style name = "myStyle"＞
　　　　　＜item name = "android：layout_width"＞fill_parent＜/item＞
　　＜/style＞

　　C. ＜resources＞
　　　　　＜item name = "android：layout_width"＞fill_parent＜/item＞
　　＜/resources＞

　　D. ＜resources＞
　　　　　＜style name = "android：layout_width"＞fill_parent＜/style＞
　　＜/resources＞

5. 在设计手机界面布局时，设置控件的 android：layout_toRighOf 属性为其他某个控件的 id，则该界面应该使用的布局是（　　）。

　　A. LinearLayout　　　　B. FrameLayout　　　　C. TableLayout　　　　D. RelativeLayout

6. 下列不属于 Android 布局的是（　　）。

　　A. FrameLayout　　　　B. BorderLayout　　　　C. TableLayout　　　　D. RelativeLayout

7. 下列说法错误的是（　　）。

　　A. Button 是普通按钮组件，除此外还有其他的按钮组件

　　B. TextView 是显示文本的组件，TextView 是 EditText 的父类

　　C. EditText 是编辑文本的组件，可以使用 EditText 输入特定的字符

D. ImageView 是显示图片的组件，可以通过设置显示局部图片

8. 下列哪个属性可以设置 EditText 编辑框只能输入数字？（　　）。

A. android:inputType　　　　　　B. android:text

C. android:digits　　　　　　　　D. android:hint

9. 下列关于进度条对话框（ProgressDialog）说法，错误的是（　　）。

A. ProgressDialog 通过 setProgress()方法设置当前进度

B. ProgressDialog 的 setProgressStyle()方法设置进度条显示的样式，可以设置为长形或圆形

C. ProgressDialog 上不仅可以设置文字，还可以设置图标和按钮

D. 调用 ProgressDialog 中的 start()方法，开始增加进度条的进度，直到到达 setMax()方法设置的最大值停止

10. 下列关于布局管理器的叙述，错误的是（　　）。

A. LinerLayout 是一种最常用的布局方式，在此布局中如果组件的宽度和高度超过了屏幕的宽度和高度，超出的组件不会显示在屏幕上

B. RelativeLayout 可以让 App 在不同大小、不同分辨率的屏幕上使用合适的大小显示组件

C. FrameLayout 在显示组件时只能从屏幕的左上角显示，不能设置组件在布局中居中显示

D. TableLayout 由多个 TableRow 组件组成，每个 TableRow 对应表格中的一行

二、填空题

1. 如果需要使用 TextView、EditText、Button、CheckBox 等控件，需要导入_____ 包。

2. 多个 RadioButton 控件默认情况下不会形成互斥，需要把 RadioButton 放入_____控件中。

3. 定义 LinearLayout 水平方向布局时至少要设置三个属性：android:layout_width、android:layout_height和_____。

4. 代码 android:layout_above = "@id/tvMsg"，使用的是_____布局。

5. Android 中 TabHost 控件是整个 Tab 的容器，包括_____和 FrameLayout 两个部分。

6. LinearLayout 中设置权重的属性为_____。

7. RadioButton 组件的_____属性用于设置其显示的文本内容。

三、判断题

1. 在实现普通按钮单击事件的监听器接口 OnClickListener 时，需要导入 Android.content.DialogInterface.OnClickListener。　　　　　　　　　　　　　　　　　　　　　（　　）

2. ProgressDialog 用于定义进度条对话框，其中 setProgressStyle 属性用于设置进度条的风格，可以为圆形或长形。　　　　　　　　　　　　　　　　　　　　　　　　　（　　）

3. TabHost 控件用于在界面中显示多个分页，每个分页可以显示不同的内容，因此需要为每个页建立一个布局文件。　　　　　　　　　　　　　　　　　　　　　　　（　　）

4. 一个用户界面上有多个 RadioButton 控件，如果没有放在 RadioGroup 控件中，则它们可以同时被选中。　　　　　　　　　　　　　　　　　　　　　　（　　）

5. 开发 Android App 时，由开发环境最后将应用程序编译生成.exe 文件。　　　　　　　　　　　　　　　　　　　　　　　　　　　　　　　　　　　　　（　　）

【第3章参考答案】

6. EditText 只能接收用户输入信息，它没有 setText()方法。（　　）

第 4 章
高级界面组件与布局优化

用户界面作为用户和系统交互的基础,是人(用户)机(Android 终端)交互的核心。在现在的软件开发过程中,用户界面开发的效率和质量已经成为影响整个软件产品质量的一个重要因素,Android 移动终端的应用软件也不例外。通过前面章节的学习,读者应该已经掌握了 Android 项目开发中涉及的基本组件和基本布局管理器的使用方法。这些组件和布局管理器也基本上可以满足一般的用户界面设计需求,但是在一些特定的应用场景中,还需要通过 ListView、Fragment、ViewPager、ViewFlipper 和 RecyclerView 等组件及与它们配合使用的各种各样的适配器(Adapter)来实现不同需求的用户界面和特殊的应用效果。本章将结合项目案例介绍它们的应用方法。

掌握 Android 中 Adapter、ListView、Fragment、ViewPager、ViewFlipper 和 RecyclerView 等应用场景和使用方法。

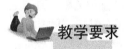

知识要点	能力要求	相关知识
Adapter	理解 Adapter 的基本概念和常用子类	接口和类
通讯录的设计与实现	(1)掌握 ListView 的常用属性和使用方法 (2)掌握 Intent、IntentFilter 的基本概念 (3)掌握利用 Intent 启动和关闭 Activity 的方法、不同 Activity 中传递数据的方法 (4)掌握利用 ListView 与不同的 Adapter 设计不同的用户界面效果	基本组件的监听事件
仿微信主界面的设计与实现	(1)掌握 Fragment 的基本概念和使用方法 (2)掌握使用 ListView 和 Fragment 相结合设计特殊应用效果的用户界面	基本布局

（续）

知识要点	能力要求	相关知识
仿今日头条的设计与实现	（1）掌握 ViewPager、PagerTitleStrip 和 PagerTabStrip 的常用属性和使用方法 （2）掌握 TabLayout 的使用方法 （3）掌握使用 ViewPager、Fragment 和 TabLayout 相结合设计特殊应用效果的用户界面	FragmentPagerAdapter
轮播效果的设计与实现	（1）掌握 ViewFlipper 类的常用属性和方法 （2）掌握使用 ViewFlipper 和基本组件实现轮播效果的步骤	ViewAnimator
商品列表布局切换效果的设计与实现	（1）掌握 RecyclerView 类的使用步骤 （2）熟悉 RecyclerView.LayoutManager 的三个子类的使用方法 （3）掌握使用 RecyclerView.Adapter、RecyclerView.ViewHolder 和普通布局文件相结合产生商品列表切换效果的实现步骤	基本布局

4.1 Adapter

4.1.1 基本概念

Adapter 是 android.widget 包中定义的一个接口。在 Android App 开发中进行较复杂的用户界面设计时，通常需要使用 Adapter 将数据源绑定到指定的 View 上。也就是说，Adapter 是连接数据和 AdapterView（ListView 是一个典型的 AdapterView）的桥梁，通过它能够有效地实现数据与 AdapterView 的分离设置，使 AdapterView 与数据的绑定更加简便、修改更加方便。高级界面设计组件 ListView、GridView 及 ViewPage 等都需要使用 Adapter 来为其设置数据源。Adapter 的继承关系如图 4.1 所示。

4.1.2 Adapter 的常用子类

1. ListAdapter

ListAdapter 是一个直接继承于 Adapter 的接口类，它是 ListView 和数据之间的桥梁，通常数据来源于一个 Cursor（游标），用于显示包装在 ListAdapter 中的数据。

2. BaseAdapter

BaseAdapter 是一个实现了 ListAdapter 和 SpinnerAdapter 接口的抽象类，通过重写 View getView（int position, View convertView, ViewGroup parent）、int getCount（）、long getItemId（int position）和 Object getItem（int position）四个方法，可以实现复杂的用户界面布局。BaseAdapter 是实际开发中用得最多的一个 Adapter。

第4章 高级界面组件与布局优化

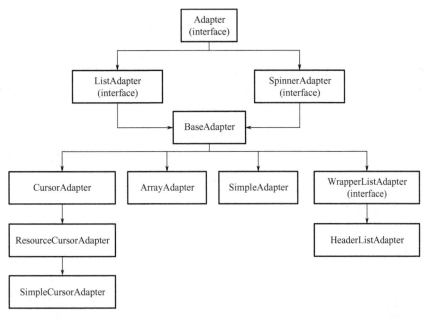

图 4.1　Adapter 的继承关系

3. ArrayAdapter

ArrayAdapter 是继承于 BaseAdapter 的一个具体类。默认情况下，ArrayAdapter 绑定每个对象的字符串值到系统默认布局定义的 TextView 或开发者自定义布局的 TextView 上，也就是说每一行只能显示一个文本。

4. SimpleAdapter

SimpleAdapter 是继承于 BaseAdapter 的一个具体类，它可以根据开发者的要求在每一行上显示图片、文本等复杂的布局对象，具有很好的扩充性，可以自定义出各种效果显示在 ListView 的每一行上。

5. SimpleCursorAdapter

SimpleCursorAdapter 用来绑定数据库里面的数据，它有以下两种构造方法。

（1）public SimpleCursorAdapter（Context context，int layout，Cursor c，String[ ] from，int[ ] to）。

（2）public SimpleCursorAdapter（Context context，int layout，Cursor c，String[ ] from，int[ ] to，int flags）。

第 2 种构造方法的最后一个参数 flags 有 CursorAdapter.FLAG_AUTO_REQUERY 和 CursorAdapter.FLAG_REGISTER_CONTENT_OBSERVER 两个值，而第 1 种构造方法就是少了最后一个参数，默认被设置为 CursorAdapter.FLAG_AUTO_REQUERY。在 Android 3.0 及以上版本中，没有 flags 参数或者 flags 参数被设置为 CursorAdapter.FLAG_AUTO_REQUERY 的使用方法已经废弃，不建议读者在开发中使用。

4.2　通讯录的设计与实现

移动终端App经常需要将要显示的内容以列表的方式显示出来。例如，在Android自带的联系人应用中显示联系人列表；在新闻浏览软件中显示的新闻条目；在微信客户端中显示微信信息；在淘宝客户端显示商品信息等。为了实现这样的用户界面效果，Android提供了一个比较常用的用户界面组件——ListView，专门用于以列表方式展示信息的具体内容，并且能够根据数据的长度自适应显示。本节以一个通讯录的实现过程来介绍ListView的用法。

4.2.1　预备知识

1. ListView

【ListView、ArrayAdapter、SimpleAdapter】

ListView 是 android.widget.AbsListView 的子类，主要用来以列表方式显示一些内容。使用 ListView 开发 App，通常有两个功能：①将数据填充到布局；②处理用户的选择单击操作。

开发者使用 ListView 组件开发 App 时必须包含三个关键要素：①ListView 中每一行的 View；②填入 View 中的数据（被映射的字符串、图片或基本组件）；③连接数据与 ListView 的 Adapter。

4.1节已经详细介绍了多个不同的 Adapter，下面分别介绍不同 Adapter 与 ListView 结合使用的方法。

（1）ArrayAdapter 与 ListView。
- 在要显示列表的界面布局文件中添加 ListView 组件，代码如下。

```
<ListView
android:layout_width="fill_parent"
android:layout_height="fill_parent"
android:id="@+id/listview" />
```

- 为 ArrayAdapter 装配数据，代码如下。

```
String[]items = {"周一","周二","周三","周四","周五","周六","周日"};
ArrayAdapter adapter = new ArrayAdapter(this, android.R.layout.simple_list_item_1, items);
```

ArrayAdapter 有多个构造方法，而最常用的一种构造方法原型如下。
public ArrayAdapter(Context context, int resource, T[] objects)

第1个参数为上下文；第2个参数为一个包含 TextView、用来填充 ListView 的每一行的布局资源 ID；第3个参数为 ListView 的每一行上具体要显示的数据。其中，第2个参数可以自定义一个布局文件，但是该布局文件中必须要有一个 TextView 组件，通常可以使用 Android 提供的布局资源。Android 提供的布局资源及功能说明见表4-1。

表4-1 Android 提供的布局资源及功能说明

布 局 资 源	功 能 说 明
android. R. layout. simple_list_item_1	每一项只有一个 TextView
android. R. layout. simple_list_item_multiple_choice	每一项有一个复选框,需要用 setChoiceMode( )方法设置选择模式
android. R. layout. simple_list_item_checked	每一项有一个选择项,需要用 setChoiceMode( )方法设置选择模式
android. R. layout. simpte. list_item_single_choice	每一项有一个单选按钮,需要用 setChoiceMode( )方法设置选择模式

- 将 Adapter 与 ListView 相关联,代码如下。

```
lv.setAdapter(adapter);
```

使用 setChoiceMode( )方法设置的选择模式主要有两种:①多选模式——ListView. CHOICE_MODE_MULTIPLE;②单选模式——ListView. CHOICE_MODE_SINGLE。以上代码执行效果如图 4.2 所示。如果要显示如图 4.3 所示效果,需要将以上代码修改如下。

```
ArrayAdapter adapter = new ArrayAdapter(this,
android.R.layout.simple_list_item_checked, items);
   lv.setChoiceMode (ListView.CHOICE_MODE_MULTIPLE);  //设置选择框可以实现选择
```

图 4.2 ListView 显示效果(1)

图 4.3 ListView 显示效果(2)

如果要实现如图 4.4 所示效果,需要将以上代码修改如下。

```
ArrayAdapter adapter = new ArrayAdapter(this, android.R.layout.simple_list_
item_single_choice, items);
   lv.setChoiceMode (ListView.CHOICE_MODE_SINGLE);  //设置单选按钮可以实现选择
```

如果要实现如图 4.5 所示效果,需要将以上代码修改如下。

```
ArrayAdapter adapter = new ArrayAdapter(this, android.R.layout.simple_list_
item_multiple_choice, items);
   lv.setChoiceMode (ListView.CHOICE_MODE_MULTIPLE);
```

图 4.4　ListView 显示效果（3）　　　　　图 4.5　ListView 显示效果（4）

- 设置监听事件

ListView 可以通过调用 setOnItemClickListener( ) 接口方法设置 "单击" ListView 某一项的监听事件，也可以通过调用 setOnItemLongClickListener( ) 接口方法设置 "长按" ListView 某一项的监听事件。这两个接口方法的原型如下。

public boolean onItemLongClick(AdapterView <?> parent, View view, int position, long id)
public void onItemClick(AdapterView <?> parent, View view, int position, long id)

这两个接口方法的参数个数和参数表示的意义完全一样：第 1 个参数表示事件发生的 AdapterView；第 2 参数表示单击或长按的 ListView 的某一项 View；第 3 个参数表示单击或长按的这一项在 Adapter 中的位置（从 0 开始计数）；第 4 个参数表示单击或长按的这一项在 ListView 中对应的位置（从 0 开始计数）。

用匿名内部类实现 Toast 显示单击 ListView 中某一项内容的代码如下。

```
1    lv.setOnItemClickListener(new AdapterView.OnItemClickListener() {
2      @Override
3      public void onItemClick(AdapterView<?> parent, View view, int position, long id) {
4        TextView tv = (TextView) view;
5        Toast.makeText(CostomListActivity.this,tv.getText().toString() + ":" + items[position],Toast.LENGTH_SHORT).show();
6      }
7    });
```

上述第 4 行代码表示将单击的某一项 View 转换为 TextView 类型（由于 Adapter 中装配数据时的默认布局中显示文本使用的是 TextView 组件）；第 5 行代码中的 tv.getText( ).toString( ) 表示获得 TextView 上的文本信息，items[position] 表示获得 items 数组中下标为 position 的元素。

在实例化 Adapter 时如果使用了 android.R.layout.simple_list_item_multiple_choice 和 android.R.layout.simple_list_item_checked 的系统布局，并且选择模式设置为 ListView.CHOICE_MODE_MULTIPLE，那么在实现多个选项被选中的事件时，就需要使用下列代码。

第4章 高级界面组件与布局优化

```
1    lv.setOnItemClickListener(new AdapterView.OnItemClickListener() {
2       @Override
3       public void onItemClick(AdapterView <? > parent, View view, int position, long id) {
4          SparseBooleanArray checked = lv.getCheckedItemPositions();
5          String str = "";
6          for(int i = 0;i < items.length;i ++ ){
7             if(checked.get(i)){//如果该位置的 checkbox 被选中
8                str + = items[i];
9             }}
10         Toast.makeText(CostomListActivity.this, + str,Toast.LENGTH_SHORT).show();
11      }
12   });
```

以上第4行代码使用了 SparseBooleanArray 类，它是一个 Map <键，值> 映射类，键为选择位置，值是被选择的 boolean 值。详细的实现代码读者可参见 sampleapp 文件夹中的 CostomListActivity.java 文件。

（2）SimpleAdapter 与 ListView。
- 在要显示列表的界面布局文件中添加 ListView 组件（lxrlistview.xml），代码如下。

```
<ListView
    android:layout_width = "fill_parent"
    android:layout_height = "fill_parent"
    android:id = "@ + id/listview" />
```

- 根据需要定义 ListView 每行所显示内容的布局文件（lxr_layout.xml），代码如下。

```
<?xml version = "1.0" encoding = "utf-8"? >
<LinearLayout xmlns:android = "http://schemas.android.com/apk/res/android"
    android:layout_width = "match_parent"
    android:layout_height = "match_parent"
    android:orientation = "horizontal" >
    <ImageView
        android:id = "@ + id/lxrimg"
        android:layout_width = "40dp"
        android:layout_height = "40dp"
        android:src = "@mipmap/keu" />
    <TextView
        android:id = "@ + id/lxrname"
        android:layout_width = "wrap_content"
        android:layout_height = "wrap_content"
        android:layout_marginLeft = "15dp"
        android:layout_marginTop = "10dp"
        android:text = "张三丰"
        android:textSize = "20sp" />
</LinearLayout>
```

- 在功能代码块中（LxrActivity.java）定义一个 HashMap 构成的 List（列表），将数据以键值对的方式存放在里面，然后构造 SimpleAdapter 对象，并将数据装配到该 Adapter

中,代码如下。

```
1    int    imgId[] = {R.mipmap.keu,R.mipmap.kev,R.mipmap.kew,R.mipmap.kex};
2    String lxrName[] = {"范小萤","王大江","吴海波","李晶晶"};
3    ArrayList<HashMap<String,Object>> lxrList = new ArrayList<HashMap<String,Object>>();
4    for(int i=0;i<imgId.length;i++){
5        HashMap<String,Object> map = new HashMap<String,Object>();
6        map.put("itemImg",imgId[i]);
7        map.put("itemname",lxrName[i]);
8        lxrList.add(map);
9    }
10   String [] from = {"itemImg","itemname"};
11   int [] to = {R.id.lxrimg,R.id.lxrname};
12   SimpleAdapter simpleAdapter = new SimpleAdapter(this,lxrList,R.layout.lxr_layout,from,to);
13   lvLxr.setAdapter(simpleAdapter);
```

上述第3～9行代码用于为SimpleAdapter准备HashMap构成的List数据集;第10行代码用于指定数据源与ListView的每一行中每个显示项对应的键;第11行代码用于指定ListView的每一行中每个显示项对应键的值填充到对应项组件的ID;第12行代码为SimpleAdapter装配数据、指定每行布局资源及将HashMap的每个键值数据映射到每行布局文件中对应ID的组件上。SimpleAdapter的原型如下。

public SimpleAdapter(Context context, List <? extends Map <String, ? >> data, int resource, String[] from, int[] to)

图4.6 ListView显示效果(5)

其中,第1个参数为上下文;第2个参数为由HashMap(String, Object)构成的List;第3个参数为ListView上显示的每一行布局文件的ID;第4个参数为HashMap中所有键构成的字符串数组;第5个参数为ListView上每一行布局文件中对应组件ID构成的int型数组。

以上代码运行后的显示效果如图4.6所示。

(3) SimpleCursorAdapter与ListView。

由于SimpleCursorAdapter需要与数据库结合使用,本书的数据库使用将在第7章介绍,本节以调用Android通讯录并显示在自定义的用户界面上为例介绍它的用法。与其他Adapter的使用一样,SimpleCursorAdapter需要首先设计一个包含ListView组件的布局文件,然后使用如下功能代码实现。

```
1    ListView listView = (ListView) this.findViewById(R.id.lvlxr);
2    ContentResolver contentResolver = getContentResolver();
3    //获得一个指向系统通讯录数据库的Cursor对象获得数据来源
4    Cursor cur = contentResolver.query(Contacts.People.CONTENT_URI, null, null, null, null);
```

```
 5      startManagingCursor(cur);
 6      String[] from = {Contacts.People.NAME};
 7      int[] to = {android.R.id.text1};
 8      //实例化适配器
 9      SimpleCursorAdapter adapter = new SimpleCursorAdapter(this,
android.R.layout.simple_list_item_1, cur, from, to,
CursorAdapter.FLAG_REGISTER_CONTENT_OBSERVER);
 10     listView.setAdapter(adapter);
```

Android App 间的数据一般是不可以共享访问的，如果需要共享访问，就需要通过 Provider/Resolver 进行。Provider 用于提供数据（内容），Resolver 用于提供接口对这些数据（内容）进行解读。本例中需要读取系统通讯录中的联系人信息，而这个联系人信息由 Android 提供给 Provider，上述第 2 行代码构建了一个 Resolver 用于读取联系人的信息，第 4 行代码使用 Resolver 到系统提供的 Provider——Contacts.People.CONTENT_URI 中获取联系人信息。关于 Provider/Resolver 的详细内容会在以后章节中进行介绍，此处内容读者只要稍做了解即可。

2. Intent

【Intent】

Android 中任何 App 都可以启动其他程序中的组件。例如，想要开发一个包含相机拍照功能的 App，如果已经有其他开发者开发出照相机功能，这个时候就可以直接调用其他开发者开发的照相机功能，而不需要再独立开发。但是由于 Android 中所有 App 都分别运行在每个独立的进程中，并且被权限配置限制对其他程序进行访问，所以用户程序不能从其他程序中直接激活这个拍照组件。如果要激活这个拍照组件，用户程序就必须向 Android 发送一个消息，这个消息通过用户程序的 Intent 对象来启动拍照组件，然后 Android 才会为用户程序激活拍照功能。

在 Android 的四个核心组件中，除 ContentProvider 以外，其他三个核心组件（Activity、Service、BroadcastReceiver）实际上都是被一个叫作 Intent 的异步消息激活的。激活的组件可以是 Android 自身提供的，也可以是由开发者自定义的。其激活方式有显式 Intent 和隐式 Intent 两种。

显式 Intent：通过指定目标组件名称来启动组件，并且每次启动的组件只能有一个。一般情况下，由于开发者不知道其他 App 的组件名称，所以显式 Intent 通常用于启动 App 的内部组件。

隐式 Intent：不指定要启动的目标组件名称，而是指定 Intent 的 Action、Data 或 Category 等，通常用隐式 Intent 激活其他 App 中的组件；在启动组件时，会去匹配 AndroidManifest.xml 相关组件的 IntentFilter，并逐一匹配出满足属性的组件。当不止一个满足时，就会弹出一个让用户选择启动哪个目标组件的对话框。

一个 Intent 对象可以包含以下六个信息，组成结构如图 4.7 所示。

• ComponentName（组件名称）：指明 Intent 对象的组件名称，是一个可选项，如果被设置，Intent 对象就显式指定了要转向的组件。

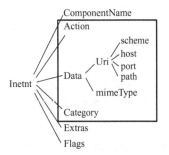

图 4.7 Intent 对象的组成结构

- Action（动作）：指明 Intent 要完成的一个动作，是在 Intent-Filter 中指定的字符串常量；该字符串常量可以在 App 的配置文件中定义，也可以是系统自带的。Intent 不用直接指明这个组件，交给 Android 去根据 IntentFilter 做匹配筛选，选择符合的组件去启动。表 4-2 所示为 Action 常用常量和功能说明。

表 4-2 Action 常用常量和功能说明

常 量 名	常 量 值	功 能 说 明
ACTION_MAIN	android.intent.action.MAIN	应用程序入口
ACTION_VIEW	android.intent.action.VIEW	显示数据给用户
ACTION_DIAL	android.intent.action.DIAL	电话拨号并显示拨号用户界面
ACTION_CALL	android.intent.action.CALL	直接打电话
ACTION_SEND	android.intent.action.SEND	直接发短信
ACTION_SENDTO	android.intent.action.SENDTO	选择对象发短信

- Data（数据）：通常用于向系统提供数据，即 Action 需要的数据，一般用 URI 和对应的 MIME 类型来表示，也是在 IntentFilter 中定义。例如，如果动作是 ACTION_CALL，则需要提供拨出的电话号码（用 tel：URI 表示）数据；如果动作是 ACTION_VIEW，则需要提供访问的目标 URI（用 http：//URI 表示）数据。
- Category（类别）：用于为 Action 提供额外的附加类别信息，也是 IntentFilter 中的字符串常量；一个 Intent 对象只能设置一个 Action，但是可以设置多个 Category。Category 和 Action 通常在 IntentFilter 中使用，用于实现隐式 Intent。Category 可以直接使用系统定义的，也可以由开发者自己定义。表 4-3 所示为 Categroy 常用常量和功能说明。

表 4-3 Category 常用常量和功能说明

常 量 名	常 量 值	功 能 说 明
CATEGORY_LAUNCHER	android.intent.category.LAUNCHER	程序最优先启动的 Activity
CATEGORY_HOME	android.intent.category.HOME	程序启动后显示的第 1 个界面

- Extras（扩展域）：用于为 Intent 对象附加一个或多个键值对信息，该键值对是一个 Bundle 对象。
- Flags（标志）：用于告诉 Android 如何启动 Activity，启动后如何处理。

3. IntentFilter

在启动组件前，Android 必须知道该组件具有哪些功能，而组件所具有的功能就是由 IntentFilter 来实现的，即 IntentFilter 负责过滤所提供的组件功能。IntentFilter 是由动作（Action）、类别（Category）和数据（Data）构成的一种过滤筛选机制。Android 通过解析 IntentFilter，就可以把不满足条件的组件过滤掉，然后筛选出满足条件的组件，供应用程序调用。其匹配原则如下。

第4章 高级界面组件与布局优化

- 任何不匹配的信息都将被过滤。
- 没有指定 Action 的过滤器可匹配任何 Intent，但是没有指定 Category 的过滤器只能匹配没有类别的 Intent。
- Intent 数据 URI 的每一部分都将与 Data 过滤器中的每个属性（协议、主机名、路径名、MIME 类型）进行匹配，只要有一个不匹配，就会被过滤掉。
- 如果匹配了多个结果，则先根据＜intent-filter＞中定义的优先级进行排序，然后再选择优先级最高的那个匹配结果。

在实际应用开发中，可以在配置文件 Androidmanifest. xml 中设置 IntentFilter，也可以直接在功能代码中设置。以下代码是在配置文件中分别设置 action、category 和 data 过滤器。

在应用程序配置文件 Androidmanifest. xml 中设置。

通过 action android:name 属性指定 action 过滤器的代码如下。

```
1    <intent-filter … >
2        <action android:name ="cn. edu. nnutc. edu. SHOW_Name" />
3        <action android:name =" cn. edu. nnutc. edu. SHOW_Tel" />
4        ……
5    </intent-filter>
```

通过 category android:name 属性指定 category 过滤器的代码如下。

```
1    <intent-filter... >
2        <category android:name ="android. intent. category. DEFAULT" />
3        <category android:name ="android. intent. category. BROWSABLE" />
4        ……
5    </intent-filter>
```

通过 data android:mimeTyp 属性指定 data 过滤器的代码如下。

```
1    <intent-filter... >
2        <data android:mimeType ="video/mpeg"
3            android:scheme ="http"  android:host ="www. nnutc. edu. cn"
4            android:port = "8080" android:path ="web "/>
5        <data android:mimeType =" audio/mpeg "
6            android:scheme ="http"  android:host ="www. nnutc. edu. cn"
7            android:port = "8080" android:path ="web "/>
8        ……
9    </intent-filter>
```

每个＜data＞元素可以指定一个 URI 和一个数据类型（MIME 类型）。构成 URI 的每个部分是通过各种属性（scheme、host、port、path）来组合的，即 scheme://host:port/path。

例如，上面 Data 过滤器的 URI 可以写成"http://www.nnutc.edu.cn:8080/web"。也就是说，该代码中的 URI 协议是 http，主机是 www.nnutc.edu.cn，端口是 8080，路径是 web/audio 或 web/video。主机和端口构成 URI 的 authority，如果没有指定主机则忽略端口。

4. 利用 Intent 启动和关闭 Activity

Android App 中一般都有多个 Activity，在 Activity 中通过调用 startActivity()或 start-

ActivityForResult()方法，并在方法的参数中传递 Intent 对象，就可以实现不同 Activity 之间的切换和数据传递。

（1）显式 Intent 启动 Activity。

显式 Intent 启动 Activity 即直接指定需要打开的 Activity 对应类。下面以执行 App 时单击 Button 实现 AActivity 跳转到 BActivity 为例说明显式 Intent 启动 Activity 的方法。

首先，在 App 中创建 AActivity.java、BActivity.java 和对应的布局文件 aactivity_layout.xml、bactivity_layout.xml。该 App 的配置清单文件代码如下。

```xml
1  <?xml version="1.0" encoding="utf-8"?>
2  <manifest xmlns:android="http://schemas.android.com/apk/res/android"
3     package="com.example.sampleapp">
4    <application
5       android:allowBackup="true"
6       android:icon="@mipmap/ic_launcher"
7       android:label="@string/app_name"
8       android:roundIcon="@mipmap/ic_launcher_round"
9       android:supportsRtl="true"
10      android:theme="@style/AppTheme">
11      <activity android:name=".AActivity">
12        <intent-filter>
13          <action android:name="android.intent.action.MAIN" />
14          <category android:name="android.intent.category.LAUNCHER"/>
15        </intent-filter>
16      </activity>
17      <activity android:name=".BActivity"></activity>
18    </application>
19  </manifest>
```

从配置文件清单可以看出，BActivity 并没有设置 intent filter 选项，同时也没有设置 action 和 category。如果要启动 BActivity，就需要在 AActivity 的 Button 单击监听事件中用显式 Intent 启动 BActivity，代码如下。

```java
1  btnA.setOnClickListener(new View.OnClickListener() {
2    @Override
3    public void onClick(View v) {
4      Intent intent = new Intent(AActivity.this,BActivity.class);
5      AActivity.this.startActivity(intent);
6    }
7  });
```

以上第 4 行代码直接使用 Intent 构造方法传入 ComponentName，这种方法在显式 Intent 中使用最多，除了这种方法外，还有以下两种方法，但用得较少。

● setComponent 方法，代码如下。

```java
1  ComponentName componentName = new ComponentName(AActivity.this, BActivity.class);
2  Intent intent = new Intent();
3  intent.setComponent(componentName);
4  startActivity(intent);
```

- setClass/setClassName 方法，代码如下。

```
1    Intent intent = new Intent();
2    intent.setClass(AActivity.this,BActivity.class);
3    //intent.setClassName(AActivity.this, "cn.edu.nnutc.edu.ie.BActivity");
4    startActivity(intent);
```

（2）隐式 Intent 启动 Activity。

隐式 Intent 启动 Activity 是指 Android 根据过滤规则自动去匹配对应的 Intent，即不需要在 Intent 对象中明确指明启动的是哪个 Activity，而是让 Android 来决定应该启动谁。在这种情况下，Android 会自动匹配最适合处理 Intent 的一个或多个 Activity。匹配的 Activity 可能是 App 自身的，也可能是 Android 内置的，还可能是第三方 App 提供的。下面仍以显式 Intent 的示例来介绍隐式 Intent 启动 Activity 的方法。

首先，修改配置清单文件代码，添加 BActivity 的 intent filter 属性，并且 category 属性值设置为 "android.intent.category.DEFAULT"。详细代码如下。

```
1        <activity android:name=".BActivity">
2            <intent-filter>
3                <action android:name="cn.edu.nnutc.ie.MYACTION"/>
4                <category android:name = "android.intent.category.DEFAULT" />
5            </intent-filter>
6        </activity>
```

其次，在 AActivity 的 Button 单击监听事件中设置 Intent 发送的 Action 为 "cn.edu.nnutc.ie.MYACTION"，这样与清单文件中配置的 BActivity 的 Action 能进行匹配。实现代码如下。

```
1        btnA.setOnClickListener(new View.OnClickListener() {
2            @Override
3            public void onClick(View v) {
4                Intent intent = new Intent();
5                intent.setAction("cn.edu.nnutc.ie.MYACTION");
6                AActivity.this.startActivity(intent);
7            }
8        });
```

综上所述，Intent 到底发给哪个 Activity，需要进行三个匹配，即 action、category 和 data。理论上来说，如果 Intent 不指定 category，那么无论 intent filter 的内容是什么都应该是匹配的。但是，如果是隐式 Intent，Android 默认给 Intent 加上一个 CATEGORY_DEFAULT，这样如果配置清单文件中的 intent filter 项中没有设置 category 属性值为 "android.intent.category.DEFAULT"，匹配就会失败。所以，如果 Activity 支持接收隐式 Intent，就一定要在 intent filter 中加入 "android.intent.category.DEFAULT"。如果开发的 App 要明确拒绝被别的 App 启动某一个 Actvity，就需要在配置清单文件的 activity 标签中设置 exported 的属性值为 false。

（3）Intent 调用常用的 Android 组件。

如果希望 App 通过 Activity 展示打电话、上网、发短信等动作，可以调用系统提供的功能去实现，即只要在某个监听事件用如下代码实现。

- 打电话

```
1   //直接拨打电话
2   Intent intent = new Intent(Intent.ACTION_CALL);
3   intent.SetData(Uri.Parse("tel:18811112220"));
4   startActivity(call);
```

- 拨电话

```
1   //打开拨打电话的用户界面
2   Intent intent = new Intent(Intent.ACTION_DIAL);
3   intent.setData(Uri.parse("tel:10086"));
4   startActivity(intent);
```

- 打开指定网页

```
1   //使用浏览器打开百度的首页
2   Intent intent = new Intent(Intent.ACTION_VIEW);
3   intent.setData(Uri.parse("http://www.baidu.com"));
4   startActivity(intent);
```

- 发短信

```
1   //向10086发送Hello内容短信
2   Intent intent =new Intent(Intent.ACTION_SENDTO);
3   Intent.setData(Uri.parse("smsto:10086"));
4   intent.putExtra("sms_body", "Hello");
5   startActivity(intent);
```

（4）Intent 传递数据。

若想使用 Intent 传递一个数据，可以直接通过调用 Intent 的 putExtra( )方法存入数据，然后在获得 Intent 后调用 getXXXExtra( )方法获得对应类型的数据。若想传递多个数据，可以使用 Bundle 对象作为容器，通过调用 Bundle 的 putXXX( )方法先将数据存储到 Bundle 中，接着调用 Intent 的 putExtras( )方法将 Bundle 存入 Intent 中，然后在获得 Intent 以后调用 getExtras( )获得 Bundle 容器，最后调用其 getXXX( )方法获取对应的数据。例如，传递一个数据的步骤如下。

- 在 AActivity 中存入数据，代码如下。

```
1   Intent it = new Intent(AActivity.this,BActivity.class);
2   It.putExtra("key",value);//value 值可以是很多简单类型数据
3   startActivity(it);
```

- 在 BActivity 中取数据，代码如下。

```
1   Intent it = this.getIntent();
2   String s = getStringExtra("key");//不同类型的数据修改 String
```

如果要一次传递多个数据，就需要使用下列步骤。

- 在 AActivity 中存入数据，代码如下。

```
1    Intent it = new Intent(AActivity.this,BActivity.class);
2    Bundle bundle = new Bundle();
3    bundle.putInt("age",12);
4    bundle.putString("name",Jake);
5    It.putExtra(bundle);
6    startActivity(it);
```

- 在 BActivity 中取数据，代码如下。

```
1    Intent it = this.getIntent();
2    Bundle bundle = it.getExtras();
3    int age = bundle.getInt("age");
4    String name = bundle.getString ("name");
```

4.2.2 通讯录的实现

【通讯录的实现】

1. 通讯录界面设计（图 4.8，图 4.9）

图 4.8　通讯录的主界面

图 4.9　添加联系人信息界面

从图 4.8 和图 4.9 可以看出本案例需要设计两个布局文件，图 4.8 的布局关键代码如下。

```
1    <?xml version ="1.0" encoding ="utf-8"?>
2    <LinearLayout xmlns:android ="http://schemas.android.com/apk/res/android"
3        android:layout_width ="match_parent"
4        android:layout_height ="match_parent"
5        android:orientation ="vertical" >
6        <RelativeLayout
7            android:layout_width ="match_parent"
```

```
8        android:layout_height="35dp"
9        android:background="#000000">
10       <TextView
11            android:layout_width="wrap_content"
12            android:layout_height="wrap_content"
13            android:layout_centerInParent="true"
14            android:text="通讯录"
15            android:textColor="#ffffff"
16            android:textSize="20sp"/>
17       <TextView
18            android:id="@+id/tvadd"
19            android:layout_width="wrap_content"
20            android:layout_height="wrap_content"
21            android:layout_alignParentRight="true"
22            android:layout_marginRight="20dp"
23            android:layout_marginTop="6dp"
24            android:text="添加"
25            android:textColor="#ffffff"
26            android:textSize="18sp"/>
27       </RelativeLayout>
28       <ListView
29            android:id="@+id/lvall"
30            android:dividerHeight="2dp"
31            android:layout_width="match_parent"
32            android:layout_height="wrap_content">
33       </ListView>
34   </LinearLayout>
```

其中第6～27行代码用相对布局设置界面中"通讯录"和"添加"文字所在行的显示效果，第28～33行代码用ListView显示联系人的信息。由于联系人信息显示包括头像、姓名和电话号码，所以需要给ListView的行信息定义一个布局文件，代码如下。

```
1    <?xml version="1.0" encoding="utf-8"?>
2    <LinearLayout xmlns:android="http://schemas.android.com/apk/res/android"
3        android:orientation="horizontal" android:layout_width="match_parent"
4        android:layout_height="match_parent">
5        <ImageView
6            android:id="@+id/lxrimg"
7            android:layout_width="60dp"
8            android:layout_height="80dp"
9            android:src="@mipmap/man1"/>
10       <LinearLayout
11            android:orientation="vertical"
12            android:layout_width="match_parent"
13            android:layout_height="match_parent">
14            <TextView
15                android:id="@+id/lxrname"
16                android:layout_width="wrap_content"
17                android:layout_height="wrap_content"
18                android:layout_marginLeft="15dp"
```

```
19          android:layout_marginTop="10dp"
20          android:text="jack"
21          android:textSize="20sp" />
22      <TextView
23          android:id="@+id/lxrtel"
24          android:layout_width="wrap_content"
25          android:layout_height="wrap_content"
26          android:layout_marginLeft="15dp"
27          android:layout_marginTop="10dp"
28          android:text="119"
29          android:textSize="20sp" />
30     </LinearLayout>
31 </LinearLayout>
```

从图4.9可以看出，布局界面上可以输入姓名、电话号码和选择头像，姓名和电话号码直接使用 EditText 组件，头像使用 ImageView 组件，由于要让头像可以向左(右)翻出不同的头像，所以布局时一共放了三个 ImageView 组件，中间的组件用于显示头像，左右两边的组件分别用于加载向左和向右的箭头图标。详细代码如下。

```
 1 <?xml version="1.0" encoding="utf-8"?>
 2 <LinearLayout xmlns:android="http://schemas.android.com/apk/res/android"
 3     android:layout_width="match_parent"
 4     android:layout_height="match_parent"
 5     android:orientation="vertical" >
 6     <RelativeLayout
 7         android:layout_width="match_parent"
 8         android:layout_height="35dp"
 9         android:background="#000000" >
10         <ImageView
11             android:id="@+id/imgreturn"
12             android:layout_width="30dp"
13             android:layout_height="30dp"
14             android:layout_marginLeft="10dp"
15             android:src="@mipmap/row_left" />
16         <TextView
17             android:layout_width="wrap_content"
18             android:layout_height="wrap_content"
19             android:layout_marginTop="5dp"
20             android:layout_toRightOf="@id/imgreturn"
21             android:text="返回"
22             android:textColor="#ffffff"
23             android:textSize="20sp" />
24     </RelativeLayout>
25     <LinearLayout
26         android:layout_width="match_parent"
27         android:layout_height="wrap_content"
28         android:layout_margin="10dp"
29         android:orientation="vertical" >
30         <EditText
31             android:id="@+id/edtname"
```

```
32          android:layout_width ="match_parent"
33          android:layout_height ="wrap_content"
34          android:hint ="请输入姓名" />
35        <EditText
36          android:inputType ="phone"
37          android:id ="@ +id/edttel"
38          android:layout_width ="match_parent"
39          android:layout_height ="wrap_content"
40          android:hint ="请输入电话号码" />
41      </LinearLayout >
42      <RelativeLayout
43          android:layout_width ="match_parent"
44          android:layout_height ="120dp"
45          android:layout_margin ="10dp"
46          android:orientation ="horizontal" >
47        <ImageView
48          android:id ="@ +id/imgleft"
49          android:layout_width ="40dp"
50          android:layout_height ="120dp"
51          android:src ="@ mipmap/row_left" />
52        <ImageView
53          android:id ="@ +id/img"
54          android:layout_width ="120dp"
55          android:layout_height ="wrap_content"
56          android:layout_centerInParent ="true"
57          android:src ="@ mipmap/man1" />
58        <ImageView
59          android:id ="@ +id/imgright"
60          android:layout_width ="40dp"
61          android:layout_height ="120dp"
62          android:layout_alignParentRight ="true"
63          android:src ="@ mipmap/row_right" />
64      </RelativeLayout >
65      <LinearLayout
66          <! --用于放置"重置"和"添加"按钮,实现比较简单,此处略 -->
67      </LinearLayout >
68    </LinearLayout >
```

根据上面的布局分析,每个界面是没有标题栏的,所以需要在配置清单文件中的 activity 选项中添加 android:theme =" @ android:style/Theme.Light.NoTitleBar" 属性,另外还需要预先将需要的图片复制到 res/mipmap 目录中。

2. 通讯录的功能实现

(1) 定义 Person 类。

联系人信息包含了头像、姓名和联系电话。本案例单独定义了一个 Person 类来封装联系人的所有信息,另外为了便于 Intent 传递 ArrayList <Person >类型数据,在定义时实现了 Serializable 接口,详细代码如下。

```
1   public class Person implements Serializable {
2       private String name;
3       private String tel;
4       private int imgid;
5       public Person(String name,String tel,int imgid){
6           this.name=name;
7           this.tel=tel;
8           this.imgid=imgid;
9       }
10      public String getName() {
11          return name;
12      }
13      public String getTel() {
14          return tel;
15      }
16      public int getImgid() {
17          return imgid;
18      }
19  }
```

(2) 单击"添加"按钮监听事件。

单击"添加"按钮，使用显式 Intent 启动 AddActivity，代码如下。

```
1   tvAdd.setOnClickListener(new View.OnClickListener() {
2       @Override
3       public void onClick(View v) {
4           Intent intent = new Intent(MainActivity.this, AddActivity.class);
5           MainActivity.this.startActivity(intent);
6       }
7   });
```

(3) 当 MainActivity 加载时，通过 Intent 获得 AddActivity 传递过来的 ArrayList <Person>，并使用 HashMap <String, Object> 格式数据将其显示在 ListView 的行布局对应组件上，即在 MainActivity.java 的 onCreate( )方法中写入以下代码。

```
1   Intent intent = this.getIntent();
2   //从 AddActivity 的 Intent 附加信息中取出添加的联系人信息
3   persons = (ArrayList <Person>) intent.getSerializableExtra("persons");
4   //标记已经通过 AddActivity 界面添加了联系人信息，即 persons 不为空
5   flag = intent.getBooleanExtra("flag", false);
6   if (flag) {
7       for (int i = 0; i <persons.size(); i++) {
8           HashMap <String, Object> map = new HashMap <String, Object>();
9               map.put("imgid", persons.get(i).getImgid());
10              map.put("name", persons.get(i).getName());
11              map.put("tel", persons.get(i).getTel());
12              lxrList.add(map);
13      }
14      String[] from = {"imgid", "name", "tel"};
15      int[] to = {R.id.lxrimg, R.id.lxrname, R.id.lxrtel};
```

```
16        adapter = new SimpleAdapter(this,lxrList, R.layout.person_layout, from,
to);
17        lvAll.setAdapter(adapter);
18    }
```

(4) 单击 ListView 时使用 Intent 实现直接拨号功能，代码如下。

```
1     lvAll.setOnItemClickListener(new AdapterView.OnItemClickListener() {
2       @Override
3       public void onItemClick(AdapterView <?> parent, View view, int position,
long id) {
4           Intent itdail = new Intent(Intent.ACTION_CALL,
Uri.parse("tel:"+persons.get(position).getTel()));
5           MainActivity.this.startActivity(itdail);
6       }
7     });
```

(5) 在 AddActivity.java 文件中实现向后、向前翻头像，添加联系人信息和返回等功能，详细代码如下。

```
1     //头像图片向后翻
2     imgRight.setOnClickListener(new View.OnClickListener() {
3       @Override
4       public void onClick(View v) {
5           if (++index >= imgId.length) {
6               index = 0;//翻到最后一张后,再翻到第1张
7           }
8           img.setImageResource(imgId[index]);
9       }
10    });
11    //头像图片向前翻
12    imgLeft.setOnClickListener(new View.OnClickListener() {
13      @Override
14      public void onClick(View v) {
15          if (--index <0) {
16              index = imgId.length-1;//翻到第1张后,再翻到最后一张
17          }
18          img.setImageResource(imgId[index]);
19      }
20    });
21    //添加联系人信息
22    btnOk.setOnClickListener(new View.OnClickListener() {
23      @Override
24      public void onClick(View v) {
25          String name = edtName.getText().toString();
26          String tel = edtTel.getText().toString();
27          int imgid = imgId[index];
28          Person person = new Person(name,tel,imgid);
29          persons.add(person);
30          edtTel.setText("");
31          edtName.setText("");
```

```
32        }
33    });
34    //返回监听事件
35    imgReturn.setOnClickListener(new View.OnClickListener() {
36        @Override
37        public void onClick(View v) {
38            Intent intent = new Intent(AddActivity.this,MainActivity.class);
39            intent.putExtra("persons",persons);
40            intent.putExtra("flag",flag);
41            AddActivity.this.startActivity(intent);
42        }
43    });
```

由于本案例实现了拨打电话，所以需要在配置文件 AndroidManifest.xml 中添加可以拨打电话权限。且打开 App 可以使用拨打电话的权限，即设置→应用→找到 NavigationBar（案例 App 的名称）→单击后进入 App 信息→权限），代码如下所示。

```
<uses-permission android:name="android.permission.CALL_PHONE"></uses-permission>
```

全部功能实现代码请读者参见代码包中 NavigationBar 文件夹里的内容。

4.3　仿微信主界面的设计与实现

在移动互联网时代，基于移动终端开发的聊天、交友软件越来越多，典型代表就是 QQ、微信。本节将使用 Fragment 组件设计一个仿微信主界面，如图 4.10 所示，当单击界面下方的"微信""通讯录""发现"和"设置"时，界面中央的内容会根据单击的目标实现相应的变换，并且文字上方的图片也会切换成另外一种效果（图 4.11）。

图 4.10　仿微信主界面(1)

图 4.11　仿微信主界面(2)

【Fragment】

4.3.1 预备知识

为了提高代码的重用性、改善用户体验和适应大屏幕的需求，Android 自 3.0（API Level 11）开始引入 Fragment（碎片）技术。例如，要开发一个新闻展示的 App，普通 UI 设计的竖屏显示效果如图 4.12 所示、横屏显示效果如图 4.13 所示。图 4.12 的显示效果需要开发者设计两个布局文件分别对应两个不一样的 Activity，当移动终端横屏放置时，就会出现图 4.13 的效果，而这种效果并不是用户想要的。用户想要的效果是：移动终端横屏放置时，新闻标题和新闻内容分左右两栏显示，左侧显示新闻标题，右侧显示新闻内容；当移动终端竖屏放置时仅仅显示一栏，其显示效果如图 4.14 所示。想要实现这样的效果，最好将新闻标题列表界面和新闻详细内容界面分别放在两个 Fragment 中，然后在同一个 Activity 里引入这两个 Fragment，这样就可以将屏幕空间充分地利用起来。下面以设计图 4.14 所示的显示效果介绍 Fragment 的用法。

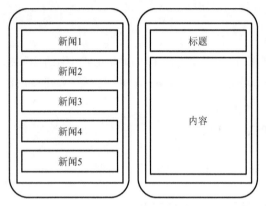

图 4.12　竖屏显示效果

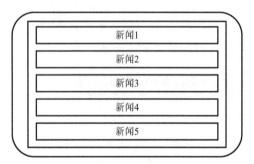

图 4.13　横屏显示效果

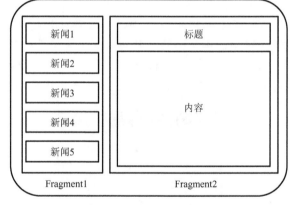

图 4.14　Fragment 横屏显示效果

1. 新建左侧 Fragment 和右侧 Fragment 的布局文件

左侧布局（本案例中文件名为 left_layout.xml）比较简单，只要在线性布局中放置一个 ListView 组件用于显示新闻条目，代码如下。

```
1    <?xml version="1.0" encoding="utf-8"?>
2    <LinearLayout xmlns:android="http://schemas.android.com/apk/res/android"
3        android:layout_width="match_parent"
4        android:layout_height="match_parent"
```

```
5       android:orientation ="vertical" >
6       <ListView
7           android:id ="@ + id/lvnews"
8           android:layout_width ="match_parent"
9           android:layout_height ="wrap_content" > </ListView >
10      </LinearLayout >
```

右侧布局（本案例中文件名为 right_layout.xml）使用线性布局，放置一个 TextView 用于显示标题，一个 EditText 用于放置新闻内容，其代码如下。

```
1   <?xml version ="1.0" encoding ="utf-8"? >
2   <LinearLayoutxmlns:android ="http://schemas.android.com/apk/res/android"
3       android:layout_width ="match_parent"
4       android:layout_height ="match_parent"
5       android:orientation ="vertical" >
6       <TextView
7           android:id ="@ + id/tvtitle"
8           android:layout_width ="match_parent"
9           android:layout_height ="wrap_content"
10          android:gravity ="center"
11          android:text ="市委书记来学院作十九大宣读报告"
12          android:textSize ="20sp" />
13      <EditText
14          android:id ="@ + id/editText"
15          android:layout_width ="match_parent"
16          android:layout_height ="wrap_content"
17          android:ems ="10"
18          android:inputType ="textMultiLine"
19          android:text ="首先介绍了党的十……" />
20      </LinearLayout >
```

2. 新建左侧布局和右侧布局的 Fragment 类

在 App 的 src/main/java/包名文件夹下，分别创建继承于 Fragment 的类 LeftFragment.java 和 RightFragment。在继承 Fragment 类时，可以导入 android.app 包，也可以导入 android.support.v4.app 包，它们的使用方法完全一样，只不过最低支持版本不同。android.app.Fragment 兼容的最低版本是 android:minSdkVersion ="11"，即 3.0 版；android.support.v4.app.Fragment 兼容的最低版本是 android:minSdkVersion ="4"，即 1.6 版。本书此处的案例程序使用了 android.app.Fragment 包。左侧布局的 Fragment 类的代码如下。

```
1   public class LeftFragment extends Fragment {//左侧布局的 Fragment 类
2       @Override
3       public View onCreateView(LayoutInflater inflater, ViewGroup container,
4           Bundle savedInstanceState) {
5           return inflater.inflate(R.layout.left_layout,container,false);
6       }
7   }
```

上述第 5 行代码中使用 inflate()方法加载布局文件，该方法在 LayoutInflater 类中定义，其常用的方法原型为 public View inflate（int resource, ViewGroup root, boolean attachToRoot）。

其中第 1 个参数为需要加载的布局文件 ID，即需要将这个布局文件加载到 Fragment 中；第 2 个参数为附加到 resource 文件的根据控件，即该方法返回的 View 对象；第 3 个参数如果为 true，就将 root 作为根对象返回，否则仅将这个 root 对象的 LayoutParams 属性附加到 resource 对象的根布局对象上，即 resource 的最外层的 View 上。右侧布局的 Fragment 类代码与此类似，限于篇幅不再赘述。

3. 在主界面的布局文件中引用左侧 Fragment 和右侧 Fragment

在主界面布局文件（本案例中文件名为 main_layout.xml）中加载自定义的 Fragment 子类的方法和引用普通组件一样，代码如下。

```
1    <?xml version="1.0" encoding="utf-8"?>
2    <LinearLayout xmlns:android="http://schemas.android.com/apk/res/android"
3      xmlns:tools="http://schemas.android.com/tools"
4      android:layout_width="match_parent"
5      android:layout_height="match_parent">
6     <fragment
7        android:id="@+id/left_fragment"
8        android:name="cn.edu.nnutc.ie.newsfragment.LeftFragment"
9        android:layout_width="0dp"
10       android:layout_height="match_parent"
11       android:layout_weight="1"
12       tools:layout="@layout/left_fragment" />
13    <fragment
14       android:id="@+id/right_fragment"
15       android:name="cn.edu.nnutc.ie.newsfragment.RightFragment"
16       android:layout_width="0dp"
17       android:layout_height="match_parent"
18       android:layout_weight="1"
19       tools:layout="@layout/right_fragment" />
20   </LinearLayout>
```

主界面中加载自定义的 Fragment 子类有静态加载和动态加载两种方法，此处使用了静态加载。而动态加载 Fragment 在仿微信主界面实现时会使用到，这里介绍它的一般使用步骤。

- 创建待添加的 Fragment 实例

```
1    RightFragment rFragment = new RightFragment();
```

- 获取到 FragmentManager

在 Activity 中可以直接调用 getFragmentManager() 方法得到，即

```
2    FragmentManager fm = getFragmentManager();
```

- 开启一个事务

```
3    FragmentTransaction transaction = fm.beginTransaction();
```

- 向容器内加入碎片

一般使用 replace() 方法实现，需要传入容器的 ID 和待添加的 Fragment 实例，即

```
4    transaction.replace(R.id.right_layout, rFragment);
```

- 提交事务

```
5    transaction.commit();
```

4. 在主界面的 Activity 文件（本案例中文件名为 MainActivity.java）中实现功能

由于本案例实现效果如图 4.15 所示，新闻标题和新闻内容直接定义了两个字符串数组 titles 和 news，左侧新闻标题列表 ListView、右侧新闻标题 TextView 和右侧新闻内容 EditText 中内容的加载实现代码如下。

```
1    private ListView lvNews;
2    private String[] titles ={"学生会成员参加义务劳动","市委书记来校作十九大宣读报告",……};
3    private String[] news = {"……", "……", "……"};
4    private ArrayAdapter arrayAdapter = null;
5    private TextView tvTitle;
6    private EditText edtContent;
7    lvNews = (ListView) this.findViewById(R.id.lvnews);
8    arrayAdapter = new ArrayAdapter(this, android.R.layout.simple_list_item_1, titles);
9    lvNews.setAdapter(arrayAdapter);
10   tvTitle = (TextView) this.findViewById(R.id.tvtitle);
11   edtContent = (EditText) this.findViewById(R.id.edtcontent);
12   lvNews.setOnItemClickListener(new AdapterView.OnItemClickListener() {
13     @Override
14     public void onItemClick(AdapterView <? > parent, View view, int position, long id) {
15       tvTitle.setText(titles[position]);//根据单击的列表项显示新闻标题
16       edtContent.setText(news[position]); //根据单击的列表项显示新闻内容
17     }
18   });
```

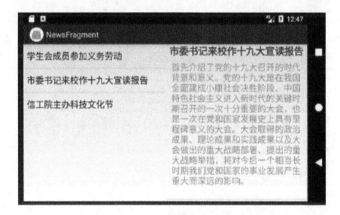

图 4.15 新闻横屏展示效果

4.3.2 仿微信主界面的实现

【仿微信主界面的实现】

1. 主界面的设计

根据图 4.10 的显示效果，可以将主界面的设计分解成如图 4.16 所示。

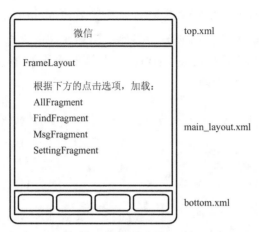

图 4.16 仿微信主界面分解效果

（1）新建顶部布局文件 top.xml。

```
1   <?xml version ="1.0" encoding ="utf-8"?>
2   <LinearLayout xmlns:android ="http://schemas.android.com/apk/res/android"
3       android:layout_width ="match_parent"
4       android:layout_height ="wrap_content"
5       android:background ="#4d4949"
6       android:gravity ="center"
7       android:orientation ="vertical">
8       <TextView
9           android:layout_width ="wrap_content"
10          android:layout_height ="wrap_content"
11          android:text ="微信"
12          android:textStyle ="bold"
13          android:textColor ="#ffffff"
14          android:textSize ="18sp" />
15  </LinearLayout>
```

（2）新建底部布局文件 bottom.xml。

```
1   <?xml version ="1.0" encoding ="utf-8"?>
2   <LinearLayout xmlns:android ="http://schemas.android.com/apk/res/android"
3       android:layout_width ="match_parent"
4       android:layout_height ="55dp"
5       android:background ="#4d4949"
6       android:orientation ="horizontal">
7       <!--底部的微信图标和文字显示效果 -->
8       <LinearLayout
9           android:id ="@+id/llmsg"
10          android:orientation ="vertical"
11          android:layout_width ="0dp"
12          android:layout_height ="match_parent"
13          android:gravity ="center"
14          android:layout_weight ="1">
15          <ImageButton
```

```
16          android:id="@+id/imgmsg"
17          android:layout_width="wrap_content"
18          android:layout_height="wrap_content"
19          android:src="@mipmap/wxmsg"
20          android:background="#00000000"/>
21      <TextView
22          android:layout_width="wrap_content"
23          android:layout_height="wrap_content"
24          android:textColor="#ffffff"
25          android:textSize="15sp"
26          android:textStyle="bold"
27          android:text="微信"/>
28      </LinearLayout>
29      <!-- 底部的通讯录图标和文字显示效果,实现代码与上面类似,此处略-->
30      <!--底部的发现图标和文字显示效果,实现代码与上面类似,此处略-->
31      <!--底部的设置图标和文字显示效果,实现代码与上面类似,此处略-->
32  </LinearLayout>
```

底部布局包括一个横向的 LinearLayout,其中又包括四个竖向的 LinearLayout,每个 LinearLayout 中包括一个 ImageButton 和一个 TextView。每个竖向的 LinearLayout 宽度设置为 0,layout_weight 为 1,表示四个共同均分屏幕宽度。在设置 Layout_weight 的时候最好将宽度或高度设置为 0,Layout_weight 平分的是屏幕剩余的宽度或高度。

(3) 新建主界面布局文件 main_layout.xml。

```
1   <?xml version="1.0" encoding="utf-8"?>
2   <LinearLayout xmlns:android="http://schemas.android.com/apk/res/android"
3       android:layout_width="match_parent"
4       android:layout_height="match_parent"
5       android:orientation="vertical">
6       <!--顶部加载-->
7       <include layout="@layout/top"></include>
8       <FrameLayout
9           android:id="@+id/id_content"
10          android:layout_weight="1"
11          android:layout_width="match_parent"
12          android:layout_height="0dp">
13      </FrameLayout>
14      <include layout="@layout/bottom"></include>
15  </LinearLayout>
```

从上述代码可以看出,代码中引入了两个布局文件(top.xml 和 bottom.xml),然后在 top 和 bottom 之间写入 FrameLayout,高度设置为 0,layout_weight 为 1,表示 FrameLayout 占据了除 top、bottom 之外屏幕的剩余空间,并可以动态加载要展示的 Fragment。

(4) 添加四个布局文件用于在主界面的 FrameLayout 位置处动态显示内容。

实现的布局代码比较简单,限于篇幅不再赘述。本案例中微信、通讯录、发现和设置分别对应的布局文件为 tab_msg.xml、tab_all.xml、tab_find.xml 和 tab_setting.xml。

2. 功能实现

（1）新建四个 Fragment 类。

MsgFragment.java、AllFragment.java、FindFragment.java 和 SettingFragment.java 分别对应 tab_msg.xml、tab_all.xml、tab_find.xml 和 tab_setting.xml 布局文件。这四个 Fragment 类文件实现代码类似，限于篇幅只列出 MsgFragment.java 的代码。

```
1    public class MsgFragment extends Fragment {
2        @Override
3        public View onCreateView(LayoutInflater inflater, ViewGroup container,
4                                Bundle savedInstanceState) {
5            return inflater.inflate(R.layout.tab_msg, container, false);
6        }
7    }
```

（2）新建继承于 FragmentActivity 的微信主界面显示的 MainActivity.java 文件。

● 定义变量

```
1    //底部的四个导航布局微信、通讯录、发现、设置
2    private LinearLayout llTabMsg, llTabAll, llTabFind, llTabSetting;
3    //底部四个导航布局中的图片按钮
4    private ImageButton imgMsg, imgAll, imgFind, imgSetting;
5    //初始化四个 Fragment
6    private Fragment tabMsg, tabALL, tabFind, tabSetting;
```

● 实现应用程序界面没有标题栏

应用程序界面没有标题栏的问题，第 3 章中是通过修改配置文件 AndroidManifest.xml 实现的，本案例中使用如下代码实现。

```
1    requestWindowFeature(Window.FEATURE_NO_TITLE);
```

● 定义 setSelect( ) 方法

```
1    private void setSelect(int i) {
2        FragmentManager fm = getSupportFragmentManager();
3        FragmentTransaction transaction = fm.beginTransaction();//创建一个事务
4        //先隐藏所有的 Fragment,然后根据单击的选项处理具体要显示的 Fragment
5        hideFragment(transaction);
6        switch (i) {
7            case 0://单击的是微信选项
8                if (tabMsg == null) {
9                    tabMsg = new MsgFragment();
10                   //将微信聊天界面的 Fragment 添加到 Activity 中
11                   transaction.add(R.id.id_content, tabMsg);
12               } else {
13                   transaction.show(tabMsg);
14               }
15               imgMsg.setImageResource(R.mipmap.bwxmsg); //将图片设置为另一种效果
16               break;
```

```
17              case 1://单击的是通讯录选项
18                  ……
19              case 2://单击的是发现选项
20                  ……
21              case 3://单击的是设置选项
22                  ……
23          }
24          transaction.commit();//提交事务
25      }
```

该方法用于切换显示内容的 Fragment 并将图片设置为另一种效果。上述第 7～16 行代码表示单击"微信"选项后,将微信对应的 MsgFragment 加载到主界面 FrameLayout 区域,即 R.id.id_content 标识的对象;单击通讯录选项、发现选项和设置选项的功能代码与单击"微信"选项类似,限于篇幅不再赘述,读者可以参阅 wxui 文件夹中的 MainActivity.java 文件。

● 定义 hideFragment( )方法

```
1       private void hideFragment(FragmentTransaction transaction) {
2           if (tabMsg != null) {
3               transaction.hide(tabMsg);
4           }
5           if (tabALL != null) {
6               transaction.hide(tabALL);
7           }
8           if (tabFind != null) {
9               transaction.hide(tabFind);
10          }
11          if (tabSetting != null) {
12              transaction.hide(tabSetting);
13          }
14      }
```

该方法用于隐藏所有的 Fragment。

● 重写四个导航选项的单击事件

```
1       @Override
2       public void onClick(View v) {
3           //将底部的导航按钮图标设置为初始状态
4           imgMsg.setImageResource(R.mipmap.wxmsg);
5           imgAll.setImageResource(R.mipmap.wxall);
6           imgFind.setImageResource(R.mipmap.wxfind);
7           imgSetting.setImageResource(R.mipmap.wxset);
8           switch (v.getId()) {
9               case R.id.llmsg:
10                  setSelect(0);
11                  break;
12              case R.id.llall:
13                  setSelect(1);
14                  break;
15              case R.id.llfind:
16                  setSelect(2);
```

```
17              break;
18          case R.id.llsetting:
19              setSelect(3);
20              break;
21      }
```

用上述代码实现单击四个导航选项时，首先将导航上的 ImageButton 设置为初始图标效果，其次根据单击的选项调用 setSelect()方法以切换显示内容的 Fragment 和改变 ImageButton 组件上加载的另一种图片效果。全部功能实现代码请读者参见 ALLAPP 代码包中 wxui 文件夹里的内容。

4.4 仿今日头条主界面的设计与实现

随着移动终端 App 越来越多，人们对 App 使用的方便性和灵活性提出了更高的要求。现在有很多 App 能实现左右滑屏来展现相应内容，如美团的"发现"界面、淘宝的"发现"界面和今日头条的主界面。本节将仿照今日头条 App 的主界面介绍左右滑屏效果的实现。

4.4.1 预备知识

【ViewPager、PagerTitleStrip、PageTabStrip】

1. ViewPager

ViewPager 是 Android 3.0 后引入的一个用户界面组件，可以通过手势滑动来完成 View 的切换，如作为 App 的引导页、实现图片轮播等功能。ViewPager 是 android-support-v4.jar 中定义的类，它直接继承了 ViewGroup 类，它也是一个容器类，可以在其中添加其他的 View 类。

（1）由于 ViewPager 类被定义在扩展包中，所以使用前需要使用如下操作过程引入 android-support-v4.jar。

① 选择 File 菜单下的 Project Structure 选项，打开 Project Structure 对话框。

② 在 Project Structure 对话框中单击 Modules 列表下正在编写的 App 名（本案例为 ViewPage），并单击对话框右侧的 Dependencies 选项卡。

③ 单击 Dependencies 选项卡右侧的"＋"标记，在弹出的选项中选择 Library Dependencies。

④ 在弹出的 Choose Library Dependencies 对话框中选择 support-v4（本案例的 ViewPage 支持库选择的是 com.android.support:support-v4:26.0.0）即可，如果要选择其他必需的支持库也可以使用此方法实现。

（2）ViewPager 虽然扩展自 ViewGroup 类，但不能使用 addView()为其添加子一级 View 对象，也不能在它的布局文件中添加子组件。为了实现在 ViewPager 中显示子一级 View 对象，就需要 PageAdapter 协调来完成。下面以实现三个 View 对象的左右滑屏为例介绍 ViewPager 的使用方法。

① 在 Activity 的布局中添加 ViewPager 组件，其代码如下。

```
1   <?xml version="1.0" encoding="utf-8"?>
2   <LinearLayout xmlns:android="http://schemas.android.com/apk/res/android"
```

```
3        android:layout_width ="match_parent"
4        android:layout_height ="match_parent"
5        android:orientation ="vertical" >
6        <android.support.v4.view.ViewPager
7            android:id ="@ +id/myviewpager"
8            android:layout_width ="wrap_content"
9            android:layout_height ="wrap_content" >
10       </android.support.v4.view.ViewPager >
11   </LinearLayout >
```

从上述代码可以看出 ViewPager 组件的声明必须指出其所在包。

② 在 res/layout 下创建各个子一级 View 对象的布局文件,本案例中创建了三个相对布局文件,居中放置了一个 TextView,其中 view1.xml 文件的代码如下。

```
1    <?xml version ="1.0" encoding ="utf-8"? >
2    <RelativeLayout xmlns:android ="http://schemas.android.com/apk/res/android"
3        android:layout_width ="match_parent"
4        android:layout_height ="match_parent" >
5        <TextView
6            android:textSize ="20sp"
7            android:layout_centerInParent ="true"
8            android:layout_width ="wrap_content"
9            android:layout_height ="wrap_content"
10           android:text ="这是第 1 个 View" / >
11   </RelativeLayout >
```

view2.xml 和 view3.xml 布局文件代码与 view1.xml 文件类似,限于篇幅不再赘述。

③ 在 java/包名下创建继承于 PagerAdapter 的子类 MyPagerAdapter,代码如下。

```
1    public class MyPagerAdapter extends PagerAdapter {
2        private ArrayList <View > viewLists;
3        public MyPagerAdapter() {      }
4        public MyPagerAdapter(ArrayList <View > viewLists) {
5            super();
6            this.viewLists = viewLists;
7        }
8        @Override
9        public int getCount() {
10           return viewLists.size();//设置为当前 list 集合中的布局个数
11       }
12       @Override
13       public boolean isViewFromObject(View view, Object object) {
14           return view == object;//保持当前 View 与目标 View 一致
15       }
16       @Override
17       public Object instantiateItem(ViewGroup container, int position) {
18           //把从 list 中加载的 View 添加到当前 ViewPager 中
19           container.addView(viewLists.get(position));
20           return viewLists.get(position);
21       }
22       @Override
```

```
23          public void destroyItem(ViewGroup container, int position,
Object object) {
24              //移除从 list 当前加载的 View
25              container.removeView(viewLists.get(position));
26          }
27      }
```

PagerAdapter 是一个抽象类，需要开发者创建重写其方法的子类，该类有下列四个需要重写的方法。

- public int getCount()：获得 ViewPager 中所有 View 对象的个数。
- public Object instantiateItem（ViewGroup container, int position）：将给定 position 位置的 View 添加到 container 中，并创建后显示出来；返回一个代表新增页面的 Object(key)，通常都是直接返回 View，也可以自定义 key，但是 key 和 View 要一一对应。
- public boolean isViewFromObject（View view, Object object）：判断 instantiateItem()方法所返回来的 key 与一个页面视图是否代表的同一个 View（即它俩是否是对应的，对应的表示同一个 View），通常直接写 "return view == object" 代码。
- public void destroyItem（ViewGroup container, int position, Object object）：移除一个给定 position 位置的页面。

④ 在 Activity 文件（本案例为 MainActivity.java 文件）中初始化子一级 View 对象，并使用 List 集合或数组进行存储，代码如下。

```
1    private ArrayList<View> aList;//ViewPager 中显示的 View 的 List 集合
2    aList = new ArrayList<View>();
3    LayoutInflater li = getLayoutInflater();
4    aList.add(li.inflate(R.layout.view1,null,false));
5    aList.add(li.inflate(R.layout.view2,null,false));
6    aList.add(li.inflate(R.layout.view3,null,false));
```

⑤ 创建 PagerAdapter 对象，并为 ViewPager 配置 PagerAdapter，代码如下。

```
1    ViewPager vpager_one = (ViewPager) findViewById(R.id.myviewpager);
2    mAdapter = new MyPagerAdapter(aList);
3    vpager_one.setAdapter(mAdapter);
```

按照以上五个步骤实现的 App，在运行后可以实现 View1～View3 的左右滑屏效果，详细代码读者可以参见 FirstApp 代码包中 viewpage 文件夹里的内容。

2. PagerTitleStrip

PagerTitleStrip 是 ViewPage 的一个关于当前页面、上一个页面和下一个页面的非交互性指示器（相当于每个页面的标题），显示效果如图 4.17 所示。使用时一般将它作为 ViewPage 的子级组件，代码如下。

```
1    <?xml version="1.0" encoding="utf-8"?>
2    <LinearLayout xmlns:android="http://schemas.android.com/apk/res/android"
3        android:layout_width="match_parent"
4        android:layout_height="match_parent"
5        android:orientation="vertical">
```

```
6      <android.support.v4.view.ViewPager
7          android:id="@+id/viewtitlepager"
8          android:layout_width="match_parent"
9          android:layout_height="wrap_content">
10         <android.support.v4.view.PagerTitleStrip
11             android:id="@+id/pagertitle"
12             android:layout_width="wrap_content"
13             android:layout_height="wrap_content"
14             android:layout_gravity="top">
15         </android.support.v4.view.PagerTitleStrip>
16     </android.support.v4.view.ViewPager>
17 </LinearLayout>
```

图 4.17　PagerTitleStrip 效果

通过将 PagerTitleStrip 的属性设置为 top 或 bottom 来将指示信息（相当于每个页面的标题）显示在 ViewPager 的顶部或底部。每个页面的标题通过 PagerAdapter 的 getPageTitle(int) 方法提供给 ViewPager。在 java/包名下创建继承于 PagerAdapter 的子类 MyTitlePagerAdapter，代码如下。

```
1  public class MyTitlePagerAdapter extends PagerAdapter {
2      private ArrayList<View> viewLists;//View 对象 List
3      private ArrayList<String> titleLists;//标题对象 List
4      public MyTitlePagerAdapter() {
5      }
6      public MyTitlePagerAdapter(ArrayList<View> viewLists, ArrayList<String> titleLists) {
7          super();
8          this.viewLists = viewLists;
9          this.titleLists = titleLists;
10     }
```

```
11      @Override
12      public int getCount() {
13          return viewLists.size();
14      }
15      @Override
16      public boolean isViewFromObject(View view, Object object) {
17          return view == object;
18      }
19      @Override
20      public CharSequence getPageTitle(int position) {
21          return titleLists.get(position);
22      }
23      @Override
24      public Object instantiateItem(ViewGroup container, int position) {
25          container.addView(viewLists.get(position));
26          return viewLists.get(position);
27      }
28      @Override
29      public void destroyItem(ViewGroup container, int position, Object object) {
30          container.removeView(viewLists.get(position));
31      }
32  }
```

初始化子一级 View 对象和 PagerTitleStrip 显示的指示信息对象（标题），并使用 List 集合或数组进行存储。初始化 View 对象的实现代码与上面一样，限于篇幅不再赘述。最后创建 PagerAdapter 对象，并为 ViewPager 配置 PagerAdapter，代码如下。

```
1   tList = new ArrayList<String>();
2   tList.add("第1页");
3   tList.add("第2页");
4   tList.add("第3页");
5   mAdapter = new MyTitlePagerAdapter(aList,tList);
6   vpager_one.setAdapter(mAdapter);
```

3. PagerTabStrip

PagerTabStrip 是 ViewPage 的一个关于当前页面、上一个页面和下一个页面的交互性指示器（相当于每个页面的标题），它与 PagerTitleStrip 的区别在于指示器的交互性。即 PagerTabStrip 在当前页面下，会有一条下划线条来提示当前页的 Tab 是哪一个，当用户单击某一个 Tab 时，当前页面就会跳转到这个页面。使用时只需要将前面 PagerTitleStrip 示例中布局代码引用的 PagerTitleStrip 修改为 PagerTabStrip，其他使用方法与 PagerTitleStrip 完全一样，功能代码不需要做任何修改。

```
1   <?xml version="1.0" encoding="utf-8"?>
2   <LinearLayout xmlns:android="http://schemas.android.com/apk/res/android"
3       android:layout_width="match_parent"
4       android:layout_height="match_parent"
5       android:orientation="vertical">
```

```
6       <android.support.v4.view.ViewPager
7           android:id="@+id/viewtitlepager"
8           android:layout_width="match_parent"
9           android:layout_height="wrap_content">
10          <android.support.v4.view.PagerTabStrip
11              android:id="@+id/pagertab"
12              android:layout_width="wrap_content"
13              android:layout_height="wrap_content"
14              android:layout_gravity="top">
15          </android.support.v4.view.PagerTabStrip>
16      </android.support.v4.view.ViewPager>
17  </LinearLayout>
```

上述代码运行后的显示效果如图 4.18 所示,从图中可以明显看出与图 4.17 的区别,即在指示器标题下有一条下划线,当用户滑动屏幕到第 2 页时下划线会自动移动到第 2 页标题下面,用户可以单击第 2 页标题来显示第 2 页对应的 View 页。为了改变指示器栏标题的显示效果,可以在功能代码中使用表 4-4 中的方法实现。

图 4.18　PagerTabStrip 效果

表 4-4　PagerTabStrip 常用方法及功能说明

方 法 名	功 能 说 明
setTextColor()	设置指示器文字颜色
setBackgroundColor()	设置指示器背景颜色
setDrawFullUnderline()	设置指示器下方是否有完整下划线颜色
setTabIndicatorColor()	设置指示器文字下方的指示颜色
setTextSpacing()	设置指示器文字的间隔

从以上两个示例（读者可以参见 FirstAPP 代码包中 viewpager 文件夹里的内容）可以看出，当左右滑屏时指示器标题会跟着滑动，而且并不能将所有指示器标题显示在页面的上端，而开发者需要的效果是所有指示器标题能够全部显示在页面的上端，并且当左右滑屏时指示器标题的显示状态（如颜色、图标等）会跟着改变。下面以仿今日头条主界面的实现为例介绍这类效果的实现方法。

4.4.2 仿今日头条主界面的实现

1. 主界面的设计

实现图 4.19 所示的带滑动的界面的具体步骤如下。

【仿今日头条主界面的实现】

图 4.19 仿今日头条主界面

（1）分别创建四个 Tab 选项内容对应的 Fragment 布局文件。

本案例一共有"推荐""热点""泰州"和"社会"四个 Tab 选项，所以需要创建四个 Fragment 布局文件，限于篇幅，本节只列出 tuijian_fragment.xml 文件的详细代码，其他文件代码类似。

```xml
1  <?xml version="1.0" encoding="utf-8"?>
2  <RelativeLayout xmlns:android="http://schemas.android.com/apk/res/android"
3      android:layout_width="match_parent"
4      android:layout_height="match_parent">
5      <!--以下就是Tab选项下显示的内容,本例以显示TextView为例-->
6      <TextView
7          android:layout_centerInParent="true"
8          android:textSize="20sp"
9          android:layout_width="wrap_content"
10         android:layout_height="wrap_content"
11         android:text="推荐"/>
12 </RelativeLayout>
```

（2）创建主布局文件 main_layout.xml 文件。

```xml
1   <?xml version="1.0" encoding="utf-8"?>
2   <LinearLayout xmlns:android="http://schemas.android.com/apk/res/android"
3       android:layout_width="match_parent"
4       android:layout_height="match_parent"
5       android:orientation="vertical">
6       <android.support.design.widget.TabLayout
7           android:id="@+id/tablayout"
8           android:layout_width="match_parent"
9           android:layout_height="50dp"
10          android:background="#cfd7d7">
11      </android.support.design.widget.TabLayout>
12      <android.support.v4.view.ViewPager
13          android:id="@+id/viewpager"
14          android:layout_width="match_parent"
15          android:layout_height="0dp"
16          android:layout_weight="1">
17      </android.support.v4.view.ViewPager>
18  </LinearLayout>
```

上述第 6～11 行代码引用了 TabLayout 组件（Android 5.0 后才可以使用），该组件可以用来与 ViewPager 进行联动，既能实现单击 Tab 选项的效果，也能实现滑动屏幕效果。Google 在 2014 年的 Google I/O 大会上推出了全新的设计语言——Material Design，并推出了一系列实现 Material Design 效果的组件库——Android Design Support Library。TabLayout 组件就包含这个组件库中。如果需要使用 Android Design Support Library，就需要在项目的 Gradle 文件中的 dependencies 中添加"compile 'com.android.support:design:25.3.1'"依赖。

2. 功能实现

（1）创建四个 Tab 选项布局文件对应的 Fragment 类。

四个 Fragment 类文件的代码类似，限于篇幅，本节只列出 tuijianFragment.java 文件的代码。

```java
1   public class tuijianFragment extends Fragment {
2       public tuijianFragment() {
3       }
4       @Override
5       public View onCreateView(LayoutInflater inflater, ViewGroup container, Bundle savedInstanceState) {
6           return inflater.inflate(R.layout.tuijian_fragment,container,false);
7       }
8   }
```

（2）创建适配器类。

为了让 Fragment 与 ViewPager 进行适配，需要创建继承于 FragmentPagerAdapter 类的适配器类，并重写相关的方法，代码如下。

```java
1   public class MyPagerAdapter extends FragmentPagerAdapter {
2       private String[] mTitles = new String[]{"推荐","热点","泰州","社会"};
3       public MyPagerAdapter(FragmentManager fm) {
4           super(fm);
```

```
5        }
6        @Override
7        public Fragment getItem(int position) {
8            switch (position) {
9                case 1:
10                   return new redianFragment();//热点
11               case 2:
12                   return new taizhouFragment();//泰州
13               case 3:
14                   return new shehuiFragment();//社会
15           }
16           return new tuijianFragment();//推荐
17       }
18       @Override
19       public int getCount() {
20           return mTitles.length;
21       }
22       @Override
23       public CharSequence getPageTitle(int position) {
24           return mTitles[position];
25       }
26   }
```

上述第2行代码定义了 Tab 选项上显示的文字，第7～17 行代码表示 Tab 选项卡位置改变时在选项区域显示对应的 Fragment。

（3）定义主布局对应的 MainActivity 类。

在类中需要为 ViewPager 设置 FragmentPagerAdapter 适配器，详细代码如下。

```
1    public class MainActivity extends AppCompatActivity {
2        private TabLayout tabLayout;
3        private ViewPager viewPager;
4        private MyPagerAdapter myPagerAdapter;
5        private TabLayout.Tab tab1,tab2,tab3,tab4;
6        @Override
7        protected void onCreate(Bundle savedInstanceState) {
8            super.onCreate(savedInstanceState);
9            setContentView(R.layout.main_layout);
10           getSupportActionBar().hide();//隐藏 ActionBar
11           initView();
12       }
13       void initView(){
14           //使用适配器将 ViewPager 与 Fragment 绑定在一起
15           viewPager = (ViewPager) this.findViewById(R.id.viewpager);
16           myPagerAdapter = new MyPagerAdapter(getSupportFragmentManager());
17           viewPager.setAdapter(myPagerAdapter);
18           //将 TabLayout 与 ViewPager 绑定在一起
19           tabLayout = (TabLayout) this.findViewById(R.id.tablayout);
20           tabLayout.setupWithViewPager(viewPager);
21           //指定 Tab 的位置
22           tab1 = tabLayout.getTabAt(0);
```

```
23            tab2 = tabLayout.getTabAt(1);
24            tab3 = tabLayout.getTabAt(2);
25            tab4 = tabLayout.getTabAt(3);
26        }
27    }
```

如果需要在 Tab 选项的文字上方增加 Tab 图标,可以使用下列代码:
tab1.setIcon(R.mipmap.ic_launcher);
全部功能实现代码请读者参见 ALLAPP 代码包中 headline 文件夹里的内容。

4.5 轮播效果的设计与实现

4.4 节使用 ViewPager、Fragment 和 Tablayout 的完美结合实现了滑屏效果,实际上 Android 还提供了一个与 ViewPager 功能相似的组件——ViewFlipper。ViewPager 可以理解为一页一页地翻转视图,而 ViewFlipper 可以理解为快速地翻转视图。ViewFlipper 一般用于广告图片的轮播效果。

4.5.1 预备知识

ViewFlipper 是系统自带组件之一,主要用于在同一个屏幕间的切换及设置动画效果、间隔时间,且可以自动播放。它直接继承于 ViewAnimator,而 ViewAnimator 继承于 FrameLayout。ViewAnimator 类的作用主要是为其中的 View 切换提供动画效果,该类的相关方法及功能说明见表 4-5;ViewFilpper 类的作用主要用来实现 View 的自动切换,该类的相关方法及功能说明见表 4-6。

表 4-5 ViewAnimator 的相关方法及功能说明

相 关 方 法	功 能 说 明
setInAnimation	设置 View 进入屏幕时使用的动画,可以直接传入 Animation 对象,也可以传入定义的 Animation 文件的 resourceID
setOutAnimation	设置 View 退出屏幕时使用的动画,使用方法和 setInAnimation 方法一样
showNext	显示 FrameLayout 里面的下一个 View
showPrevious	显示 FrameLayout 里面的上一个 View

表 4-6 ViewFilpper 的相关方法及功能说明

相 关 方 法	功 能 说 明
setFlipInterval	设置 View 切换的时间间隔,单位为毫秒
startFlipping	开始进行 View 的切换,切换会循环进行
stopFlipping	停止 View 切换
setAutoStart	设置是否自动开始,如果为 true,则当 ViewFlipper 显示的时候 View 的切换会自动开始

ViewFlipper 一般用于图片的切换,当然也可以添加用户自定义的 View 或其他 View 对象,并实现 View 的切换。

4.5.2 左右轮播的实现

【ViewFlipper、动画效果资源文件】

（1）下面以春、夏、秋、冬四幅图的左右轮播效果为例介绍 ViewFlipper 实现 ImageView 左右轮播切换的实现过程。

① 在布局文件中添加 ViewFlipper 组件和四个 ImageView 组件，代码如下。

```
1   <?xml version = "1.0" encoding = "utf-8"? >
2   <LinearLayout xmlns:android = "http://schemas.android.com/apk/res/android"
3       android:layout_width = "match_parent"
4       android:layout_height = "match_parent"
5       android:orientation = "vertical" >
6       <ViewFlipper
7           android:id = "@ + id/flipper"
8           android:layout_width = "match_parent"
9           android:layout_height = "match_parent"
10          android:flipInterval = "2000" >
11          <!--'春'图片的实现代码-->
12          <ImageView
13              android:layout_width = "match_parent"
14              android:layout_height = "match_parent"
15              android:scaleType = "fitXY"
16              android:src = "@mipmap/spring" />
17          <!--'夏''秋''冬'图片的实现代码与上面类似,此处略 -->
18      </ViewFlipper >
19  </LinearLayout >
```

上述第 10 行代码表示下面加载的图片（春、夏、秋、冬，这四幅图片事先已经存放到 mipmap 文件夹下）每隔 2s 切换一次。

② Activity 中实现的功能代码如下。

```
1   ViewFlipper viewFlipper = (ViewFlipper) findViewById(R.id.flipper);
2   viewFlipper.startFlipping();//开始进行 ImageView 图片的切换
```

（2）上面介绍的是在 ViewFlipper 中直接加入 ImageView 来实现本地图片的加载，这种实现效果的 ImageView 对象个数是固定的。在实际 App 中需要加载的 View 对象个数可能并不固定，也就是说 View 对象可能根据用户需要动态变化，实现时可以使用动态加载 View 来实现，其实现步骤如下。

① 在布局文件中添加 ViewFlipper 组件，代码如下。

```
1   <?xml version = "1.0" encoding = "utf-8"? >
2   <LinearLayout xmlns:android = "http://schemas.android.com/apk/res/android"
3       android:layout_width = "match_parent"
4       android:layout_height = "match_parent"
5       android:orientation = "vertical" >
6       <ViewFlipper
7           android:id = "@ + id/flipper"
8           android:layout_width = "match_parent"
9           android:layout_height = "match_parent"
10          android:flipInterval = "2000" >
```

```
11        </ViewFlipper>
12    </LinearLayout>
```

② Activity 中实现的功能代码如下。

```
1    ViewFlipper mFlipper = (ViewFlipper) findViewById(R.id.flipper);
2    ImageView imageView1 = new ImageView(this);//创建 ImageView 对象
3    imageView1.setImageResource(R.mipmap.spring);//设置'春'图片
4    mFlipper.addView(imageView1);//将 ImageView 对象加载到 ViewFlipper 上
5    //夏、秋、冬图片的加载与第 2～4 行代码相似,此处略
6    mFlipper.startFlipping();
```

以上介绍的图片在 App 运行后就会自动轮播,有的时候可能需要根据用户的手势来实现图片的轮播效果。即当用户在屏幕上向左滑屏时,图片向左滚动切换;当用户在屏幕上向右滑屏时,图片向右滚动切换。如果要实现这样的效果,就需要使用 Android SDK 自带的用户手势检测类——GestureDetector(Gesture:手势,Detector:识别),通过这个类可以识别很多手势,即通过它的 onTouchEvent(event)方法可以完成不同手势的识别。虽然 GestureDetector 能识别手势,但是不同的手势要怎么处理,应该是通过开发者实现的。下面介绍在前面功能的基础上通过增加手势识别功能实现左右滑屏时图片滚动切换效果,其主要功能代码如下。

```
1    public class Dong_Activity extends Activity implements View.OnTouchListener {
2        private ViewFlipper mFlipper;
3        private GestureDetector mDetector; //手势检测
4        @Override
5        protected void onCreate(Bundle savedInstanceState) {
6            super.onCreate(savedInstanceState);
7            this.setContentView(R.layout.dongtai_layout);
8            mFlipper = (ViewFlipper) findViewById(R.id.flipper);
9            ImageView imageView1 = new ImageView(this);//创建 ImageView 对象
10           imageView1.setImageResource(R.mipmap.spring);//设置春图片
11           mFlipper.addView(imageView1);//将 ImageView 对象加载到 ViewFlipper 上
12           //夏、秋、冬图片的加载与第 2～4 行代码相似,此处略
13           mFlipper.setOnTouchListener(this);
14           mDetector = new GestureDetector(new simpleGestureListener());
15       }
16
17       @Override
18       public boolean onTouch(View v, MotionEvent event) {
19           return mDetector.onTouchEvent(event);
20       }
21
22       private class simpleGestureListener extends
     GestureDetector.SimpleOnGestureListener {
23           final int FLING_MIN_DISTANCE = 100;//滑动距离
24           final int FLING_MIN_VELOCITY = 200;//滑动速度
25           //用户按下屏幕就会触发
26           @Override
27           public boolean onDown(MotionEvent e) {
28               return true;
29           }
```

```
30      /*
31      滑屏,用户按下触摸屏、快速移动后松开,由 1 个 MotionEvent ACTION_DOWN,
32      多个 ACTION_MOVE, 1 个 ACTION_UP 触发
33      参数解释:
34      e1:第 1 个 ACTION_DOWN MotionEvent
35      e2:最后 1 个 ACTION_MOVE MotionEvent
36      velocityX:X 轴上的移动速度,像素/秒
37      velocityY:Y 轴上的移动速度,像素/秒* /
38      @Override
39      public boolean onFling(MotionEvent e1, MotionEvent e2, float velocityX,
40                      float velocityY) {
41          //左滑屏
42          if (e1.getX() - e2.getX() > FLING_MIN_DISTANCE
43                  && Math.abs(velocityX) > FLING_MIN_VELOCITY) {
44              mFlipper.showNext();
45          } else if (e2.getX() - e1.getX() > FLING_MIN_DISTANCE
46                  && Math.abs(velocityX) > FLING_MIN_VELOCITY) {
47              //右滑屏
48              mFlipper.showPrevious();
49          }
50          return true;
51      }
52  }
53  }
```

以上代码实现的功能是:当用户向左滑动距离超过 100px,且滑动速度超过 100px/s 时,即判断为向左滑动;当用户向右滑动距离超过 100px,且滑动速度超过 100px/s 时,即判断为向右滑动。

4.5.3　上下轮播的实现

下面以三个中奖号码从下往上滚动的效果为例,介绍 ViewFlipper 实现 TextView 上下轮播切换的实现过程。

(1) 在布局文件中添加 ViewFlipper 组件和三个 TextView 组件,代码如下。

```
1   <?xml version ="1.0" encoding ="utf-8"? >
2   <LinearLayout xmlns:android ="http://schemas.android.com/apk/res/android"
3       android:layout_width ="match_parent"
4       android:layout_height ="match_parent"
5       android:orientation ="vertical" >
6       <ViewFlipper
7           android:id ="@ +id/flipper"
8           android:layout_width ="match_parent"
9           android:layout_height ="wrap_content"
10          android:inAnimation ="@ anim/anim_comein"
11          android:outAnimation ="@ anim/anim_getout"
12          android:flipInterval ="3000" >    //设置滚动间隔时间(毫秒)
13          <!--'第 1 个中奖号码'的实现代码-->
14          <TextView
15              android:layout_width ="match_parent"
```

```
16          android:layout_height ="wrap_content"
17          android:gravity ="center_horizontal"
18          android:text ="中奖号码   13010101100"
19          android:textSize ="26sp" />
20      <!--'其他三个中奖号码'的实现代码与上面类似,此处略 ->
21      </ViewFlipper >
22  </LinearLayout >
```

上述第10行代码、第11行代码分别表示 TextView 滑入和滑出的动画效果,此效果需要开发者在 res 文件夹下创建一个 anim 文件夹,然后在此文件夹下分别创建 anim_comein.xml 和 anim_getout.xml 文件,在这两个文件中定义动画效果,其代码如下所示。上述第12行代码表示下面加载的 TextView 内容每隔3s 按照第10行代码和第11行代码定义的动画效果切换一次。

- anim_comein.xml 文件代码

```
1   <?xml version ="1.0" encoding ="utf-8"? >
2   <set xmlns:android ="http://schemas.android.com/apk/res/android" >
3       <translate
4           android:fromYDelta ="100% p"
5           android:toYDelta ="0"
6           android:duration ="1000"/ >
7   </set >
```

其中 translate 位置属性及功能说明见表4-7。

表4-7 translate 位置属性及功能说明

位 置 属 性	功 能 说 明
fromXDelta	动画起始时 x 坐标上的位置
toXDelta	动画结束时 x 坐标上的位置
fromYDelta	动画起始时 y 坐标上的位置
toYDelta	动画结束时 y 坐标上的位置

在这些属性里面还可以加上%和 p,例如:

android:toXDelta = "100% ",表示自身的100%,也就是从 View 自己的位置开始。

android:toXDelta = "80% p",表示父层 View 的80%,是以它父层 View 为参照的,其应用分析读者可以参见图4.20。

从图4.20可以看出,以手机屏幕下边为 x 轴,屏幕左边为 y 轴,当 Activity 在 x 轴值为 -100% p 时,刚好在屏幕的左边(位置1),当 x 轴值为0% p 时,刚好在屏幕内(位置2),当 x = 100% p 时刚好在屏幕右边(位置3)。

- anim_getout.xml 文件代码

```
1   <?xml version ="1.0" encoding ="utf-8"? >
2   <set xmlns:android ="http://schemas.android.com/apk/res/android" >
3       <translate
4           android:fromYDelta ="0"
5           android:toYDelta ="-100% p"
6           android:duration ="1000"/ >
7   </set >
```

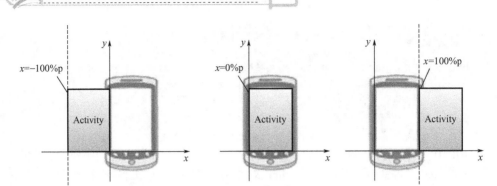

图 4.20 toXDelta 位置说明

（2）Activity 中实现的功能代码如下。

```
1   ViewFlipper viewFlipper = (ViewFlipper) findViewById(R.id.flipper);
2   viewFlipper.startFlipping();//开始进行 ImageView 图片的切换
```

本节的全部功能实现代码，读者可以参见 FirstAPP 代码包中 viewflipper 文件夹里的内容。

4.6 商品列表布局切换效果的设计与实现

随着网上购物需求越来越大，越来越多的商家都需要开发移动客户端的商品展示 App，这些 App 都有一个共同的特点：可以实现商品展示效果的切换，即默认打开时类似于图 4.21 的效果，当单击图中上面部分右侧的图标时，商品展示效果就立即切换成了图 4.22 所示的效果。为了达到这样的展现效果，Android SDK 中提供了 RecyclerView 类和 LayoutManager 类，本节将通过对这两个类的详细介绍来让读者掌握商品列表布局切换效果的设计和开发方法。

图 4.21 商品列表（1 行 1 个商品）

图 4.22 商品列表（1 行 3 个商品）

4.6.1 预备知识

1. RecyclerView

RecyclerView 是 Android 5.0 开始推出的用来替代 ListView 和 GridView 的组件,在使用 RecyclerView 前需要引入 support-v7 中

【RecycleView、RecycleView.LayoutManager】

的 RecyclerView,引入的步骤与 4.4.1 节介绍引入 support-v4 的步骤一样,本节从支持库选择了 com.android.support:recyclerview-v7:25.3.1 支持库。

下面以实现图 4.21 所示效果为例介绍 RecyclerView 组件的使用步骤。

(1) 在主布局文件中添加 RecyclerView 组件,代码如下。

```
1   <?xml version="1.0" encoding="utf-8"?>
2   <LinearLayout xmlns:android="http://schemas.android.com/apk/res/android"
3       android:layout_width="match_parent"
4       android:layout_height="match_parent"
5       android:orientation="vertical">
6       <android.support.v7.widget.RecyclerView
7           android:id="@+id/rv"
8           android:layout_width="match_parent"
9           android:layout_height="match_parent">
10      </android.support.v7.widget.RecyclerView>
11  </LinearLayout>
```

(2) 编写要展示的 item 样式布局文件,此处以布局文件中显示一个 TextView 为例,代码如下。

```
1   <?xml version="1.0" encoding="utf-8"?>
2   <LinearLayout xmlns:android="http://schemas.android.com/apk/res/android"
3       android:layout_width="match_parent"
4       android:layout_height="70dp"
5       android:layout_margin="2dp"
6       android:background="#00ff00"
7       android:orientation="vertical">
8       <TextView
9           android:id="@+id/recycle_tv"
10          android:layout_width="72dp"
11          android:layout_height="match_parent"
12          android:gravity="center"
13          android:text="test"
14          android:textSize="25dp" />
15  </LinearLayout>
```

读者需要注意,在 item 布局设计时,其 layout_height 属性值不能设置成 match_parent,而一定要给定一个固定的高度值。

(3) 自定义继承自 RecyclerView.ViewHolder 的类 MyViewHolder.java,代码如下。

```
1   public class MyViewHolder extends RecyclerView.ViewHolder {
2       TextView tv;
3       public MyViewHolder(View itemView) {
```

```
4          super(itemView);
5          tv = (TextView) itemView.findViewById(R.id.recycle_tv);
6          tv.setTextSize(25);
7      }
8  }
```

(4) 自定义继承自 RecyclerView.Adapter 的适配器类 MyAdapter.java，代码如下。

```
1  public class MyAdapter extends RecyclerView.Adapter <MyViewHolder> {
2      private LayoutInflater inflater;
3      private Context mContext;
4      private List <String> mDatas;
5      //创建构造参数
6      public MyAdapter(Context context, List <String> datas){
7          this.mContext = context;
8          this.mDatas = datas;
9          inflater = LayoutInflater.from(context);
10     }
11     //创建 ViewHolder,指定 item 并绑定
12     @Override
13     public MyViewHolder onCreateViewHolder(ViewGroup parent, int viewType) {
14         View view = inflater.inflate(R.layout.item_layout, parent, false);
15         MyViewHolder viewHolder = new MyViewHolder(view);
16         return viewHolder;
17     }
18     //绑定 ViewHolder,为绑定 item 中的组件设置相应的属性值
19     @Override
20     public void onBindViewHolder(MyViewHolder holder, int position) {
21         holder.tv.setText(mDatas.get(position));    //为 item 中的 TextView 赋值
22     }
23     //返回 item 项数目
24     @Override
25     public int getItemCount() {
26         return mDatas.size();
27     }
28     //新增 item 项
29     public void addData(int pos){
30         mDatas.add("新增");
31         notifyItemInserted(pos);
32     }
33     //移除 item 项
34     public void deleateData(int pos){
35         mDatas.remove(pos);
36         notifyItemRemoved(pos);
37     }
38 }
```

(5) 在主 Activity 中实现功能，代码如下。

```
1  mRecycleView = (RecyclerView) findViewById(R.id.rv);
2  mData = new ArrayList <String>();
3  for (int i = 0; i <33; i ++) {
```

```
4              mData.add("新闻" + i);
5          }
6      myAdapter = new MyAdapter(this, mData);//实例化适配器
7      mRecycleView.setAdapter(myAdapter);// 给 RecyclerView 对象设置适配器
8      //设置布局管理器,将布局设置成纵向(默认)
9      mRecycleView.setLayoutManager(new LinearLayoutManager(this));
```

上述第 2～5 行代码用于创建一些文字内容通过适配器填充 item,这些内容在实际开发应用中也可以是从网络取来的数据。第 9 行代码为 RecyclerView 对象设置布局管理器,以实现 RecyclerView 布局里面的内容显示方式,代码格式如下。

```
recyclerView.setLayoutManager(LayoutManager layoutManager)
```

2. RecyclerView. LayoutManager

RecyclerView. LayoutManager 是一个抽象类,它包含 LinearLayoutManager(线性布局管理器)、StaggeredGridLayoutManager(错列网格布局管理器)、GridLayoutManager(网格布局管理器)三个子类。

(1)LinearLayoutManager。

LinearLayoutManager 类有两种常用的构造方法。

第 1 种构造方法格式如下。

LinearLayoutManager(Context context),context 参数为上下文,实现的是默认的垂直布局。前面示例中使用的就是这一种构造方法。

第 2 种构造方法格式如下。

LinearLayoutManager(Context context, int orientation, boolean reverseLayout),第 1 个参数 context 为上下文,第 2 个参数 orientation 为布局显示方式(该参数值及功能说明见表 4-8),第 3 个参数 reverseLayout 为是否反转。

表 4-8 orientation 参数值及功能说明

参 数 值	功 能 说 明
LinearLayoutManager. VERTICAL	不反转的垂直布局:数据从上向下加载(新数据在底部)
LinearLayoutManager. VERTICAL	反转的垂直布局:数据从下向上加载(新数据在顶部)
LinearLayoutManager. HORIZONTAL	不反转的水平布局:数据从左向右加载(新数据在右)
LinearLayoutManager. HORIZONTAL	反转的水平布局:数据从右向左加载(新数据在左)

如果将前面第 9 行代码修改为如下代码,其显示效果为不反转的垂直布局,如图 4.23 所示。

```
1   mRecycleView.setLayoutManager(new LinearLayoutManager(this,
LinearLayoutManager.VERTICAL,false));
```

如果将前面第 9 行代码修改为如下代码,其显示效果为反转的垂直布局,如图 4.24 所示。

```
1   mRecycleView.setLayoutManager(new LinearLayoutManager(this,
LinearLayoutManager.VERTICAL,true));
```

图 4.23 不反转的垂直布局效果

图 4.24 反转的垂直布局效果

（2）StaggeredGridLayoutManager。

StaggeredGridLayoutManager 的构造方法格式如下。

StaggeredGridLayoutManager（int spanCount，int orientation），第 1 个参数 spanCount 为要显示的列数，第 2 个参数 orientation 为显示的方向。

如果将前面第 9 行代码修改为如下代码，其显示效果为水平方向，并且每列四个 item，可以通过向左滑屏翻转内容，如图 4.25 所示。

```
1    mRecycleView.setLayoutManager(new
StaggeredGridLayoutManager(4,StaggeredGridLayoutManager.HORIZONTAL));
```

如果将前面第 9 行代码修改为如下代码，其显示效果为垂直方向，并且每行四个 item，可以通过向上滑屏翻转内容，如图 4.26 所示。

```
1    mRecycleView.setLayoutManager(new
StaggeredGridLayoutManager(4,StaggeredGridLayoutManager.VERTICAL));
```

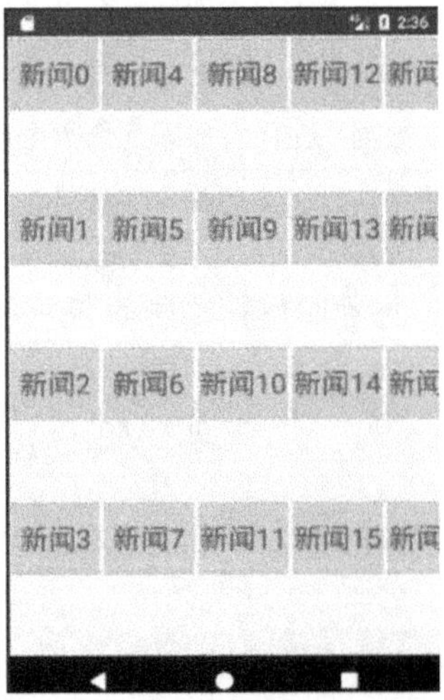

图 4.25 错列网格水平布局效果 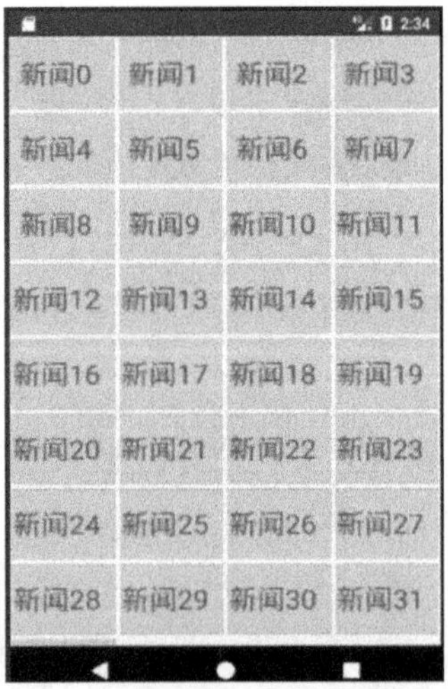　　　　图 4.26 错列网格垂直布局效果

（3）GridLayoutManager。

GridLayoutManager 既可以设置列数，也可以设置方向，还可以设置加载数据是否反转，其功能可以理解为 LinearLayoutManager 和 StaggeredGridLayoutManager 的结合体。它有两种构造方法。

① GridLayoutManager(Context context, int spanCount)，第 1 个参数 context 为上下文，第 2 个参数 spanCount 为显示列数，默认显示方向为垂直布局。

② GridLayoutManager(Context context, int spanCount, int orientation, boolean reverseLayout)，第 1 个参数 context 为上下文，第 2 个参数 spanCount 为显示列数，第 3 个参数 orientation 为显示方向，第 4 个参数 reverseLayout 为是否反转。

GridLayoutManager 的使用方法与前面两个类的使用方法一样，限于篇幅，本节不再赘述。

4.6.2 商品列表布局切换效果的实现

1. 主界面的设计

从图 4.21 和图 4.22 可以看出，整个界面的布局分为上下两个部分，上部用于显示"衣服系列"文字对象和可供使用者单击的"图片"对象，下部用于显示商品列表，代码如下。

【商品列表效果的实现】

```
1   <?xml version ="1.0" encoding ="utf-8"?>
2   <LinearLayout xmlns:android ="http://schemas.android.com/apk/res/android"
3       android:layout_width ="match_parent"
4       android:layout_height ="match_parent"
5       android:orientation ="vertical">
6       <RelativeLayout
7           android:layout_width ="match_parent"
8           android:layout_height ="40dp">
9           <TextView
10              android:layout_centerVertical ="true"
11              android:layout_marginLeft ="10dp"
12              android:text ="衣服系列"
13              android:textSize ="20sp"
14              android:layout_width ="wrap_content"
15              android:layout_height ="wrap_content" />
16          <ImageView
17              android:layout_marginRight ="10dp"
18              android:layout_centerVertical ="true"
19              android:id ="@+id/img"
20              android:src ="@mipmap/maket1"
21              android:layout_alignParentRight ="true"
22              android:layout_width ="wrap_content"
23              android:layout_height ="wrap_content" />
24      </RelativeLayout>
25      <android.support.v7.widget.RecyclerView
26          android:background ="#dfecdf"
27          android:id ="@+id/rv"
```

```
28        android:layout_width="match_parent"
29        android:layout_height="match_parent">
30    </android.support.v7.widget.RecyclerView>
31 </LinearLayout>
```

上述第 6~24 行代码定义了显示效果的上面部分（显示"衣服系列"文字对象和可供使用者单击的"图片"对象），第 25~30 行代码定义了 RecyclerView 对象用于达到切换商品列表的效果。

2. 展示商品系列的 item 布局设计

（1）每行只显示一个商品 item 的布局文件代码如下。

```
1  <?xml version="1.0" encoding="utf-8"?>
2  <LinearLayout xmlns:android="http://schemas.android.com/apk/res/android"
3      android:layout_width="match_parent"
4      android:layout_height="wrap_content"
5      android:orientation="horizontal"
6      android:background="#FFFFFF"
7      android:layout_margin="4dp">
8      <ImageView
9          android:id="@+id/image_big"
10         android:layout_width="80dp"
11         android:layout_height="80dp"
12         android:scaleType="centerCrop"
13         android:src="@mipmap/y1"/>
14     <LinearLayout
15         android:layout_width="match_parent"
16         android:layout_height="match_parent"
17         android:orientation="vertical"
18         android:gravity="center_vertical"
19         android:padding="16dp">
20         <TextView
21             android:id="@+id/title_big"
22             android:textAppearance="@style/TextAppearance.AppCompat.Medium"
23             android:layout_width="match_parent"
24             android:layout_height="wrap_content"
25             android:layout_marginBottom="4dp"
26             android:text="Image 1"/>
27         <TextView
28             android:id="@+id/tv_info"
29             android:textAppearance="@style/TextAppearance.AppCompat.Small"
30             android:layout_width="match_parent"
31             android:layout_height="wrap_content"
32             android:text="20 人感兴趣 · 33 个评论"/>
33     </LinearLayout>
34 </LinearLayout>
```

当切换成每行显示一个商品的效果时，需要显示商品图片、商品名、该商品感兴趣人数和对该商品的评论人数。

(2) 每行显示多个商品 item 的布局文件代码如下。

```xml
1   <?xml version ="1.0" encoding ="utf-8"? >
2   <LinearLayout xmlns:android ="http://schemas.android.com/apk/res/android"
3       android:layout_width ="match_parent"
4       android:layout_height ="wrap_content"
5       android:layout_margin ="4dp"
6       android:background ="#FFFFFF"
7       android:gravity ="center_horizontal"
8       android:orientation ="vertical" >
9       <ImageView
10          android:layout_marginTop ="10dp"
11          android:id ="@ +id/image_small"
12          android:layout_width ="wrap_content"
13          android:layout_height ="wrap_content"
14          android:scaleType ="centerCrop"
15          android:src ="@mipmap/y1" />
16      <TextView
17          android:id ="@ +id/title_small"
18          android:layout_width ="wrap_content"
19          android:layout_height ="50dp"
20          android:padding ="8dp"
21          android:text ="Image 1"
22          android:textAppearance ="@style/TextAppearance.AppCompat.Medium" />
23  </LinearLayout>
```

当切换成每行显示多个商品时,需要显示商品图片和商品名。

3. 功能实现

(1) 定义商品类 ItemInfo。

从图 4.21 和每行显示一个商品 item 的布局文件可以看出,每个商品包含商品图片、商品名、感兴趣人数和评论人数四个属性,所以该类的代码如下。

```java
1   public class ItemInfo {
2       private int imgResId;
3       private String title;
4       private int likes;
5       private int comments;
6       public ItemInfo(int imgResId, String title, int likes, int comments) {
7           this.imgResId = imgResId;//商品图片
8           this.title = title;//商品名
9           this.likes = likes;//感兴趣人数
10          this.comments = comments;//评论人数
11      }
12      public int getImgResId() {
13          return imgResId;
14      }
15      public String getTitle() {
16          return title;
17      }
```

```
18      public int getLikes()
19          return likes;
20      }
21      public int getComments() {
22          return comments;
23      }
24  }
```

(2) 自定义继承自 RecyclerView.ViewHolder 的类 ItemViewHolder.java，代码如下。

```
1   public class ItemViewHolder extends RecyclerView.ViewHolder {
2       ImageView iv;
3       TextView title;
4       TextView info;
5       ItemViewHolder(View itemView, int viewType) {
6           super(itemView);
7           if (viewType == 1) {//一行显示一个商品 Item
8               iv = (ImageView) itemView.findViewById(R.id.image_big);
9               title = (TextView) itemView.findViewById(R.id.title_big);
10              info = (TextView) itemView.findViewById(R.id.tv_info);
11          } else {//一行显示多个商品 Item
12              iv = (ImageView) itemView.findViewById(R.id.image_small);
13              title = (TextView) itemView.findViewById(R.id.title_small);
14          }
15      }
16  }
```

在一行显示一个商品时，需要视图列表项包括商品图片对象（ImageView）、商品名对象（TextView）、商品感兴趣人数和评论人数对象（TextView）；在一行显示多个商品时，需要视图列表项包括商品图片对象（ImageView）、商品名对象（TextView）。

(3) 自定义继承自 RecyclerView.Adapter 的适配器类 ItemAdapter.java，代码如下。

```
1   public class ItemAdapter extends RecyclerView.Adapter<ItemViewHolder> {
2       private List<ItemInfo> mItems;
3       private GridLayoutManager mLayoutManager;
4       public ItemAdapter(List<ItemInfo> items, GridLayoutManager layoutManager) {
5           mItems = items;
6           mLayoutManager = layoutManager;
7       }
8       @Override
9       public int getItemViewType(int position) {
10          int spanCount = mLayoutManager.getSpanCount();
11          if (spanCount == 1) {//显示一列
12              return 1;
13          } else {//显示三列
14              return 3;
15          }
16      }
17      @Override
18      public ItemViewHolder onCreateViewHolder(ViewGroup parent, int viewType) {
```

```
19            View view;
20            if (viewType == 1) {//一列时加载 item_big 布局文件
21                view =
LayoutInflater.from(parent.getContext()).inflate(R.layout.item_big, parent, false);
22            } else {//三列时加载 item_small 布局文件
23                view =
LayoutInflater.from(parent.getContext()).inflate(R.layout.item_small, parent, false);
24            }
25            return new ItemViewHolder(view, viewType);
26        }
27        @Override
28        public void onBindViewHolder(ItemViewHolder holder, int position) {
29            ItemInfo item = mItems.get(position % 4);
30            holder.title.setText(item.getTitle());
31            holder.iv.setImageResource(item.getImgResId());
32            if (getItemViewType(position) == 1) {
33                holder.info.setText(item.getLikes() + " 人感兴趣  ·  " +
item.getComments() + " 个评论");
34            }
35        }
36        @Override
37        public int getItemCount() {
38            return 30;//默认为 30 个商品,实际应用开发中可以根据服务器获取的数据确定
39        }
40    }
```

(4) 主 Activity 中的功能实现。

- 定义变量,代码如下。

```
1    private ImageView imageView;
2    private int cols = 3;
3    private RecyclerView recyclerView;
4    private ItemAdapter itemAdapter;
5    private GridLayoutManager gridLayoutManager;
6    private List<ItemInfo> items;
```

- 具体实现,代码如下。

```
1    //初始化主界面对象、给上面部分的 ImageView 对象设置监听事件
2    void initView() {
3        imageView = (ImageView) this.findViewById(R.id.img);
4        imageView.setOnClickListener(new MyClickImageView());
5        recyclerView = (RecyclerView) findViewById(R.id.rv);
6    }
7    //初始化 item 数据
8    void initItemsData() {
9        items = new ArrayList<>;
10       items.add(new ItemInfo(R.mipmap.y1, "休闲", 20, 33));
11       items.add(new ItemInfo(R.mipmap.y2, "正装", 10, 54));
12       items.add(new ItemInfo(R.mipmap.y3, "礼服", 27, 20));
13       items.add(new ItemInfo(R.mipmap.y4, "冬装", 45, 67));
14    }
```

```
15      //根据i值改变商品列表布局(i=1时一行显示一个商品,i=3时一行显示三个商品)
16      void switchLayout(int i) {
17          gridLayoutManager = new GridLayoutManager(this, i);
18          itemAdapter = new ItemAdapter(items, gridLayoutManager);
19          recyclerView.setAdapter(itemAdapter);
20          recyclerView.setLayoutManager(gridLayoutManager);
21      }
22      //单击上部的 ImageView 对象的监听事件
23      class MyClickImageView implements View.OnClickListener {
24          @Override
25          public void onClick(View v) {
26              if (cols == 3) {
27                  imageView.setImageResource(R.mipmap.maket2);
28                  cols = 1;
29                  switchLayout(3);
30              } else {
31                  imageView.setImageResource(R.mipmap.maket1);
32                  cols = 3;
33                  switchLayout(1);
34              }
35          }
36      }
```

上述第 4 行代码为显示效果图上面部分右侧的 ImageView 对象设置监听事件，默认状态下该对象加载的是 maket1.png。当单击该图片对象后，该图片对象切换成 maket2.png，并将商品列表切换成一行显示三个商品对象的布局；再单击该图片对象，该图片对象切换成 maket1.png，并将商品列表切换成一行显示一个商品对象的布局。为了达到这个效果，使用了 cols 变量作为一行三个商品对象和一行一个商品对象切换的标志。详细功能实现代码请读者参见 ALLAPP 代码包中 maket 文件夹里的内容。

本 章 小 结

本章结合通讯录、仿微信主界面、仿今日头条主界面、轮播效果和商品列表切换效果等案例项目的开发过程，介绍了 ListView、Fragment、ViewPager、ViewFilpper 和 RecyclerView 等组件的使用方法和应用场景。读者通过对本章高级界面组件与布局优化的理解和掌握，在将来的项目开发中可以设计出更令用户满意的具有特殊效果的用户界面。

习　　题

一、选择题

1. Intent 过滤器用于匹配 Intent 和接受 Intent 的 Activity，过滤器可以根据 Intent 中的以下选项进行匹配，除了（　　）。

A. content　　　　　　　　　　B. data
C. action　　　　　　　　　　　D. category

2. 下列关于适配器的说法正确的是（　　）。
   A. 它主要用来存储数据　　　　B. 它主要用来把数据绑定到组件上
   C. 它主要用来解析数据　　　　D. 它主要用来存储 XML 数据
3. 在隐式启动 Activity 时，需要设置 Intent 的动作，如果需要根据 URI 的数据类型来匹配 Activity，则将动作设置为（　　）。
   A. ACTION_PICK　　　　　　B. ACTION_VIEW
   C. ACTION_EDIT　　　　　　D. ACTION_ANSWER
4. 在 Android 中，ArrayAdapter 类用于（　　）。
   A. 把数据绑定到组件上　　　　B. 把数据显示到 Activity 上
   C. 把数据传递给广播　　　　　D. 把数据传递给服务
5. 下列关于 ListView 组件叙述错误的是（　　）。
   A. ListView 组件里装的是一行一行的数据，一行中可以显示多列
   B. ListView 组件中若选中一行，无论一行有几列数据，则该行的列都会被选中
   C. ListView 组件可以使用 SimpleAdapter 适配器来设置复杂的数据
   D. ListView 组件中的一行中只能装入最多两列数据
6. 下列属于 Intent 的作用的是（　　）。
   A. 实现 App 之间的数据共享
   B. 是一段长的生命周期，没有用户界面的程序，可以保持 App 在后台运行，而不会因为切换页面而消失
   C. 可以实现界面间的切换，可以包含动作和动作数据，是连接四大组件的纽带
   D. 处理一个 App 整体性的工作
7. 进度条中哪个属性是设置进度条大小格式的？（　　）
   A. android：secondaryProgress　　B. android：progress
   C. android：max　　　　　　　　D. style
8. 下列针对 ListView 组件叙述错误的是（　　）。
   A. ListView 自带滚动面板功能，如果数据超出屏幕范围，可以自动滚动
   B. ListView 在使用时，必须通过适配器来加入数据
   C. ListView 如果想改变显示内容，只需要调整对应的 List 集合中的数据即可
   D. ListView 中可以通过 OnItemClickListener 来完成针对某一项目的单击监听
9. Activity 对一些资源及状态的操作保存，最好是保存在生命周期的哪个函数中进行？（　　）
   A. onPause(　)　　　　　　　B. onCreate(　)
   C. onResume(　)　　　　　　D. onStart(　)
10. 在 Activity 的生命周期中，当 Activity 开始显示的时候执行生命周期中的哪个回调函数？（　　）
    A. onCreate(　)　　　　　　B. onPause(　)
    C. onResume(　)　　　　　D. onStart(　)
11. Intent 过滤器可以根据 Intent 中包含的信息对组件是否接受该 Intent 进行筛选，Intent 过滤器中不包含哪项信息？（　　）

A. <action>　　　　　　　B. <data>

C. <category>　　　　　　D. <name>

12. 如果希望自定义TabHost标题部分的显示内容需要使用下列哪个方法？（　　）

A. tabHost.addTab（tabHost.newTabSpec（"tab1"））

B. setIndicator( )

C. setContent( )

D. setView( )

二、填空题

1. 关闭Activity时需要调用_____函数。

2. 当启动一个Activity并且新的Activity执行完后需要返回到启动它的Activity时执行的回调函数是_____。

3. Android中表示下拉列表的组件是_____。

4. TabHost组件可以实现一个屏幕上多个页面间的切换，该组件可以使用_____方法创建一个选项卡。

5. 如果Activity A被Activity B完全遮住，那么该Activity A处于生命周期中的_____。

三、判断题

1. Activity的onResume( )是在onStart( )方法之前被调用，用于Activity的重新启动。（　　）

2. 如果一个Activity显示内容给用户看，并且接收用户的输入，那么此Activity所在的进程为可见进程。（　　）

3. Android App中如果包含多个Activity，则每个Activity都需要在AndroidManifest.xml中进行注册，否则不能使用。（　　）

4. 在创建Notification时，需要在AndroidManifest.xml文件中注册。（　　）

5. SimpleAdapter是继承于BaseAdapter的一个具体类，它可以根据开发者的要求在每一行上显示图片、文本等复杂的布局对象。（　　）

【第4章参考答案】

第 5 章 菜单和对话框

近年来，Android 平台正在变得越来越完善，尤其是用户体验方面显著提升。前面章节已经介绍了用户界面组件和界面布局，本章将详细介绍常用菜单和对话框的使用。菜单和对话框是各类 App 中非常重要的组成部分，能够在不占用界面空间的前提下，为应用程序提供统一的功能和交互界面。Android 提供了四种类型的菜单：ContextMenu（上下文菜单）、OptionsMenu（选项菜单）、SubMenu（子菜单）和 PopupMenu（弹出式菜单）。Android 也提供了丰富的对话框：AlertDialog（提示对话框）、DatePickerDialog（日期选择对话框）、TimePickerDialog（时间选择对话框）和 ProgressDialog（进度对话框）。另外，开发者还可以根据需要自定义菜单和对话框。

教学目标

掌握 ContextMenu、OptionsMenu 和 SubMenu 的使用方法。
了解 PopupMenu 的基本使用方法。
理解 ActionBar（活动栏）和传统标题栏的区别，并掌握其创建方法。
掌握 AlertDialog 的常用属性和方法，以及几种不同风格对话框的实现。
掌握 DatePickerDialog、TimePickerDialog 和 ProgressDialog 的使用。
掌握自定义菜单和自定义对话框的使用方法。

教学要求

知识要点	能力要求	相关知识
概述	（1）理解四种菜单的作用和应用场景 （2）了解对话框的种类和作用	快捷菜单

（续）

知识要点	能力要求	相关知识
满意度调查表的设计与实现	（1）掌握创建 ContextMenu 的方法 （2）掌握菜单的常用方法和属性 （3）掌握 AlertDialog、ProgressDialog 及自定义对话框的实现方法	菜单文件设计与保存 菜单项的监听事件
其他常用菜单	（1）掌握创建 OptionMenu、SubMenu 和 PopupMenu 的方法 （2）掌握 Action Bar 的设计和实现方法	菜单选项选中事件监听
宾馆预订 App 用户界面的设计与实现	（1）掌握 DatePickerDialog 和 TimePickerDialog 的创建和使用方法 （2）了解 DatePickerDialog 和 TimePickerDialog 的样式	Calendar

5.1 概　　述

5.1.1 菜单（Menu）

　　从 Android 3.0（API Level 11）开始，Android 的设备不再要求提供一个专门的菜单（Menu）按钮，而是开始推荐使用 ActionBar，所以现在市面上的 Android 设备均使用三个虚拟按键，而不再提供额外的 Menu 按钮。带 Menu 按钮的手机外观如图 5.1 所示（小米 1，Android 2.3），不带 Menu 按钮的手机外观如图 5.2 所示（小米 4，基于 Android 4.4）。

图 5.1　带 Menu 按钮的手机外观

图 5.2　不带 Menu 按钮的手机外观

随着 Android 的发展，不同版本的系统对菜单的支持也不尽相同。Android 中的菜单有如下四种。

1. ContextMenu

Android 中的 ContextMenu 与 Windows 中的快捷菜单一样，都是与某个组件相关的菜单，只是弹出的方式不一样。Windows 中的快捷菜单是右击鼠标弹出，而 Android 中的 ContextMenu 是在某个组件（视图）上长按（超过 2s）后弹出。

2. OptionsMenu

OptionsMenu 是 Android 中最常见的菜单，Android 3.0 版本之前需要通过硬件上的 Menu 键调用，因为屏幕的限制，屏幕上最多只能展示六个菜单项，如果定义的菜单项超过六个，其他菜单项会被隐藏，并在第 6 个菜单显示"更多"。由于 Android 3.0 版本之后的硬件设备上已经没有 Menu 按钮，所以就不能再依靠单击 Menu 按钮弹出 OptionsMenu，而此时 OptionsMenu 默认放到了 ActionBar 上。ActionBar 位于 Activity 顶部，用于显示 Activity 的图标、标题及菜单，也可用于导航等功能，广泛应用于 View 的交互，它通常可分为三部分，分别是 Icon 按钮、Item 按钮、overflow 按钮。

3. SubMenu

SubMenu 是将功能相同或相似的分组进行多级显示的一种菜单，选择子菜单将弹出悬浮窗口显示子菜单项，但是 Android 中的子菜单不支持嵌套，即子菜单中不能再包括其他子菜单。

4. PopupMenu

PopupMenu 可以非常方便地在指定组件（视图）的下方显示一个弹出菜单，其显示效果类似于 ContextMenu，它必须在 Android 3.0 或更高的版本上才有效。

5.1.2 对话框（Dialog）

Android 开发中经常需要在 App 用户界面上弹出一些对话框，如询问用户或者让用户选择。在实际应用开发中可以根据不同的需求选择不同样式的对话框，Android 提供了普通（包含提示消息和按钮）、列表、单选、多选、等待、进度条、编辑和自定义等多种形式的对话框。Dialog 的直接子类有 AlertDialog、CharacterPickerDialog；间接子类有 DatePickerDialog、ProgressDialog 和 TimePickerDialog，它们的具体使用方法将在后面的章节中进行详细介绍。

5.2　满意度调查表的设计与实现

我国每年都有非常多的大学毕业生步入社会，为了能及时了解毕业生的就业情况，各高校都会开展满意度情况调查。本节以一个简单的满意度调查表为例介绍菜单（Menu）和对话框（Dialog）在 Android 应用开发中的基本用法。

5.2.1 预备知识

1. ContextMenu

Android 提供了 Java 代码和标准的 XML 菜单文件两种方式来定义菜单及菜单项,如果要在 Android 的 App 中使用 ContextMenu,就需要使用以下步骤实现。

【ContextMenu、AlertDialog】

(1) 重写 onCreateContextMenu()方法。

通过重写每个 Activity 类中的 onCreateContextMenu()方法来加载菜单,该方法的原型如下。

```
1    @Override
2      public void onCreateContextMenu(ContextMenu menu, View v, Context-
Menu.ContextMenuInfo menuInfo) {
3         super.onCreateContextMenu(menu, v, menuInfo);
4         //实现菜单项代码
5      }
```

该方法的第 1 个参数 menu 为要加载的上下文菜单,第 2 个参数 v 为与菜单相关的组件,第 3 个参数 menuInfo 是菜单的附加信息。加载菜单的方式有以下两种。

方式一:XML 菜单文件

首先,在 res/menu 文件夹下创建 XML 菜单资源文件(以 main_menu.xml 为例),并定义菜单项。如果没有 menu 文件夹,就需要开发者创建。main_menu.xml 的代码如下。

```
1    <menu xmlns:android = "http://schemas.android.com/apk/res/android" >
2        <group android:id = "@ + id/gbackcolor" >
3            <item android:id = "@ + id/bgreen"
4                android:title = "背景绿色"/ >
5            <item android:id = "@ + id/bblue"
6                android:title = "背景蓝色"/ >"
7        </group >
8        <group android:id = "@ + id/gfontcolor" >
9            <item android:id = "@ + id/fgreen"
10               android:title = "前景绿色"/ >
11           <item android:id = "@ + id/fblue"
12               android:title = "前景蓝色"/ >
13       </group >
14   </menu >
```

从上述代码可以看出,在 XML 文件中包含 menu、item 和 group 等三个元素。

● menu:定义一个菜单(Menu),是菜单资源文件的根节点,里面可以包含一个或者多个 <item > 和 <group > 元素。

● item:创建一个菜单项(MenuItem),代表了菜单中一个选项。item 的常用属性及说明见表 5-1。item 元素除了常规的 id、icon 和 title 等属性外,还有一个在 Android 3.0 以后的高版本中的重要属性——showAsAction(见 5.3.3 节),该属性起向低版本的兼容性作用,即菜单项在何时以何种方式加入 ActionBar 中(在 OptionMenu 菜单中详细介绍)。

表 5-1　item 的常用属性及说明

属　性　名	说　　明
android:id	设置菜单项(item)的 id
android:title	设置菜单项标题
android:icon	设置菜单项图标
android:alphabeticShortcut	设置菜单项字母快捷键
android:numericShortcut	设置菜单项数字快捷键
android:checkable	是否为可选，值为 true 或 false
android:checked	是否为选中状态，值为 true 或 false
android:visible	设置是否可见，值为 true 或 false

● group：对菜单项进行分组，可以组的形式操作菜单项，也可省略。group 的常用属性及说明见表 5-2。group 是对菜单项进行分组，分组的菜单与不分组的菜单显示效果没有区别，但是针对菜单组可以进行统一操作，从而让开发者可以根据实际的菜单项进行分类，Android 中提供的统一操作方法如下。

Menu. setGroupCheckable()：菜单组内的菜单是否都可选。

Menu. setGroupVisible()：菜单组内的菜单是否都可见。

Menu. setGroupEnabled()：菜单组内的菜单是否都可用。

如果菜单项需要单选或者多选，可以使用 android:checkableBehavior 属性设置，它可以对单个 item 或者 group 设置一个组，该属性值可以是 single（单选）、all（多选）、none（没有 Checked 的选项，默认值）。

表 5-2　group 的常用属性及说明

属　性　名	说　　明
android:id	设置菜单组(group)的 id
android:orderInCategory	定义这组菜单在菜单中的默认次序，为整数值
android:checkableBehavior	定义这组菜单是否 checkable，有效值：none、all（复选框）、single（单选按钮）
android:visible	设置菜单是否可见，值为 true 或 false
android:enabled	设置菜单是否可用，值为 true 或 false

接着，在 onCreateContextMenu()方法中使用 MenuInflater. inflate()方法填充菜单资源，即将 XML 菜单资源转换成一个可编程的对象，代码如下。

```
1       @Override
2       public void onCreateContextMenu(ContextMenu menu, View v,
ContextMenu. ContextMenuInfo menuInfo) {
3           MenuInflater inflater = MainActivity. this. getMenuInflater();
4           //填充菜单（读取 XML 文件,解析,加载到 Menu 组件上）
5           inflater. inflate(R. menu. main_menu, menu);
6           super. onCreateContextMenu(menu, v, menuInfo);
7       }
```

图 5.3 上下文菜单

上述第 5 行代码使用了 inflate（int menuRes，Menu menu）方法实现 XML 菜单资源文件转换为 Menu 对象，其中第 1 个参数表示待转换的菜单资源文件，第 2 个参数表示菜单对象。运行效果如图 5.3 所示。

方式二：Java 代码

通过代码动态添加菜单项可以直接使用 onCreateOptionsMenu( )方法的 menu 参数，并调用 add( )方法添加菜单项，即 menu.add（菜单项的组号，菜单项的ID，菜单项的排序号，菜单项标题），其中菜单项的排序号如果是按照菜单项的添加顺序排序，该参数的值可以都为 0。

直接在 onCreateContextMenu( )方法中用 Java 代码实现，实现代码如下。

```
1     @Override
2     public void onCreateContextMenu(ContextMenu menu, View v, Context-
Menu.ContextMenuInfo menuInfo) {
3         int bgroup = 0;
4         int fgroup = 1;
5         menu.add(0, 0, 0, "背景绿色");
6         menu.add(0, 1, 1, "背景蓝色");
7         menu.add(1, 2, 2, "前景绿色");
8         menu.add(1, 3, 3, "前景蓝色");
9         super.onCreateContextMenu(menu, v, menuInfo);
10    }
```

（2）为 View 组件注册 ContextMenu。

```
1     TextViewtvinfo = (TextView) this.findViewById(R.id.tvinfo);
2     this.registerForContextMenu(tvinfo);
```

（3）重写 onContextItemSelected( )方法。

ContextMenu 菜单建成后，需要给 ContextMenu 指定监听器为每个菜单项添加执行功能，即重写 onContextItemSelected( )方法，实现代码如下。

```
1     @Override
2     public boolean onContextItemSelected(MenuItem item) {
3         switch (item.getItemId()){
4             case R.id.bgreen:
5                 tvinfo.setBackgroundColor(0xff00ff00);
6                 break;
7             case R.id.bblue:
8                 tvinfo.setBackgroundColor(0xff0000ff);
9                 break;
10            case R.id.fgreen:
11                tvinfo.setTextColor(0xff00ff00);
```

```
12                break;
13            case R.id.fblue:
14                tvinfo.setTextColor(0xff0000ff);
15                break;
16        }
17        return super.onContextItemSelected(item);
18    }
19
```

上述代码适用于 XML 菜单资源文件创建的菜单。如果是 Java 代码创建的菜单，就需要使用如下代码实现菜单项监听事件。

```
1    @Override
2    public boolean onContextItemSelected(MenuItem item) {
3        switch (item.getItemId()){
4            case 0:
5                tvinfo.setBackgroundColor(0xff00ff00);
6                break;
7            case 1:
8                tvinfo.setBackgroundColor(0xff0000ff);
9                break;
10           case 2:
11               tvinfo.setTextColor(0xff00ff00);
12               break;
13           case 3:
14               tvinfo.setTextColor(0xff0000ff);
15               break;
16       }
17       return super.onContextItemSelected(item);
18   }
```

2. AlertDialog

AlertDialog 的功能比较强大，可以生成以下四种预定义的对话框。
- 带消息和按钮的提示对话框。
- 带列表和按钮的列表对话框。
- 带单选列表和按钮的单选列表对话框。
- 带复选列表和按钮的复选列表对话框。

AlertDialog 类继承自 Dialog，AlertDialog 的构造方法全部都是 protected 的，所以不能直接通过 AlertDialog 类创建一个 AlertDialog 对象，但是可以通过其内部类 AlterDialog.Builder 来创建，AlertDialog.Builder 的常用方法及功能见表 5-3。

表 5-3　AlertDialog.Builder 的常用方法及功能

方　法　名	功　　能
create()	创建对话框
setIcon(Drawable icon)	为对话框设置图标
setTitle(CharSequence title)	为对话框设置标题

(续)

方 法 名	功 能
setItems(CharSequence[] items, DialogInterface.OnClickListener listener)	设置对话框显示的列表项
setMessage(CharSequence message)	为对话框设置消息内容
setNegativeButton(CharSequence text, DialogInterface.OnClickListener listener)	给对话框添加"取消"按钮
setNeutralButton(CharSequence text, DialogInterface.OnClickListener listener)	给对话框添加"中立"按钮
setPositiveButton(CharSequence text, DialogInterface.OnClickListener listener)	给对话框添加"确定"按钮
setMultiChoiceItems(CharSequence[] items, boolean[] checkedItems, DialogInterface.OnMultiChoiceClickListener listener)	设置对话框显示的复选列表项
setSingleChoiceItems(CharSequence[] items, int checkedItem, DialogInterface.OnClickListener listener)	设置对话框显示的单选列表项
setView(View view)	为对话框设置自定义样式
show()	显示对话框

创建对话框通常由下列五个步骤实现。

● 创建 AlertDialog.Builder 对象。

● 调用 setIcon()、setTitle()或 setCustomTitle()等设置对话框标题栏的图标、标题等。

● 调用 setMessage()方法设置对话框中显示的消息内容或调用 setItems()、setSingle-ChoiceItems()、setMultiChoiceItems()方法设置不同类别的对话框。

● 调用 setPositive/Negative/NeutralButton()方法设置确定按钮、取消按钮或中立按钮。

● 调用 create()、show()方法创建对话框对象并显示出来。

(1) 带消息和按钮的提示对话框。

带消息和按钮的提示对话框主要由标题、图标、提示信息和按钮等几个部分组成,实际开发中不需要开发者自己设计界面布局,只需要直接用 Java 代码实现,具体代码如下。

```
1        AlertDialog.Builder alertDialog = new AlertDialog.Builder(this);
2        alertDialog.setTitle("请认真选择");// 设置标题
3        alertDialog.setMessage("你喜欢智能手机开发这门课吗?");// 设置提示信息
4        alertDialog.setIcon(R.drawable.ic_launcher); // 设置图标
5        //添加 PositiveButton
6        alertDialog.setPositiveButton("非常喜欢",new DialogInterface.OnClickListener(){
7            @Override
8            public void onClick(DialogInterface dialog, int which){
```

```
         9                     //设置选择PositivieButton时所执行的操作
        10                 }
        11             });
        12             //添加NegativeButton
        13             alertDialog.setNegativeButton("不喜欢",new
DialogInterface.OnClickListener(){
        14                 @Override
        15                 public void onClick(DialogInterface dialog, int which){
        16                     //设置选择NegativeButton时所执行的操作
        17                 }
        18             });
        19             //添加NeutralButton
        20             alertDialog.setNeutralButton("一般般",new
DialogInterface.OnClickListener(){
        21                 @Override
        22                 public void onClick(DialogInterface dialog, int which){
        23                     //设置选择NeutralButton时所执行的操作
        24                 }
        25             });
        26             AlertDialog dialog = alertDialog.create();//创建对话框
        27             dialog.show();//显示对话框
        28         }
```

程序中使用了setPositiveButton( )、setNegativeButton ( )和setNeutralButton等方法直接添加按钮，这些方法的第1个参数表示按钮上显示的文本，第2个参数是对该按钮单击事件的监听。运行效果如图5.4所示。

（2）带列表和按钮的列表对话框。

带列表和按钮的列表对话框不需要界面布局，但与带消息和按钮的提示对话框不同的是，实现时需要先定义在对话框中列表显示的内容，列表显示的内容通常定义为字符串数组，如表示课程的字符串数组course[ ]，用于存放一组课程名。然后调用setItems（CharSequence [ ] items, DialogInterface. OnClickListener listener）方法实现，该方法的第1个参数表示对话框中显示的列表项数组，第2个参数表示对应选项的监听事件。运行效果如图5.5所示，详细代码如下。

图5.4　提示对话框

```
        1    String course[]=new String[]{"语文","数学","英语","化学","物理"};
        2    AlertDialog.Builder alertDialog=new AlertDialog.Builder(this);
        3    alertDialog.setTitle("你喜欢哪门功课?");//设置标题
        4    alertDialog.setIcon(R.drawable.ic_launcher);//设置图标
        5    //设置列表项对话框
        6    alertDialog.setItems(course,new DialogInterface.OnClickListener(){
        7        @Override
        8        public void onClick(DialogInterface dialog, int which){
        9            Toast.makeText(MainActivity.this,course[which],Toast.LENGTH_SHORT).show();
```

```
10          }
11      });
12      //设置 PositivieButton
13      alertDialog.setPositiveButton("确定",new DialogInterface.OnClickListener
(){
14          @Override
15          public void onClick(DialogInterface dialog, int which){
16              //设置选择 PositivieButton 时所执行的操作
17          }
18      });
19      //设置 NegativeButton 与 NeutralButton,此处略
20      AlertDialog dialog = alertDialog.create();//创建对话框
21      dialog.setCanceledOnTouchOutside(false);//使得单击对话框外部时对话框不消失
22      dialog.show(); //显示对话框
23  }
24 }
```

上述第 8～10 行代码表示单击列表项时执行的功能，此处代码表示单击某列表项后，用 Toast 将列表项的内容显示出来，即第 8 行的 which 参数表示返回单击列表中某行行号（从 0 开始）；第 12～19 行代码表示在列表下面显示"确定"和"取消"按钮，也可以不设置这两个按钮；默认状态下弹出的对话框在单击对话框外部时会自动消失，这一般不符合应用程序的使用逻辑。如果需要单击对话框的外部后对话框也不会消失的功能，就需要使用第 21 行代码。

（3）带单选列表和按钮的单选列表对话框。

单选列表对话框的实现方法与列表对话框的几乎一样，实现时只需要将列表对话框中的 setItems（CharSequence [ ] items, DialogInterface.OnClickListener listener）方法修改为 setSingleChoiceItems（CharSequence [ ] items, int checkedItem, DialogInterface.OnClickListener listener）方法，setSingleChoiceItems 方法比 setItems 方法多一个参数，该参数表示列表项中默认选中列表项的下标（从 0 开始）。程序运行效果如图 5.6 所示。

图 5.5 列表对话框

图 5.6 单选列表对话框

(4) 带复选列表项和按钮的复选列表对话框。

复选列表对话框的实现方法与单选列表对话框几乎一样，实现时只需要将单选列表对话框中的 setSingleChoiceItems（）方法修改为 setMultiChoiceItems（CharSequence［］ items，boolean［］ checkedItems，DialogInterface.OnMultiChoice ClickListener listener）方法。setMultiChoiceItems（）方法中的第 2 个参数是一个布尔型的数组，用于表示复选框中被选中列表项。如果列表项选中，则其对应的数组元素值为 true，否则为 false。图 5.7 所示效果中默认选中"语文"和"物理"复选框，所以定义的布尔型数组为

```
private  boolean[] courseSelected = new boolean[]{true,false,false,false,true};
```

3. 自定义对话框

系统提供的预定义 AlertDialog 有时不能够满足实际应用的需要。例如，要实现一个登录对话框，需要在对话框中输入用户名和密码，如图 5.8 所示，这时就需要用到自定义 View 的 AlertDialog。

【自定义对话框、进度条对话框】

图 5.7　复选列表对话框

图 5.8　自定义对话框

实现一个自定义对话框的步骤如下。

（1）在 res/layout 文件夹下自定义对话框的界面布局，本案例中用到了 TextView 组件和 EditText 组件。布局文件 dialog_login.xml 代码如下。

```xml
<LinearLayout
    android:layout_margin="15dp"
    android:layout_width="match_parent"
    android:layout_height="wrap_content"
    android:orientation="horizontal" >
    <TextView
        android:layout_width="0dp"
        android:layout_height="wrap_content"
        android:layout_weight="1"
        android:text="用户名:" />
    <EditText
        android:id="@+id/edtname"
        android:layout_width="0dp"
        android:layout_height="wrap_content"
        android:layout_weight="1" />
</LinearLayout>
<LinearLayout
    android:layout_margin="15dp"
    android:layout_width="match_parent"
    android:layout_height="wrap_content"
    android:orientation="horizontal" >
    <TextView
        android:layout_width="0dp"
        android:layout_height="wrap_content"
        android:layout_weight="1"
        android:text="密码:" />
    <EditText
        android:id="@+id/edtpwd"
        android:layout_width="0dp"
        android:layout_height="wrap_content"
        android:layout_weight="1" />
</LinearLayout>
```

(2) 把 dialog_login.xml 布局文件添加到 AlertDialog 上,即使用 LayoutInflater 类中的 inflate(int resource, ViewGroup root) 方法取得自定义的界面布局,其中第 1 个参数表示自定义的界面布局文件(本例中为 dialog_login),第 2 个参数表示将这个布局文件放在哪个父类视图对象中(若没有则为 null)。详细代码如下。

```
1      LayoutInflater layoutInflater = LayoutInflater.from(MainActivity.this);
2      View loginView = layoutInflater.inflate(R.layout.dialog_login, null);
3      //从自定义 View 获得 edtname、edtpwd 对象
4      EditText edtName = (EditText)loginView.findViewById(R.id.edtname);
5      EditText edtPwd = (EditText)loginView.findViewById(R.id.edtpwd);
6      AlertDialog.Builder loginDialog = new AlertDialog.Builder(this);
7      alertDialog.setTitle("用户登录");
8      alertDialog.setIcon(R.mipmap.ic_launcher);
9      alertDialog.setView(loginView); // 为对话框设置自定义 View
10     alertDialog.setPositiveButton("登录",
```

```
11              new DialogInterface.OnClickListener(){
12                  @Override
13                  public void onClick(DialogInterface dialog, int which){
14                      //登录按钮功能代码
15                  }
16              });
17      alertDialog.setNegativeButton("取消",
18              new DialogInterface.OnClickListener(){
19                  @Override
20                  public void onClick(DialogInterface dialog, int which){
21                      //取消按钮功能代码
22                  }
23              });
24      AlertDialog dialog = loginDialog.create();
25      dialog.setCanceledOnTouchOutside(false);
26      dialog.show();
27      }
28  }
```

4. ProgressDialog

ProgressDialog 类继承自 AlertDialog，它和 ProgressBar 有着异曲同工之处，都是用于显示执行进度；不同的是 ProgressDialog 以对话框的形式展示出来。ProgressDialog 对话框也可以通过相应方法设置对话框上显示的文字、图标和进度条的样式等，当然进度的改变同样与 ProgressBar 一样，也需要使用线程来控制。

ProgressDialog 有圆形不明确状态和水平进度条状态两种。图形进度条如图 5.9 所示，长形进度条如图 5.10 所示。其创建方式也有以下两种。

图 5.9　圆形进度条

图 5.10　长形进度条

(1) 直接使用 new ProgressDialog( ) 语句，代码如下。

```
ProgressDialog dialog = new ProgressDialog(this);
dialog.show();
```

(2) 调用 ProgressDialog 的静态方法 show( ) 创建并显示，但这种进度条只能是圆形条样式，代码如下。

```
ProgressDialogdialog1 = ProgressDialog.show(this,"提示","正在登录中");//只能是圆形条,可以通过show()方法的参数设置Title和Message提示内容
ProgressDialog dialog2 = ProgressDialog.show(this,"提示","正在登录中",false);//使用静态方式创建并显示,只能是圆形条,最后一个参数是boolean型,用于设置是否是不明确的状态
ProgressDialog dialog3 = ProgressDialog.show(this,"提示","正在登录中",false,true);//使用静态方式创建并显示,只能是圆形条,最后一个参数是boolean型,用于设置进度条是否可以取消
ProgressDialog dialog4 = ProgressDialog.show(this,"提示","正在登录中",true,true,cancelListener);//使用静态方式创建并显示,只能是圆形条,最后一个参数Dia-logInterface.OnCancelListener表示取消对话框的监听事件
```

ProgressDialog 的常用方法及功能说明见表 5-4。

表 5-4 ProgressDialog 的常用方法及功能说明

方 法 名	功 能 说 明
getMax( )	获取对话框进度条最大值
getProgress( )	获取对话框当前进度值
getSecondaryProgress( )	获取对话框第 2 进度条的值
setMax(int max)	设置对话框进度条的最大值
setMessage(CharSequence message)	设置对话框提示信息文字
setIcon(Drawable Icon)	设置对话框进度条图标
setIndeterminate (boolean indeterminate)	设置对话框进度条是否明确
setProgress(int value)	设置对话框进度条当前进度值
setProgressStyle(int style)	设置对话框进度条的样式
setSecondaryProgress(int secondaryProgress)	设置对话框第 2 进度条的值
setTitle(CharSequence message)	设置对话框进度条对话框标题
setCancelable(Boolean flag)	设置是否可以按返回键取消对话框进度条
setButton(CharSequence text,DialogInterface,OnClickListenerlistener)	设置对话框进度条上的按钮及提示信息和事件
show( )	显示进度条对话框
cancel( )	取消进度条对话框
dismiss( )	取消进度条对话框

创建完 ProgressDialog 后，通过调用 setProgressStyle （ProgressDialog.STYLE_SPINNER）方法设置圆形进度条，调用 setProgressStyle （ProgressDialog.STYLE_HORIZONTAL）方法设

置长形的进度条，然后使用表 5-4 中提供的方法设置进度条的相关属性。在使用 setButton( )方法为对话框设置按钮时，必须设置按钮的文本和按钮单击的监听事件，最后使用线程更新进度；另外，setIndeterminate（boolean indeterminate）方法的参数应为 false，否则不能显示进度条的进度，而是在进度条最小值和最大值之间循环移动，默认的值为 true。创建长形进度条和圆形进度条很相似，下面给出创建长形进度条对话框的代码。

```java
        ProgressDialog pgDialog;
        count = 0;
        pgDialog = new ProgressDialog(MainActivity.this);
        pgDialog.setProgressStyle(ProgressDialog.STYLE_HORIZONTAL); // 设置长形进度条
        pgDialog.setTitle(" 提示"); // 设置进度条标题
        pgDialog.setIcon (R.drawable.ic_launcher);    // 设置进度条图标
        //设置进度条提示信息
        pgDialog.setMessage (" 这是一个长形进度条对话框");
        //设置进度条为不明确才可以在进度条上显示具体进度
        pgDialog.setIndeterminate (false);
        pgDialog.setProgress (0);    // 设置当前进度值
        pgDialog.setSecondaryProgress (0); // 设置第 2 进度条的值
        pgDialog.setMax (1000); // 设置进度条最大值
        pgDialog.setCancelable (true); // 设置按" 返回" 键取消进度条
        pgDialog.setButton (" 取消", new CancelBtnListener()); // 添加 Button
        pgDialog.show(); // 显示对话框
            // 创建线程更新进度
            new Thread() {
            public void run() {
                try {
                    while (count < = pgDialog.getMax()) {
                        pgDialog.setProgress (count ++);
                        Thread.sleep (200); // 暂停 0.2s
                    }
                        pgDialog.cancel(); // 取消对话框
                    } catch (Exception e) {
                        pgDialog.cancel(); // 取消对话框
                    }
                }
            }.start();
          }
    });
}
// "取消" 按钮的监听器
class CancelBtnListener implements DialogInterface.OnClickListener {
    public void onClick (DialogInterface dialog, int which) {
        dialog.cancel();
    }
}
```

5.2.2 满意度调查表的实现

1. 主界面的设计

满意度调查表的主界面如图 5.11 所示,其实现方法比较简单,限于篇幅,不再赘述,详细代码请读者参见 ALLAPP 代码包中 investigationsystem 文件夹里的内容。

【满意度调查表的实现】

图 5.11 满意度调查表的主界面

2. 功能实现

(1) 实现性别选择功能。

在本案例中,性别选择功能使用 ContextMenu 实现,当长按"您的性别"右边的 EditText 后,显示如图 5.12 所示效果,然后在"性别"菜单中单击"男"或"女"后,会将单击的内容填入 EditText 中,其功能代码如下。

```
1       //创建上下文菜单
2       @Override
3       public void onCreateContextMenu(ContextMenu menu, View v,
ContextMenu.ContextMenuInfo menuInfo) {
4           super.onCreateContextMenu(menu, v, menuInfo);
5           menu.setHeaderTitle("性别");
6           menu.add(0, 0, 1, "男");
7           menu.add(0, 1, 1, "女");
8       }
9       //单击上下文菜单后的实现
10      @Override
11      public boolean onContextItemSelected(MenuItem item) {
12          switch (item.getItemId()) {
13              case 0:
```

```
14                    edtSex.setText("男");
15                    break;
16               case 1:
17                    edtSex.setText("女");
18                    break;
19           }
20           return super.onContextItemSelected(item);
21      }
```

由上述代码可以看出，在 Activity 的 onCreate()方法中重写 onCreateContextMenu()方法用于创建"性别"的上下文菜单，重写 onContextItemSelected()方法用于实现单击菜单项时的操作功能。重写完以上两个方法后，还需要在 onCreate()方法中为"性别"的 EditText 注册该菜单，即使用如下语句实现菜单的注册。

```
1      this.registerForContextMenu(edtSex);// edtSex 为用于显示性别的 EditText
```

（2）实现专业选择功能。

本案例中专业选择功能使用列表对话框实现，当单击"您的专业"右边的 EditText 时弹出"选择专业"对话框，显示如图 5.13 所示的效果，其功能代码如下。

```
1      //用列表对话框列出专业
2      private void showSpecDialog() {
3          final String[] specs = {"计算机科学与技术","通信工程","电子信息工程","电气及其自动化"};
4          AlertDialog.Builder specsDialog =
5                  new AlertDialog.Builder(MainActivity.this);
6          specsDialog.setTitle("选择专业").setIcon(R.mipmap.spec);
7          specsDialog.setItems(specs, new DialogInterface.OnClickListener() {
8              @Override
9              public void onClick(DialogInterface dialog, int which) {
10                 edtSpec.setText(specs[which]);
11             }
12         });
13         specsDialog.show();
14     }
```

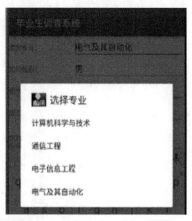

图 5.12 性别选择(上下文菜单)　　　　图 5.13 选择专业(列表对话框)

由上述代码可以看出，第2行代码表示初始化专业名称；第6～11行代码表示给列表对话框设置数据源specs，并实现列表项的单击事件监听代码。

（3）实现毕业年份选择功能。

本案例中毕业年份选择功能使用单选列表对话框实现，单击"您毕业的年份"右边的EditText时弹出"选择入学年份"对话框，显示如图5.14所示的效果图，其功能代码如下。

```
1     //用单选对话框列出入学年份
2     inttimeChoice;
3     private void showTimeChoiceDialog() {
4         final String[] times = {"1995-2000", "2001-2005", "2006-2010", "2011-2015", "2016-2020"};
5         timeChoice = -1;//没有选择
6         AlertDialog.Builder timeDialog =
7             new AlertDialog.Builder(MainActivity.this);
8         timeDialog.setTitle("选择入学年份");
9         timeDialog.setSingleChoiceItems(times, 0,
10            new DialogInterface.OnClickListener() {
11                @Override
12                public void onClick(DialogInterface dialog, int which) {
13                    timeChoice = which;
14                }
15            });
16        timeDialog.setPositiveButton("确定",
17            new DialogInterface.OnClickListener() {
18                @Override
19                public void onClick(DialogInterface dialog, int which) {
20                    if (timeChoice ! = -1) {
21                        edtTime.setText(times[yourChoice]);
22                    }
23                }
24            });
25        timeDialog.show();
26    }
```

上述第9行代码的setSingleChoiceItems( )方法的第2个参数是默认选中的选项，此处设置为0，表示弹出的单选列表项对话框中"1995—2000"项处于选中状态。

图5.14 选择入学年份(列表对话框)

第5章 菜单和对话框

(4) 实现工作过的单位性质选择功能。

本案例中工作过的单位性质选择功能使用复选列表对话框实现,当单击"您工作过的单位性质"右边的 EditText 时弹出"选择单位性质"对话框,显示如图 5.15 所示的效果,其功能代码如下。

```
1       //用复选列表对话框列出单位性质
2       ArrayList <Integer> worksChoices = new ArrayList<>();
3       private void showWorkChoiceDialog() {
4           final String[] works = {"国家行政企业","公私合作企业","中外合资企业",
"社会组织机构","国际组织机构","外资企业","私营企业","集体企业","国防军事企业"};
5           final boolean initChoiceSets[] = {false, false, false, false,
false, false, false, false, false };
6           worksChoices.clear();
7           AlertDialog.Builder worksDialog =
8               new AlertDialog.Builder(MainActivity.this);
9           worksDialog.setTitle("选择单位性质");
10          worksDialog.setMultiChoiceItems(works, initChoiceSets,
11              new DialogInterface.OnMultiChoiceClickListener() {
12                  @Override
13                  public void onClick(DialogInterface dialog, int which,
14                                      boolean isChecked) {
15                      if (isChecked) {
16                          worksChoices.add(which);
17                      } else {
18                          worksChoices.remove(which);
19                      }
20                  }
21              });
22          worksDialog.setPositiveButton("确定",
23              new DialogInterface.OnClickListener() {
24                  @Override
25                  public void onClick(DialogInterface dialog, int which) {
26                      String result = "";
27                      for (int i = 0; i <worksChoices.size(); i++) {
28                          result += works[worksChoices.get(i)] + ",";
29                      }
30                      edtFactory.setText(result);
31                  }
32              });
33          worksDialog.show();
34      }
```

上述第 5 行代码用于初始化复选列表选项中默认选中的选项,此处全为 false,表示默认均未选中。第 10 行代码 setMultiChoiceItems() 方法的第 2 个参数用于表示复选列表项中默认选中的项。

(5) 实现母校学习期间的收获功能。

由图 5.16 可以看出,这种类型的对话框需要使用自定义对话框的方法实现,所以需要在 layout 文件夹下创建一个自定义对话框的布局。本案例的自定义布局代码如下。

```xml
1   <?xml version="1.0" encoding="utf-8"?>
2   <LinearLayout xmlns:android="http://schemas.android.com/apk/res/android"
3       android:layout_width="match_parent"
4       android:layout_height="match_parent"
5       android:orientation="vertical">
6       <TextView
7           android:layout_width="match_parent"
8           android:layout_height="wrap_content"
9           android:text="请选择您的学习收获或在其他栏填入"
10          android:textSize="18sp" />
11      <RadioGroup
12          android:layout_width="wrap_content"
13          android:layout_height="wrap_content"
14          android:orientation="horizontal">
15          <RadioButton
16              android:id="@+id/radknow"
17              android:layout_width="wrap_content"
18              android:layout_height="wrap_content"
19              android:text="知识"
20              android:textSize="18sp" />
21          <!-- 技能、经验代码类似,此处省略 -->
22      </RadioGroup>
23      <EditText
24          android:id="@+id/edtother"
25          android:layout_width="match_parent"
26          android:layout_height="wrap_content"
27          android:hint="其他收获"
28          android:textSize="18sp" />
29  </LinearLayout>
```

图5.15 选择单位性质(复选列表对话框)

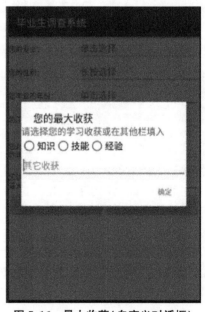

图5.16 最大收获(自定义对话框)

上述代码在本案例中以 gain_layout.xml 文件名保存在 layout 文件夹中。为了能够让这个布局文件通过 setView 装入对话框中，就需要使用下列功能代码的第 5～6 行语句将布局文件转化为 View 类型，详细的功能代码如下。

```
1       //自定义对话框
2       private void showCustomizeDialog() {
3           AlertDialog.Builder customizeDialog =
4                   new AlertDialog.Builder(MainActivity.this);
5           final View dialogView = LayoutInflater.from(MainActivity.this)
6                   .inflate(R.layout.gain_layout, null);
7           customizeDialog.setTitle("您的最大收获");
8           customizeDialog.setView(dialogView);
9           customizeDialog.setPositiveButton("确定",
10                  new DialogInterface.OnClickListener() {
11                      @Override
12                      public void onClick(DialogInterface dialog, int which) {
13                          String gain ="";
14                          RadioButton radioKnow = (RadioButton)
dialogView.findViewById(R.id.radknow);
15          //radioSkill、radioPractice 和 edtOther 的实例化代码与 radioKnow 一样,此处略
16                          if(radioKnow.isChecked()){
17                              gain ="知识";
18                          }else if(radioSkill.isChecked()){
19                              gain ="技能";
20                          }else if(radioPractice.isChecked()){
21                              gain ="经验";
22                          }else {
23                              gain = edtOther.getText().toString();
24                          }
25                          edtSchool.setText(gain);
26                      }
27              });
28          customizeDialog.show();
29      }
```

(6) 实现提交功能。

由图 5.17 可以看出，当单击"提交"按钮后会弹出一个进度条对话框，其实现代码如下。

```
1       //提交进度条对话框
2       private void showSubmitDialog() {
3           final int MAX_PROGRESS = 100;
4           final ProgressDialog progressDialog =
5                   new ProgressDialog(MainActivity.this);
6           progressDialog.setProgress(0);
7           progressDialog.setTitle("正在提交");
8           progressDialog.setProgressStyle(ProgressDialog.STYLE_HORIZONTAL);
9           progressDialog.setMax(MAX_PROGRESS);
10          progressDialog.show();
11          new Thread(new Runnable() {
```

```
12          @Override
13          public void run() {
14              int progress = 0;
15              while (progress <MAX_PROGRESS){
16                  try {
17                      Thread.sleep(100);
18                      progress ++;
19                      progressDialog.setProgress(progress);
20                  } catch (InterruptedException e){
21                      e.printStackTrace();
22                  }
23              }
24              progressDialog.cancel();// 进度达到最大值后,窗口消失
25          }
26      }).start();
27  }
```

上述第 3 行代码定义了一个进度最大值 MAX_PROGRESS,第 9 行代码中使用 setMax()方法对模拟调查表提交时需要的时间进行设置,通常这个值与实际应用环境有关,可能是提交到本地,也可能是提交到网络服务器中;第 11~26 行代码新开一个线程(每个 100ms,进度增加 1)用于模拟进度增加的过程。

(7) 实现退出功能。

由图 5.18 可以看出,当单击 "退出" 按钮后会弹出消息提示对话框,其实现代码如下。

```
1   //退出普通提示对话框
2   private void showQuitDialog(){
3       final AlertDialog.Builder normalDialog =
4           new AlertDialog.Builder(MainActivity.this);
5       normalDialog.setIcon(R.mipmap.help);
6       normalDialog.setTitle("退出");
7       normalDialog.setMessage("你确定要退出吗?");
8       normalDialog.setPositiveButton("确定",
9           new DialogInterface.OnClickListener() {
10              @Override
11              public void onClick(DialogInterface dialog, int which) {
12                  MainActivity.this.finish();//Activity 销毁
13              }
14          });
15      normalDialog.setNegativeButton("取消",
16          new DialogInterface.OnClickListener() {
17              @Override
18              public void onClick(DialogInterface dialog, int which) {
19                  //其他功能代码
20              }
21          });
22      normalDialog.show();
23  }
```

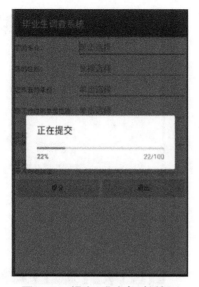

图 5.17 提交(进度条对话框)

图 5.18 退出(消息提示对话框)

5.3 其他常用菜单

5.3.1 选项菜单(OptionsMenu)

本章一开始已经介绍过 OptionsMenu 的显示效果，Android 3.0 版本以前的显示效果如图 5.19 所示（需要单击 Menu 按钮弹出菜单），Android 3.0 版本以后的显示效果如图 5.20 所示（需要单击标题栏的⁝）。在 Android App 中使用选项菜单，需要重写 Activity 中的两个方法。

【OptionsMenu、ActionBar、PopupMenu、DatePickerDialog、TimePickerDialog】

- public boolean onCreateOptionsMenu(Menu menu)：调用 OptionsMenu，在该方法中完成菜单初始化。
- public boolean onOptionsItemSelected(MenuItem item)：菜单项被选中时触发，在该方法中完成事件处理。

创建 Android App 中的选项菜单方法与 5.2 节介绍的上下文菜单的方法几乎一样，既可以使用 XML 菜单资源文件创建，也可以使用 Java 代码创建。例如，要创建如图 5.20 所示的选项菜单，就可以使用如下代码实现。

```
public boolean onCreateOptionsMenu(Menu menu){
    menu.add(0,1, 0, "向左");
    menu.add(0,2, 0, "向右");
    menu.add(0,3, 0, "向上");
    menu.add(0,4, 0, "向下");
    menu.add(0,5, 0, "快进");
    menu.add(0,6, 0, "快退");
    return true;
}
```

图 5.19　Android 3.0 版本以前的
Options Menu 显示效果

图 5.20　Android 3.0 版本以后的
Options Menu 显示效果

add()方法的返回值为一个 MenuItem 对象，可以定义一个 MenuItem 类型的变量引用它，代码如下。

```
MenuItem item1 = menu.add(0,M1,0,"向左");
item1.setIcon(R.drawable.left);//给"向左"菜单项设置一个 left.png 图片
item1.setShortcut('1','f'); //给"向左"菜单项设置快捷键
```

获得 MenuItem 对象后，就可以调用 setIcon()方法设置选项菜单的图标，图标要预先保存在 res/drawable 目录中；也可以调用 setShortcut()方法为菜单项设置快捷键，该方法的第 1 个参数是数字快捷键，第 2 个参数是全键盘快捷键。需要说明的是：Android 3.0 版本后设置的图标默认已不再显示，开发者如果仍然需要在菜单项的左侧显示图标，就需要查看 Android API 帮助文档，此处限于篇幅不再赘述。

onCreateOptionsMenu()方法仅在第 1 次使用菜单时调用，如果想每次使用菜单时都能动态地改变菜单项的某些属性，就需要重写 Activity 的 onPrepareOptionsMenu()方法，每次用户使用选项菜单时都会调用该方法，代码如下。

```
public boolean onPrepareOptionsMenu(Menu menu){
    MenuItem item = menu.findItem(0);//对应上述代码的"向左"菜单项
    item.setTitle("重新设置标题");
    return true;
}
```

该方法的参数和 onCreateOptionsMenu()方法一样，都表示 Activity 默认的菜单。如果需要重新设置每个菜单项的属性，首先需要调用 menu.findItem（int id）方法找到相应的菜单项，参数就是菜单项的 ID，方法的返回值是一个 MenuItem 对象，然后可以通过相应方法设置菜单项的属性。

如果希望选择菜单的某个选项，执行相应动作，需要覆写 onOptionsItemSelected

（MenuItem item）方法，方法的参数为用户所选择的菜单项，代码如下。

```
public boolean onOptionsItemSelected(MenuItem item){
    switch(item.getItemId()){
    case0:
        tvmsg.setText("向左");
        return true;
    case1:
        tvmsg.setText("向右");
        return true;
    }
    //其他功能代码类似,此处略
    return false;
}
```

可以通过 onOptionsItemSelected（）方法设置参数，获取被选择的菜单项，再调用 getItemId（）方法获取菜单项的 ID，最后通过 switch 语句判断所选择的是哪个菜单项，进而执行相应的操作。onOptionsItemSelected（）返回值为布尔型，值为 true 表示该事件已经得到处理，值为 false 表示该事件未处理。

5.3.2 子菜单（SubMenu）

Android 提供的 SubMenu 类似于 Windows 中的一级子菜单下的二级子菜单，只不过是用弹出的悬浮窗口来显示子菜单项。但是 Android 中的子菜单下不能再有下一级子菜单。SubMenu 继承于 Menu，其常用方法及功能说明见表 5-5。

表 5-5 SubMenu 常用方法及功能说明

方 法 名	功 能 说 明
SubMenu setHeaderIcon(Drawable icon)	设置菜单头的图标
SubMenu setHeaderIcon(int iconRes)	设置菜单头的图标
SubMenu setHeaderTitle(CharSequence title)	设置菜单头的标题
SubMenu setHeaderTitle(int titleRes)	设置菜单头的标题
SubMenu.setIcon(Drawable icon)	设置子菜单图标
SubMenu setIcon(int iconRes)	设置子菜单图标

SubMenu 既可以通过重写 onCreateOptionsMenu（）方法创建，也可以通过重写 onCreateContextMenu（）方法创建，不同的是 SubMenu 项需要调用 menu.addSubMenu（）方法创建，而不是 menu.add（）方法，具体代码如下。

```
int MENU_area =1;
SubMenu sbmenu = menu.addSubMenu(0, MENU_area, 1, "选择地区");
sbmenu.setIcon(R.mipmap.ic_launcher);
sbmenu.setHeaderTitle("选择居住的城市");
sbmenu.add(0,MENU_area +1,0,"北京");
sbmenu.add(0,MENU_area +2,1,"上海");
```

SubMenu 的单击事件也是使用 onOptionsItemSelected() 方法和 onContextItemSelected() 方法创建，在实际的 Android App 开发中，SubMenu 应用得较少。

5.3.3 活动栏（ActionBar）

在 Android 3.0 推出之前，Google 开发者网站声明 Android App 应该停止对 Android 硬件 Menu 按钮的依赖，也就是提醒开发者尽量不要再有通过 Menu 按钮实现各类操作的开发思路。基于此，在 Android 3.0 版本发布时，推出了一个新的特性——活动栏（ActionBar，也有翻译成操作栏），它主要代替了传统的标题栏，让 ActionBar 可以展示更丰富的内容，方便用户操控。

要将传统的 OptionsMenu 显示在 ActionBar 上，仍然是重写 onCreateOptionsMenu(Menu menu)方法，只不过是需要在 Java 代码实现菜单时使用如下代码。

```
1    MenuItem actionItem = menu.add("向左");
2    actionItem.setShowAsAction(MenuItem.SHOW_AS_ACTION_IF_ROOM);
```

setShowAsAction() 方法用于设置 ActionBar 中 menu 的显示方式。显示方式一共有五种，它们可以混合使用。表 5-6 所示为 ActionBar 中 Menu 的显示属性值及功能说明。

表 5-6　ActionBar 中 menu 的显示属性值及功能说明

属性值	功能说明
ifRoom	会显示在 Item 中，但是如果已经有四个或者四个以上的 Item 时会隐藏在溢出列表中。当然个数并不仅仅局限于四个，而是依据屏幕的宽窄而定
never	永远不会显示。只会在溢出列表中显示，而且只显示标题，所以在定义 Item 时，最好把标题都带上
always	无论是否溢出，总会显示
withText	withText 值表示 ActionBar 要显示文本标题。ActionBar 会尽可能地显示这个标题，但是，如果图标有效并且受到 ActionBar 空间的限制，文本标题有可能显示不全
collapseActionView	声明了这个操作视窗应该被折叠到一个按钮中，当用户选择这个按钮时，这个操作视窗展开；否则，这个操作视窗在默认的情况下是可见的，并且即便在不适用时应用，也要占据操作栏的有效空间

也可以在 XML 菜单资源文件中指定菜单项的属性值让菜单显示在 ActionBar 上，代码如下。

```
1        <group android:id="@+id/gfangxiang">
2            <item
3                android:showAsAction="ifRoom"
4                android:id="@+id/bleft"
5                android:title="向左" />
6            <item
```

```
7            android:showAsAction ="ifRoom"
8            android:id ="@ +id/bright"
9            android:title ="向右" />"
10     </group >
```

菜单的显示方式设置为 ifRoom 时,如果菜单项少于四个,其显示效果如图 5.21 所示,否则显示图 5.22 所示的效果。菜单的显示方式设置为 always 时,在屏幕宽度允许的情况下,会将所有的菜单项都显示在 ActionBar 上。

图 5.21　ActionBar 显示效果(1)

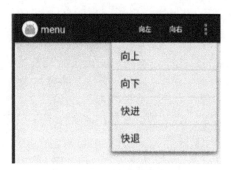

图 5.22　ActionBar 显示效果(2)

5.3.4　弹出式菜单（PopupMenu）

PopupMenu 用于在指定的 View 下显示一个弹出菜单,其菜单选项通常由 XML 菜单资源文件生成。PopupMenu 的相关方法及功能说明见表 5－7。

表 5－7　PopupMenu 的相关方法及功能说明

方　法　名	功　能　说　明
PopupMenu(Context context, View anchor)	第 1 个参数表示 Activity,第 2 个参数表示弹出菜单显示在指定 anchor 控件的下方或上方
MenuInflater.inflate(int menuResId, PopupMenu.getMenu( ))	加载 XML 菜单资源到弹出菜单对象中
setOnMenuItemClickListener( )	设置弹出菜单项的单击事件
show( )	弹出菜单
dismiss( )	关闭菜单

下面以单击 TextView 弹出菜单为例介绍弹出式菜单的使用步骤。
（1）在 res/menu 文件下创建 XML 菜单资源文件（pop_menu.xml）,代码如下。

```
1    <? xml version ="1.0" encoding ="utf-8"? >
2    <menu xmlns:android ="http://schemas.android.com/apk/res/android" >
3        <item
4            android:id ="@ +id/special_topic"
5            android:title ="专题" />
```

```
6          <item
7              android:id = "@ +id/elite"
8              android:title = "精选"/ >
9      </menu>
```

（2）在 Activity 中绑定 TextView 的单击监听事件，代码如下。

```
1          tvapp.setOnClickListener(new View.OnClickListener() {
2              @Override
3              public void onClick(View v) {
4                  PopupMenu popup = new PopupMenu(MainActivity.this, tvapp);
5                  popup.getMenuInflater().inflate(R.menu.pop_menu, popup.getMenu());
6                  popup.setOnMenuItemClickListener(new
PopupMenu.OnMenuItemClickListener() {
7                      public boolean onMenuItemClick(MenuItem item) {
8                          //单击菜单项执行
9                          return true;
10                     }
11                 });
12                 popup.show();
13             }
14         });
```

5.4 宾馆预订 App 界面的设计与实现

随着人们生活水平的不断提高，商务出行和旅游出行的需求越来越大，宾馆预订也成为人们出行之前必须重点考虑的事项之一，所以关于宾馆预订的 App 也越来越多。本节以宾馆预订界面的实现为例，介绍 Android 中的 DatePickerDialog 组件和 TimePickerDialog 组件的使用方法。

5.4.1 预备知识

1. DatePickerDialog

DatePickerDialog 类中提供了下列四种构造方法。
- public DatePickerDialog（Context context）
- public DatePickerDialog（Context context, int themeResId）
- public DatePickerDialog（Context context, OnDateSetListener listener, int year, int month, int dayOfMonth）
- public DatePickerDialog（Context context, int themeResId, OnDateSetListener listener, int year, int monthOfYear, int dayOfMonth）

以上四种构造方法中的参数说明：context 表示上下文；themeResId 表示 DatePickerDialog 的样式，见表 5-8；listener 表示日期选择时的监听事件；year 表示日期选择器默认年份；month 表示日期选择器默认月份；dayOfMonth 表示日期选择器默认天数。

第5章 菜单和对话框

表 5-8 DatePickerDialog 的样式及显示效果

DatePickerDialog 的样式	显 示 效 果
DatePickerDialog. THEME_DEVICE_DEFAULT_DARK	图 5.23
DatePickerDialog. THEME_TRADITIONAL	图 5.24
DatePickerDialog. THEME_DEVICE_DEFAULT_LIGHT	图 5.25
DatePickerDialog. THEME_HOLO_DARK	图 5.26
DatePickerDialog. THEME_HOLO_LIGHT	图 5.27

图 5.23 DatePickerDialog 显示效果(1)

图 5.24 DatePickerDialog 显示效果(2)

图 5.25 DatePickerDialog 显示效果(3)

图 5.26 DatePickerDialog 效果(4)

下面以在图 5.28 的"日期"后面的 EditText 中填入选择的日期为例介绍 DatePickerDialog 的使用步骤。

图 5.27　DatePickerDialog 显示效果(5)

图 5.28　日期时间填入效果

（1）创建 DatePickerDialog 选择日期后的监听事件类。

```
1    class MyDatePickerListener implements DatePickerDialog.OnDateSetListener{
2        @Override
3        public void onDateSet(DatePicker view, int year, int month, int dayOfMonth) {
4            Calendar ca = Calendar.getInstance();
5            ca.set(year, month, dayOfMonth);
6            //创建日期格式器对象
7            SimpleDateFormat SDformat = new SimpleDateFormat("yyyy-MM-dd");
8            //将当前日期按照格式显示在 TextView 上
9            edtDate.setText(SDformat.format(ca.getTime()));
10       }
11   }
```

上述第 7 行代码的功能是让日期以"××××年××月××日"的格式表示。

（2）创建 DatePickerDialog 对象，并实现相关功能。

```
1    //获取当前当时的年、月、日
2    Calendar ca = Calendar.getInstance();
3    int mYear = ca.get(Calendar.YEAR);
4    int mMonth = ca.get(Calendar.MONTH);
5    int mDay = ca.get(Calendar.DAY_OF_MONTH);
6    DatePickerDialog datePickerDialog = new DatePickerDialog(this,
DatePickerDialog.THEME_HOLO_LIGHT, new MyDatePickerListener() , mYear, mMonth,
mDay);
7    datePickerDialog.setTitle("日期选择");//设置日期选择对话框的标题
8    datePickerDialog.show();
```

2. TimePickerDialog

TimePickerDialog 类中提供了下列两种构造方法。

• public TimePickerDialog（Context context，OnTimeSetListener listener，int hourOfDay，int minute，boolean is24HourView）

• public TimePickerDialog（Context context，int themeResId，OnTimeSetListener listener，int hourOfDay，int minute，boolean is24HourView）

以上两种构造方法中的参数说明：context 表示上下文；themeResId 表示 TimePickerDialog 的样式，见表 5-9；listener 表示时间选择时的监听事件；hourOfDay 表示时间选择器默认小时；minute 表示时间选择器默认分钟；is24HourView 表示时间选择器是否使用 24 小时制。

表 5-9 TimePickerDialog 的样式及显示效果

TimePickerDialog 的样式	显 示 效 果
TimePickerDialog. THEME_DEVICE_DEFAULT_DARK	图 5.29
TimePickerDialog. THEME_TRADITIONAL	图 5.30
TimePickerDialog. THEME_DEVICE_DEFAULT_LIGHT	图 5.31
TimePickerDialog. THEME_HOLO_DARK	图 5.32
TimePickerDialog. THEME_HOLO_LIGHT	图 5.33

图 5.29 TimePickerDialog 显示效果（1）

图 5.30 TimePickerDialog 显示效果（2）

下面以在图 5.28 的"时间"后面的 EditText 中填入选择的时间为例介绍 TimePickerDialog 的使用步骤。

图 5.31 TimePickerDialog 显示效果(3)

图 5.32 TimePickerDialog 显示效果(4)

图 5.33 TimePickerDialog 显示效果(4)

(1) 创建 TimePickerDialog 选择时间后的监听事件类。

```
1    class MyTimePickerListener implements TimePickerDialog.OnTimeSetListener{
2        @Override
3        public void onTimeSet(TimePicker view, int hourOfDay, int minute) {
4            Calendar ca = Calendar.getInstance();
5            ca.set(Calendar.HOUR, hourOfDay);
```

```
6                ca.set(Calendar.MINUTE, minute);
7                edtTime.setText(hourOfDay + ":" + minute);
8            }
9        }
```

上述第 7 行代码的功能是让时间用"时:分"格式表示。

（2）创建 DatePickerDialog 对象，并实现相关功能。

```
1      TimePickerDialog timePickerDialog = new
TimePickerDialog(this,TimePickerDialog.THEME_HOLO_LIGHT,new
MyTimePickerListener(),12,12,true);
2        timePickerDialog.setTitle("时间选择");
3        timePickerDialog.show();
```

5.4.2 宾馆预订 App 界面的实现

1. 主界面的设计

宾馆预订 App 的主界面如图 5.34 所示，其中与前面章节介绍的 App 不一样的地方在于 ActionBar 的设计。从图 5.34 可以看出，ActionBar 的左侧显示的是图片，右侧显示的是三个供选择的选项。本案例中整个 ActionBar 都是用 Java 代码实现的，要显示左侧图片的效果，可以直接在 Activity 的 onCreate()方法中使用如下代码。

```
1    ActionBar actionBar = getSupportActionBar();//获得 ActionBar 对象
2    Drawable drawable = getApplicationContext().getDrawable(R.mipmap.hotel);
3    actionBar.setBackgroundDrawable(drawable);
4    actionBar.setTitle("");//ActionBar 的标题为空字符串
```

【宾馆预订App的实现】

图 5.34 宾馆预订 App 的主界面

上述第 2～3 行代码表示将 hotel.png 图片作为 ActionBar 的背景图片；如果没有第 4 行代码，ActionBar 上会默认显示 App 的名称。

ActionBar 右侧的"天气""公交""自驾"用 OptionsMenu 实现，其实现代码如下。

```
1   @Override
2   public boolean onCreateOptionsMenu(Menu menu) {
3       menu.add(0,0,0,"天气").setShowAsAction(MenuItem.SHOW_AS_ACTION_ALWAYS);
4       menu.add(0,1,1,"公交").setShowAsAction(MenuItem.SHOW_AS_ACTION_ALWAYS);
5       menu.add(0,2,2,"自驾").setShowAsAction(MenuItem.SHOW_AS_ACTION_ALWAYS);
6       return super.onCreateOptionsMenu(menu);
7   }
```

其他部分的布局设计比较简单，限于篇幅，不再赘述，读者可以参见 ALLAPP 代码包中 prohotel 文件夹里的内容。

2. 功能实现

（1）实现城市选择功能。

要实现选择城市时先选择省份再选择地区，需要使用子菜单。本案例使用弹出式菜单和子菜单实现，如图 5.35 所示。具体实现步骤如下。

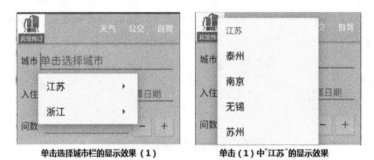

图 5.35　城市选择显示效果

- 创建菜单文件（menu_city.xml）

```
1   <?xml version="1.0" encoding="utf-8"?>
2   <menu xmlns:android="http://schemas.android.com/apk/res/android">
3       <item
4           android:id="@+id/itemjs"
5           android:title="江苏">
6           <menu>
7               <item
8                   android:id="@+id/itemtz"
9                   android:title="泰州" />
10              <!-- 其他城市代码类似,此处略 -->
11          </menu>
12      </item>
13      <item
14          android:id="@+id/itemzj"
15          android:title="浙江">
```

```
16        <menu >
17           <item
18               android:id ="@ + id/itemhz"
19               android:title ="湖州" />
20               <!--其他城市代码类似,此处略 -->
21        </menu >
22    </item >
23 </menu >
```

- 设置"单击选择城市"EditText 的监听事件

```
1    edtCity.setOnClickListener(new View.OnClickListener() {
2        @Override
3        public void onClick(View v) {
4            //在 MainActivity 的 edtCity 上生成弹出式菜单对象 popupMenu
5            PopupMenu popupMenu = new PopupMenu(MainActivity.this, edtCity);
6            //为 popupMenu 加载弹出式菜单项
7            popupMenu.getMenuInflater().inflate(R.menu.menu_city, popupMenu.getMenu());
8            //设置弹出式菜单选项的单击事件
9            popupMenu.setOnMenuItemClickListener(new PopupMenu.OnMenuItemClickListener() {
10               @Override
11               public boolean onMenuItemClick(MenuItem item) {
12                   switch (item.getItemId()){
13                       case R.id.itemtz:
14                           edtCity.setText("泰州");
15                           break;
16                       case R.id.itemnj:
17                           <!--其他城市代码类似,此处略 -->
18                   }
19                   return true;
20               }
21           });
22           popupMenu.show();
23       }
24   });
```

(2) 实现住店日期和离店日期选择功能。

住店日期和离店日期可以使用 DatePickerDialog 实现,本案例中由于对话框出现了两次,而且对话框上显示的标题信息会改变,所以将 DatePickerDialog 定义了一个方法来实现。

- DatePickerDialog 中日期改变的监听事件

```
1    class MyDatePickerListener implements DatePickerDialog.OnDateSetListener {
2        @Override
3        public void onDateSet(DatePicker view, int year, int month, int dayOfMonth) {
4            SimpleDateFormat SDformat = new SimpleDateFormat("yyyy-MM-dd");
```

```
5        Calendar calendar = Calendar.getInstance();
6        int mYear = year;
7        int mMonth = month;
8        int mDay = dayOfMonth;
9        calendar.set(year, month, dayOfMonth);
10       tempDate = SDformat.format(calendar.getTime());
11       if (flag == 1) {//单击的是到店日期 EditText
12          edtComeDate.setText(SDformat.format(calendar.getTime()));
13       } else {//单击的是离店日期 EditText
14          edtOutDate.setText(SDformat.format(calendar.getTime()));
15       }
16    }
17 }
```

- 显示 DatePickerDialog 方法

```
1    //info 就是要在对话框标题上显示的内容
2    void showDateDialog(String info) {
3       Calendar ca = Calendar.getInstance();
4       int mYear = ca.get(Calendar.YEAR);
5       int mMonth = ca.get(Calendar.MONTH);
6       int mDay = ca.get(Calendar.DAY_OF_MONTH);
7       datePickerDialog = new DatePickerDialog(MainActivity.this,
DatePickerDialog.THEME_HOLO_LIGHT, new MyDatePickerListener(), mYear, mMonth, mDay);
8       datePickerDialog.setTitle(info);
9       datePickerDialog.show();
10   }
```

- 单击入住日期和离店日期监听事件

```
1    //入住日期 EditText 监听事件
2    edtComeDate.setOnClickListener(new View.OnClickListener() {
3       @Override
4       public void onClick(View v) {
5          flag = 1;
6          showDateDialog("到店日期");
7       }
8    });
9    //离店日期 EditText 监听事件
10   edtOutDate.setOnClickListener(new View.OnClickListener() {
11      @Override
12      public void onClick(View v) {
13         flag = 2;
14         showDateDialog("离店日期");
15      }
16   });
```

(3) 实现价格/星级选择功能。

价格/星级选择对话框显示效果如图 5.36 所示。从图 5.36 的显示效果可以看出，该对话框从界面的底部开始弹出，并且要使用自定义对话框实现。

第5章 菜单和对话框

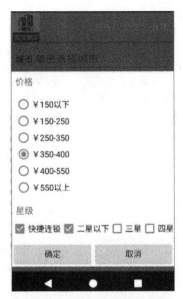

图 5.36 价格/星级选择对话框

- 自定义对话框布局文件（price_layout.xml）

```
1    <?xml version="1.0" encoding="utf-8"?>
2    <LinearLayout xmlns:android="http://schemas.android.com/apk/res/android"
3      android:layout_width="match_parent"
4      android:layout_height="match_parent"
5    <!--将对话框的背景设置为白色-->
6      android:background="#ffffff"
7      android:orientation="vertical">
8      <TextView
9        style="@style/LayoutMargin"
10       android:layout_width="match_parent"
11       android:layout_height="wrap_content"
12       android:text="价格"
13       android:textSize="15sp" />
14     <LinearLayout
15       <!--LayoutMargin 自定义的样式-->
16       style="@style/LayoutMargin"
17       android:layout_width="match_parent"
18       android:layout_height="wrap_content">
19       <RadioGroup
20         android:layout_width="match_parent"
21         android:layout_height="wrap_content">
22         <RadioButton
23           android:id="@+id/rbp1"
24           android:layout_width="wrap_content"
25           android:layout_height="wrap_content"
26           android:text="￥150以下" />
27         <!--其他 RadioButton 代码类似,此处略-->
28       </RadioGroup>
```

```
29        </LinearLayout>
30        <TextView
31            style="@style/LayoutMargin"
32            android:layout_width="match_parent"
33            android:layout_height="wrap_content"
34            android:text="星级"
35            android:textSize="15sp"/>
36        <LinearLayout
37            android:layout_width="match_parent"
38            android:layout_height="wrap_content">
39            <CheckBox
40                android:id="@+id/chb1"
41                android:layout_width="wrap_content"
42                android:layout_height="wrap_content"
43                android:text="快捷连锁"/>
44            <!--其他CheckBox代码类似,此处略-->
45        </LinearLayout>
46        <LinearLayout
47            android:layout_width="match_parent"
48            android:layout_height="wrap_content">
49            <Button
50                android:id="@+id/btnOk"
51                android:layout_width="0dp"
52                android:layout_height="wrap_content"
53                android:layout_weight="1"
54                android:text="确定"/>
55            <Button
56                android:id="@+id/btnCancel"
57                android:layout_width="0dp"
58                android:layout_height="wrap_content"
59                android:layout_weight="1"
60                android:text="取消"/>
61        </LinearLayout>
62    </LinearLayout>
```

- 定义对话框样式

修改 values/style.xml 文件，并加入如下代码。

```
1   <style name="ActionSheetDialogStyle" parent="@android:style/Theme.Dialog">
2       <!--背景透明 -->
3       <item name="android:windowBackground">@android:color/transparent</item>
4       <item name="android:windowContentOverlay">@null</item>
5       <!--浮于Activity之上 -->
6       <item name="android:windowIsFloating">true</item>
7       <!--边框 -->
8       <item name="android:windowFrame">@null</item>
9       <!--Dialog以外的区域模糊效果 -->
10      <item name="android:backgroundDimEnabled">true</item>
11      <!--无标题 -->
12      <item name="android:windowNoTitle">true</item>
```

第5章 菜单和对话框

```
13          <!--半透明-->
14          <item name="android:windowIsTranslucent">true</item>
15          <!-- Dialog 进入及退出动画 -->
16      </style>
```

- 根据自定义对话框布局和样式创建自定义对话框对象

本案例中通过自定义的方法来实现这一功能，详细代码如下。

```
1   void showPriceDialog() {
2       Dialog dialog = new Dialog(this, R.style.ActionSheetDialogStyle);
3       View view=LayoutInflater.from(this).inflate(R.layout.price_layout,null);
4       dialog.setContentView(view);//将布局设置给 Dialog
5       Window dialogWindow = dialog.getWindow();//获取当前 Activity 所在窗体
6       dialogWindow.setGravity(Gravity.BOTTOM);//设置 Dialog 从窗体底部弹出
7       WindowManager.LayoutParams lp = dialogWindow.getAttributes();//获得窗体的属性
8       lp.y = 20;//设置 Dialog 离底部的距离
9       dialogWindow.setAttributes(lp);//将属性设置给窗体
10      dialog.show();
11  }
```

其他功能的代码可以使用前面章节介绍的内容实现，限于篇幅，不再赘述，读者可以参见 ALLAPP 代码包中 prohotel 文件夹里的内容。

本 章 小 结

本章通过理论与实践相结合的方法介绍了 OptionsMenu、SubMenu、ContextMenu 和 PopupMenu 等的使用方法，介绍了功能强大的 AlertDialog、DatePickerDialog、TimePickerDialog 和不同风格的 ProgressDialog 的使用及应用场景。读者通过本章的学习，可以设计出更好的用户界面，为 Android App 提供更完备的功能。

习 题

一、选择题

1. Android 支持三种菜单，以下选项中不是 Android 支持的菜单的是（　　）。
  A. OptionsMenu   B. SubMenu
  C. ContextMenu   D. MainMenu
2. 下列关于线程的说法正确的是（　　）。
  A. 实现 Runnable 接口的类，就是线程类
  B. 调用线程类中的 run( )方法，即可启动线程
  C. 线程 sleep 的时候调用 interrupted( )方法可能会出现异常
  D. 被 interrupted 的线程调用 sleep( )方法不会出现异常
3. 下列关于选项菜单的说法错误的是（　　）。
  A. Menu 既可以在 XML 文件中定义，也可以在 Java 代码中创建

B. 在创建菜单时会执行 onCreateOptionsMenu( )方法，并且每次单击时都会再次执行
C. 当单击某个菜单项时会执行 onOptionsItemSelected( )方法
D. 选项菜单可以是图标菜单和扩展菜单，且图标菜单最多显示六个

4. 在 Android 中使用 Menu 时不需要重写的方法有（   ）。

A. onCreateOptionsMenu( )　　　　B. onCreateMenu( )
C. onOptionsItemSelected( )　　　　D. onPrepareOptionsMenu( )

5. 处理菜单项"单击事件"的方法不包含（   ）。

A. 使用 onOptionsItemSelected（MenuItem item）响应
B. 使用 onMenuItemSelected（int featureId，MenuItem item）响应
C. 使用 onMenuItemClick（MenuItem item）响应
D. 使用 onCreateOptionsMenu（Menu menu）响应

6. 下列关于 AlertDialog 的说法错误的是（   ）。

A. 要想使用对话框，首先要使用 new 关键字创建 AlertDialog 的实例
B. 对话框的显示需要调用 show 方法
C. setPositiveButton 方法是用来加确定按钮的
D. setNegativeButton 方法是用来加取消按钮的

7. 下面关于 Android 中 ContextMenu 说法错误的是（   ）。

A. 上下文菜单的菜单项需要在 onCreateContextMenu( )方法中创建，每次使用菜单时该方法都会被调用
B. 上下文菜单需要和某个界面控件进行注册才能使用，使用 registerForContextMenu( )方法进行注册，该方法的参数为控件对象
C. 在上下文菜单中选择某个菜单项后会执行 onContextItemSelected( )方法
D. 在某个控件上单击右键，会弹出与该控件相关联的上下文菜单

二、填空题

1. _____是用户界面中一种经常使用的组件，通常用于向用户显示某个耗时操作的完成百分比。
2. 创建子菜单的方法是_____。
3. _____类继承自 AlertDialog，它和 ProgressBar 都是用于显示执行进度。
4. Android 中输入日期的选择对话框是_____。
5. Android 中输入时间的选择对话框是_____。

三．判断题

1. 每次单击上下文菜单时 OnCreateContextMenu( )函数都会被调用一次。　　（   ）
2. 在定义 Menu 时使用 Menu.FIRST 可以返回第 1 个菜单项。　　（   ）
3. 在定义菜单项时只能使用 Java 代码创建，不能使用 XML 文件创建。　　（   ）
4. 选项菜单可以是图标菜单和扩展菜单，且图标菜单最多显示 10 个。　　（   ）
5. TimePickerDialog 是用来选择时间的对话框，不可以设置时间为 12 小时制。　　（   ）

【第5章参考答案】

第 6 章
服务和消息广播

Service 和 BroadcastReceiver 是 Android 的两个重要组件。Service 是 Android 的后台服务组件,适用于开发无界面且长期在后台运行的程序,如下载程序或播放背景音乐等;BroadcastReceiver 是对广播进行过滤并响应的组件,如电池电量不足或定时激发某个事件等。本章将结合实际应用案例介绍它们的用法。

教学目标

理解 Service 和 BroadcastReceiver 组件的工作机制。
掌握 Service 和 BroadcastReceiver 的使用方法。
掌握 TelephonyManager、AlarmManager、PendingIntent、Notification 和 SmsMamager 的使用方法。
了解 MediaRecorder 和 MediaPlayer 的基本使用步骤。

教学要求

知识要点	能力要求	相关知识
概述	(1) 了解进程和线程的概念及区别 (2) 了解 Service 的含义 (3) 了解 BroadcastReceiver 的含义	前台任务和后台任务
电话监听器的设计与实现	(1) 掌握 Service 的工作机制和使用方法 (2) 掌握 TelephonyManager 的使用方法	MediaRecorder 的使用
短信拦截器的设计与实现	(1) 掌握 BroadcastReceiver 的工作机制和使用方法 (2) 掌握 Android 自定义权限的含义和使用方法	系统权限和用户权限
闹钟的设计与实现	(1) 掌握 AlarmManager 的使用方法 (2) 掌握 PendingIntent 的使用方法	MediaPlayer 的使用
定时短信发送器的设计与实现	(1) 掌握 SmsManager 的使用方法 (2) 掌握 Notification 的使用方法	AlarmManager

6.1 概　　述

1. 进程与线程

进程是一个运行中的程序，是操作系统调度与资源分配的一个独立单位，在内存中有完备的数据空间和代码空间。线程是一个进程中一路单独运行的程序，是比进程更小的执行单元。每一个进程由一个或多个线程构成，线程需要放在一个进程中才能执行。

Android 会为每一个 App 分配一个单独的 Linux 用户。当一个 App 第 1 次启动时，Android 会自动启动一个 Linux 进程和一个主线程，在默认情况下，这个程序的组件都将在这个进程和线程中运行。只有当内存资源出现不足时，Android 才会尝试停止一些进程从而释放足够的资源给其他新的进程使用，也能保证用户正在访问的当前进程有足够的资源去及时响应用户的事件。

Android 会根据进程中运行的组件类别及组件的状态来判断该进程的重要性。Android 的进程可以按照重要性分为以下五种。

（1）前台进程：是指目前正在屏幕上显示的进程和一些系统进程，也就是正在和用户交互的进程。这个进程是最重要的，也是最后被销毁的。例如，正在 Android 平台上使用微信与别人聊天，此时微信程序所在的进程就是前台进程。

（2）可见进程：是指程序界面能够被用户看见，但却不能像前台进程一样与用户交互的进程。例如，正在使用的 Android App 弹出一个对话框，此时在对话框后可以看到 App 的部分界面，但是并没有与用户进行交互，那么可见界面的 App 就是可见进程。

（3）服务进程：是指通过 startService( ) 方法启动 Service 而开启的进程，Service 所在的进程虽然对用户不是直接可见的，但是它会执行用户启动的任务，只要前台进程和可见进程有足够的内存，Android 就不会回收它们。例如，在后台播放音乐或者在后台下载文件。

（4）后台进程：是指目前对用户不可见的进程，通常对用户体验没有直接影响，它可以在前台进程、可见进程和服务进程需要内存的时候被 Android 回收。例如，用户正在用微信与别人聊天时微信程序所在的进程是前台进程，当用户按下 Home 键后微信界面消失，此时该进程就转换成了后台进程。

（5）空进程：是指在这些进程内部没有任何组件在运行。保留这种进程的唯一目的是用作缓存，以缩短该应用下次在其中运行组件所需的启动时间。Android 经常会中止这些进程，这样可以调节程序缓存和系统缓存的平衡。

在进行系统资源回收时，按照空进程、后台进程、服务进程、可见进程、前台进程的顺序进行回收，以释放进程占用的系统资源。

2. Service

Android 的 Service 是一种类似于 Activity 的组件，但 Service 没有用户操作界面，也不能自己启动，其主要作用是提供后台服务调用，所以一般用于在后台处理一些耗时的逻辑或者去执行某些需要长期运行的任务。例如，在 Android 终端下载某个软件时，即使退出下载界面也不会中断下载，而是转到后台继续下载。

Service 可由 Actvity 等其他组件启动，Service 一旦被启动将在后台一直运行，即使启动该 Service 的组件已经销毁也不受影响。Service 比 Activity 具有更高的优先级，在系统资源紧张的时候，系统也不会轻易地终止 Service，即使 Service 被终止，当系统资源恢复时，系统也将自动恢复 Service 的运行。

3. 广播消息

Android 提供了一种称为广播的机制，Android App 可以使用 Intent 发送广播消息，App 可以注册接收某种广播消息，广播消息的内容可以是系统的一些状态，如电量过低、收到短消息，或者系统设置的变化，也可以是 App 自定义的一些消息。接收广播消息需要使用 BroadcastReceiver 组件，且注册 BroadcastReceiver 组件时需要说明该 BroadcastReceiver 组件能接收哪种类型的广播消息。

6.2 电话监听器的设计与实现

Service 是 Android 的四大组件之一，在 Android 开发中起着非常重要的作用。它是一个没有用户界面的、不能直接与用户交互的、通常在后台执行耗时操作的应用组件。例如，用户打开音乐播放器听音乐，在听音乐的同时又要上网聊 QQ、看新闻等，这时就需要用 Service 来实现音乐播放。本节通过电话监听器的实现案例来讲述 Service 的原理及使用方法。

6.2.1 预备知识

1. Service 简介

【Service】

Service 与 Activity 一样也是 Context 的子类，只不过 Service 没有用户界面，而是在后台运行。Service 的主要用途有以下两个。

（1）后台运行：通过启动一个服务，可以在不显示界面的情况下在后台运行指定的任务，这样可以不影响用户做其他事情。

（2）跨进程访问：通过 AIDL（Android Interface Definition Language）可以实现不同进程之间的通信。

2. Service 的生命周期

Service 的启动和关闭有以下两种方式。

（1）启动式：是指通过其他组件调用 startService()方法启动 Service，通过其他组件调用 stopService()方法或 Service 本身调用 stopSelf()方法关闭 Service。当 Service 被停止时，系统会销毁它。这种方式的优点是调用简单、控制方便；缺点是一旦启动了 Service，除了再次调用或关闭 Service 外就再没有办法对 Service 内部状态进行操作控制，缺乏灵活性。也就是说，通过该方式启动的 Service 与启动它的组件没有关联，即使启动它的组件退出了，Service 也仍然在后台运行。

（2）绑定式：是指通过其他组件（客户）调用 bindService()方法启动 Service，其他组件（客户）可以通过一个 IBinder 接口与 Service 进行通信，也可以通过调用 unbindService()方法关闭 Service。这种方式可以通过 IBinder 接口获取 Service 对 Service 状态进行检测,即通过该方式启动的 Service 与启动它的组件(客户)绑定在一起,组件(客户)一旦退出,Service 也随之关闭。

由于 Service 有以上两种完全不同的启动和关闭方式,所以其生命周期也不完全一样,如图 6.1 所示。

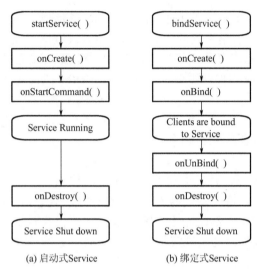

图 6.1 Service 的生命周期

从图 6.1(a)所示的启动式 Service 的生命周期可以看出,首次调用 startService()方法启动会创建一个 Service 实例,依次调用 onCreate()方法和 onStartCommand()方法,该 Service 都在后台运行。如果一个 Service 被 startService()方法多次启动,那么 onCreate()方法只会调用一次,onStartCommand()方法将会被调用多次[对应 startService()调用的次数],并且系统只会创建 Service 的一个实例,所以也只需要调用一次 stopService()方法即可停止 Service。该 Service 将会一直在后台运行,直到自身调用 stopSelf()方法或者其他 Activity 调用 stopService()方法时服务停止,onDestroy()方法将会被调用。当然在系统资源不足时,Android 也可能结束服务。

从图 6.1(b)所示的绑定式 Service 的生命周期可以看出,首次调用 bindService()方法绑定启动会创建一个 Service 实例,并调用其 onCreate()方法和 onBind()方法,然后调用者就可以通过 IBinder 和 Service 进行交互。如果再次使用 bindService()方法绑定 Service,系统不会创建新的 Service 实例,也不会再调用 onBind()方法,只会直接把 IBinder 对象传递给其他后来增加的客户端。无论调用多少次 bindService()方法,onCreate()方法都只会调用一次。当连接建立之后,Service 将会一直运行。如果要解除与服务的绑定,只需调用 unbindService(),此时 onUnbind()方法和 onDestory()方法将会被调用。这是一个客户端的情况,如果是多个客户端绑定同一个 Service 的情况,当一个客户完成和 Service 之间的互动后,它调用 unbindService()方法来解除绑定。当所有的客户端都和 Service 解除绑定后,系统会销毁 Service。即除非调用 unbindService()方法断开连接或者之前调用 bindService()方法的 Context 不存在了,如 Activity 被 finish()方法结束,系统才会自动停止 Service,对应 onDestroy()方法将被调用。

当然,如果一个 Service 既被 startService()方法启动又被 bindService()方法绑定启动,即一个 Service 又被启动又被绑定,那么该 Service 将会一直在后台运行,并且无论如何调用,onCreate()方法始终只会调用一次,而对应的 startService()方法调用多少次,Service 的 onStartCommand()方法也会调用多少次。此时调用 unbindService()方法不会停止 Service,而必须调用 stopService()方法或 stopSelf()来停止服务。

Service 生命周期中的常用方法及功能说明见表 6-1。

第6章 服务和消息广播

表6-1 Service 生命周期中的常用方法及功能说明

方 法 名	功 能 说 明
void onCreate( )	只在 Service 创建时调用一次，可以在此进行一些一次性的初始化操作
int onStartCommand(Intent i, int f, int id)	当其他组件调用 startService( )方法时此方法将会被调用，在这里实现 Service 主要的操作，其中参数 i 表示输入对象，参数 f 表示 Service 的启动方式，参数 id 表示当前启动 Service 的唯一标式符。返回值决定服务结束后的处理方式，表后作详细说明
void onStart(Intent i, int id)	Android2.0 版本前的方法，新版本已被 onStartCommand( )方法代替
IBinder onBind(Intent i)	当其他组件调用 bindService( )方法时，此方法将会被调用；如果不想让这个 Service 被绑定，直接返回 null 即可。当通过 bindService( )启动服务时触发此方法，返回一个 IBinder 对象，在远程服务时可用于对 Service 对象进行远程操控
void onRebind(Intent i)	当使用 startService( )或 bindService( )启动 Service，并且 onUnbind( )返回值为 true 时，下次再次调用 bindService( )将触发
boolean onUnbind(Intent i)	调用 unbindService( )触发此方法，默认返回 false；当返回值为 true 时，若再次调用 bindService( )将触发 onRebind( )
void onDestory( )	分三种情况： （1）以 startService( )启动 Service，调用 stopService( )时触发此方法； （2）以 bindService( )启动 Service，调用 unbindService( )时触发此方法； （3）以 startService( )启动 Service，再用 bindService( )绑定服务，Service 结束时必须先调用 unbindService( )解绑，再使用 stopService( )结束 Service 才会触发此方法

由于 Android 的移动终端设备的 RAM、内部资源有限，因此很多 Service 都会因为资源不足而被 kill，这时 onStartCommand( )的返回值就决定了 Service 被 kill 后的处理方式，onStartCommand( )的返回值一般分为以下几种。

- START_STICKY：如果 Service 进程被 kill，系统会尝试重新创建 Service，如果在此期间没有任何启动命令被传递到 Service，那么参数 Intent 为 null。
- START_NOT_STICKY：使用该返回值时，如果在执行完 onStartCommand( )后，Service 被异常 kill，系统不会自动重启该 Service。
- START_REDELIVER_INTENT：使用该返回值时，如果在执行完 onStartCommand( )后，Service 被异常 kill，系统会自动重启该 Service，并将 Intent 的值传入。

3. Service 的类型

（1）Service 根据运行地点的不同可以分为以下两种。

① 本地服务（Local Service）：这种 Service 通常与启动该服务的组件（如 Activity）在同一个进程中，当前进程结束后 Service 也会随之结束，Service 可以随时与 Activity 等多个组件进行信息交换。Service 不会自动启动线程，如果没有人工调用多线程方式启动 Service，那么 Service 将依附于主线程运行。

②远程服务（Remote Service）：这种 Service 通常与启动该服务的组件不在同一个进程中，可通过 AIDL 接口定义语言实现两个进程间通信。AIDL 的 IPC 机制是基于远程过程调用（Remote Proceduce Call，RPC）协议建立的，用于约束两个进程间的通信规则，供编译器生成代码。进程之间的通信信息首先会被转换成 AIDL 协议消息，然后发送给对方，对方收到 AIDL 协议消息后再转换成相应的对象。其使用方法将在后面的章节进行详细介绍。

（2）Service 根据运行类型的不同可以分为以下两种。

①前台服务：用户可以看到的服务（如在通知栏显示通知）。服务被终止时通知栏的通知也会消失。这类服务使用时需要让用户知道并进行相关的操作，如音乐播放服务。

②后台服务：用户无法看到的在后台运行的服务。即使服务被终止，用户也无法知道。这类服务使用时不需要用户知道并进行相关的操作，如天气预报更新。

（3）Service 根据功能的不同可以分为以下两种。

①不可通信的后台服务：用 startService( ) 启动的服务，可以应用在不需要进行任何通信的服务。

②可通信的后台服务：用 bindService( ) 启动的服务，可以应用在需要进行通信的服务。

4. 启动式 Service 的使用

启动式 Service 是最普通、最常用的后台服务 Service，通常使用 startService( ) 启动服务和 stopService( ) 停止服务。启动式 Service 通常不需要进行任何通信。下面以图 6.2 和图 6.3 所示的运行效果为例介绍本地 Service 的使用步骤：当用户单击界面上的"启动 Service"按钮时，每隔 5s 使用 Toast 显示一次"正在服务，请稍候！"信息；当用户单击界面的"停止 Service"按钮时，不再显示上面的信息。其中，图 6.2 是单击"启动 Service"按钮后的效果，图 6.3 是在图 6.2 的基础上单击 Home 按钮的运行效果。界面布局实现比较简单，限于篇幅，不再赘述，功能实现步骤如下。

图 6.2　启动式 Service 运行效果（1）

图 6.3　启动式 Service 运行效果（2）

(1) 新建继承于 Service 类的子类（本案例为 InfoService 类），并根据需要重写 onCreate( )、onStartCommand( )、onDestroy( ) 和 onBind( ) 等方法，代码如下。

```
1   public class InfoService extends Service {
2     private boolean flag = true;
3     private Handler handler = new Handler() {
4       @Override
5       public void handleMessage(Message msg) {
6         super.handleMessage(msg);
7         Context context = getApplicationContext();
8         Toast.makeText(context, "正在服务,请稍候!",Toast.LENGTH_SHORT).show();
9       }
10    };
11    @Override
12    public void onCreate() {
13      super.onCreate();
14      new Thread(new Runnable() {
15        @Override
16        public void run() {
17          while (flag) {
18            handler.sendEmptyMessage(0);
19            try {
20              Thread.sleep(5000);
21            } catch (InterruptedException e) {
22              e.printStackTrace();
23            }
24          }
25        }
26      }).start();
27    }
28    @Override
29    public void onDestroy() {//调用 stopService()后触发
30      super.onDestroy();
31      flag = false;
32    }
33    @Override
34    public IBinder onBind(Intent intent) {
35      return null;
36    }
37  }
```

(2) 构建用于启动 Service 和停止 Service 的 Intent，并调用 startService( ) 启动 Service、调用 stopService( ) 停止 Service，代码如下。

```
1       //启动 InfoService
2       btnStart.setOnClickListener(new View.OnClickListener() {
3         @Override
4         public void onClick(View v) {
5           Intent startIntent = new Intent(MainActivity.this,InfoService.class);
6           MainActivity.this.startService(startIntent);
7         }
```

```
8         });
9         //停止 InfoService
10        btnStop.setOnClickListener(new View.OnClickListener() {
11          @Override
12          public void onClick(View v) {
13            Intent startIntent = new Intent(MainActivity.this,InfoService.class);
14            MainActivity.this.stopService(startIntent);
15          }
16        });
```

(3) 在 AndroidManifest.xml 里注册 InfoService,代码如下。

```
1    <service
2        android:name=".InfoService" >
3    </service>
```

在 AndroidManifest.xml 里注册 Service 时,android:name 属性是必须要有的,还可以使用如下属性。

```
1    <service
2        android:enabled=["true" |"false"]
3        android:exported=["true" |"false"]
4        android:icon="drawable resource"
5        android:isolatedProcess=["true" |"false"]
6        android:label="string resource"
7        android:name="string"
8        android:permission="string"
9        android:process="string" >
10   </service>
```

● android:enabled:如果为 true,则这个 Service 可以被系统实例化;如果为 false,则 Service 不可以被系统实例化,默认为 true。

● android:exported:如果为 true,则其他应用的组件也可以调用这个 Service 并且可以与它进行互动;如果为 false,则只有与 Service 同一个应用或者相同 UserID 的应用可以开启或绑定此 Service。它的默认值取决于 Service 是否有 intent filters。如果一个 filter 都没有,就意味着只有指定了 Service 的准确的类名才能调用,也就是说这个 Service 只能在应用内部使用(因为其他的应用不知道它的类名)。这种情况下 exported 的默认值就为 false。反之,如果有了一个 filter,就意味着 Service 可以被外界使用,即 exported 的默认值就为 true。

● android:icon:指明 Service 的 icon。

● android:isolatedProcess:如果设置为 true,则这个 Service 将运行在一个从系统中其他部分分离出来的特殊进程中,此时只能通过 Service API 来与它进行交流。默认为 false。

● android:label:显示给用户的 Service 的名字。如果不设置,将默认使用 <application> 的 label 属性。

● android:name:Service 的包名,如 "cn.edu.nnutc.ie.MyService"。该属性是唯一一个必须要有的属性。

● android:permission:其他组件必须具有该属性值的权限才能启动这个 Service。

第6章 服务和消息广播

- android:process：Service 运行地所在进程的 name。默认启动的 Service 是运行在主进程中的。

5. 绑定式 Service 的使用

绑定式 Service 的使用比较复杂，这种服务通常可用于与其他组件通信。下面以前面的案例为基础介绍使用绑定式启动 Service，用 Service 与调用组件间的通信机制将 Service 中 Count 值传递给调用者。

（1）新建继承于 Service 类的子类（本案例为 InfoBindService 类），在 InfoBindService 类中新建一个继承于 Binder 类的内部类（本案例为 InfoBinder），代码如下。

```
1   public class InfoBindService extends Service {
2       private boolean flag = true;
3       private int count = 0;
4       private Handler handler = new Handler() {
5           @Override
6           public void handleMessage(Message msg) {
7               super.handleMessage(msg);
8               Context context = getApplicationContext();
9               Toast.makeText(context, "正在服务,请稍候!", Toast.LENGTH_SHORT).show();
10          }
11      };
12      //定义 onBind 方法所返回的 Binder 对象
13      private InfoBinder infoBinder = new InfoBinder();
14      class InfoBinder extends Binder {
15          public int getCount() {
16              return count; //通过 Binder 与 Activity 通信
17          }
18      }
19      //必须实现的方法,绑定该 Service 时调用该方法
20      @Override
21      public IBinder onBind(Intent intent) {
22          return infoBinder;
23      }
24      //Service 创建时调用
25      @Override
26      public void onCreate() {
27          super.onCreate();
28          new Thread(new Runnable() {
29              @Override
30              public void run() {
31                  while (flag) {
32                      handler.sendEmptyMessage(0);
33                      try {
34                          Thread.sleep(5000);
35                      } catch (InterruptedException e) {
36                          e.printStackTrace();
37                      }
38                      count ++;//动态修改 count 的值
39                  }
```

```
40              }
41          }).start();
42      }
43      //service 被关闭时调用
44      @Override
45      public void onDestroy() {
46          super.onDestroy();
47          flag = false;
48      }
49      //Servie 断开连接时调用
50      @Override
51      public boolean onUnbind(Intent intent) {
52          count = -100;
53          return super.onUnbind(intent);
54      }
55  }
```

上述第 12～23 行代码表示创建一个 InfoBinder 类的实例化对象 infoBinder，通过重写 InfoBindService 类中的 onBind（）方法返回 InfoBinder 类型的 infoBinder 对象，以便传入 onServiceConnected(ComponentName name, IBinder bind) 的 bind 参数中。

为了实现在 Service 与 Activity 间的通信，如本案例中将 Service 对 count 的处理结果传送给 Activity 并通过 Toast 展现出来，通常需要按照以下步骤完成。

第 1 步：在自定义的 Service 中（如本案例的 InfoBindService）自定义一个 Binder 类（如本案例的 InfoBinder），然后将需要对外暴露的方法都写到该类中（如本案例的getCount）。

第 2 步：在自定义的 Service 中实例化自定义的 Binder 类（如本案例的 infoBinder），然后重写 onBind（）方法，将实例化的 Binder 对象返回（如上述第 20～23 行代码）。

第 3 步：在 Activity 类中实例化一个 ServiceConnection 对象，重写 onServiceConnected（）方法，然后获取 Binder 对象（如下文第 4～7 行代码），最后调用相关方法（如下文第 17 行代码）。

（2）在绑定该 Service 的类中定义一个 ServiceConnection 对象，重写该对象的 onServiceConnected（）方法和 onDisconnected（）方法，然后直接读取 IBinder 传递过来的参数即可，代码如下。

```
1   private InfoBindService.InfoBinder infoBinder;
2   private ServiceConnection conn = new ServiceConnection() {
3       //Activity 与 Service 成功连接时调用该方法
4       @Override
5       public void onServiceConnected(ComponentName name, IBinder binder) {
6           infoBinder = (InfoBindService.InfoBinder) binder;
7       }
8       //Activity 与 Service 断开连接时调用该方法
9       @Override
10      public void onServiceDisconnected(ComponentName name) {
11          //断开连接时的功能代码
12      }
13  };
```

第6章　服务和消息广播

（3）绑定 Service、解绑 Service 及读取 Service 的 Count 对象值的代码如下。

```
1      btnBind.setOnClickListener(new View.OnClickListener() {
2          @Override
3          public void onClick(View v) {
4              Intent intent = new Intent(MainActivity.this, InfoBindService.class);
5              bindService(intent, conn, BIND_AUTO_CREATE);
6          }
7      });
8      btnunBind.setOnClickListener(new View.OnClickListener() {
9          @Override
10         public void onClick(View v) {
11             unbindService(conn);
12         }
13     });
14     btngetStatus.setOnClickListener(new View.OnClickListener() {
15         @Override
16         public void onClick(View v) {
17             Toast.makeText(getApplicationContext(), "当前 Service 的 Count:"
   + infoBinder.getCount(), Toast.LENGTH_SHORT).show();
18         }
19     });
```

上述第 5 行代码表示绑定 Service。bindService( )方法有三个参数：第 1 个参数是绑定了 Service 的 Intent 对象；第 2 个参数是 ServiceConnection 对象；第 3 个参数通常为 BIND_AUTO_CREATE，表示如果 Service 不存在，那么创建一个 Service，该参数的其他值读者可以自行查阅相关资料。

最后也需要在 AndroidManifest.xml 里注册 InfoBindService，注册方法与启动式 Service 一样，限于篇幅，不再赘述，详细代码读者可以参见 FIRSTAPP 代码包中 servicedemo 文件夹里的内容。

6. TelephonyManager 的使用

在 Android 平台中，通过 TelephonyManager 可以访问与手机通信相关的信息，如设备信息、SIM 卡信息及网络信息等，也可以监听电话的相关状态（如呼叫服务状态、信号强度状态等）。下面以几个方面的应用来介绍 Android 平台中使用电话技术的方法和步骤。

（1）获取 TelephonyManager 对象。

```
1      TelephonyManager tm = (TelephonyManager)getSystemService(Context.
TELEPHONY_SERVICE);
```

（2）通过 TelephonyManager 获取设备信息。

```
1      String[]phoneType = {"未知","2G","3G","4G"};
2      String phoneCategory = phoneType[tm.getPhoneType()];//电话类型
3      String phoneIMEI = tm.getDeviceId();//设备编号
4      String phoneSoft = tm.getDeviceSoftwareVersion()! = null?
tm.getDeviceSoftwareVersion():"未知";//软件版本
5      String phoneNumber = tm.getLine1Number();//电话号码
```

(3) 通过 TelephonyManager 获取网络信息。

```
1    String phoneISOid = tm.getNetworkCountryIso()//国家 ISO 代码
2    String phoneOperid = tm.getNetworkOperator();//网络运营商编号
3    String phoneOperName = tm.getNetworkOperatorName();//网络运营商名称
4    String phoneType = tm.getNetworkType()//网络类型
```

(4) 通过 TelephonyManager 获取 SIM 卡信息。

```
1    String simCountry = tm.getSimCountryIso();//SIM 卡国家 ISO
2    String[] simStates = {"状态未知","无 SIM 卡","被 PIN 加锁","被 PUK 加锁",
"被 NetWork PIN 加锁","已准备好"}
3    String[] simState = simState[tm.getSimState()];//SIM 卡状态
4    String simNo = tm.getSimSerialNumber();//SIM 卡序列号
5    String phoneAddress = tm.getCellLocation() != null ?
tm.getCellLocation().toString():"未知位置";//设备当前位置
```

(5) 通过继承 PhoneStateListener，使用 TelephonyManager 注册后进行监听电话状态。

例如，通过继承自 PhoneStateListener 的 myPhoneListener 自定义类，用于处理电话状态改变时的业务，详细代码如下。

```
1    class MyPhoneListener extends  PhoneStateListener{
2        @Override
3        public void onCallStateChanged(int state, String incomingNumber) {
4            super.onCallStateChanged(state, incomingNumber);
5            switch (state){
6                case TelephonyManager.CALL_STATE_IDLE:
7                    …//空闲状态时的业务
8                    break;
9                case TelephonyManager.CALL_STATE_OFFHOOK:
10                   …//通话状态时的业务
11                   break;
12               case TelephonyManager.CALL_STATE_RINGING:
13                   …//响铃状态时的业务
14                   break;
15           }
16       }
17       @Override
18       public void onDataConnectionStateChanged(int state) {
19           super.onDataConnectionStateChanged(state);
20           switch (state){
21               case TelephonyManager.DATA_CONNECTED:
22                   …//数据已连接的业务
23                   break;
24               case TelephonyManager.DATA_CONNECTING:
25                   …//数据正在连接中的业务
26                   break;
27               case TelephonyManager.DATA_DISCONNECTED:
28                   …//数据已关闭的业务
29                   break;
30           }
```

```
31        }
32        @Override
33        public void onDataActivity(int direction) {
34            super.onDataActivity(direction);
35            switch (direction){
36                case TelephonyManager.DATA_ACTIVITY_IN:
37                    …//收到数据(下载)的业务
38                    break;
39                case TelephonyManager.DATA_ACTIVITY_OUT:
40                    …//送出数据(上传)的业务
41                    break;
42                case TelephonyManager.DATA_ACTIVITY_INOUT:
43                    …//收到(下载)送出数据(上传)的业务
44                    break;
45                case TelephonyManager.DATA_ACTIVITY_NONE:
46                    …//未知的业务
47                    break;
48            }
49        }
50    }
```

PhoneStateListener 可以通过重写 onCallStateChanged、onDataConnectionStateChanged、onDataActivity 等方法对来电状态、数据连接状态、数据传输状态进行监听。然后通过 TelephonyManager.listen() 方法进行注册，该方法有两个参数，第 1 个参数为 PhoneStateListener，第 2 个参数为监听标志，决定了监听哪些状态，详细代码如下。

```
1    listener = new MyPhoneListener ();
2    tm.listen(listener,PhoneStateListener.LISTEN_CALL_STATE
3            |PhoneStateListener.LISTEN_DATA_CONNECTION_STATE
4            |PhoneStateListener.LISTEN_DATA_ACTIVITY);
```

上述第 2 行代码同时监听了三种状态，这三种状态可以由功能需要自由组合添加。如果要注销监听，可以使用如下代码。

```
1    tm.listen(listener,PhoneStateListener.LISTEN_NONE);
```

当然，要实现以上功能还需要获得相关的使用权限，所以就需要在 App 的配置文件中添加如下代码。

```
1    <!--添加访问手机位置的权限 -->
2    <uses-permission android:name ="android.permission.ACCESS_COARSE_LOCATION"/>
3    <!--添加访问手机状态的权限 -->
4    <uses-permission android:name ="android.permission.READ_PHONE_STATE"/>
```

6.2.2 电话监听器的实现

1. 电话监听器界面的实现

电话监听器界面设计比较简单，实际应用中电话监听器应该在开机后自动启动，所以本案例中为了阐述方便，在 App 界面上只放置了两个 Button 按钮：一个用于启动 Service，一个用于关闭 Service。

【TelephonyManager、监听器的实现】

2. PhoneService 服务类的实现

建立一个继承于 Service 的 PhoneService 类，在该类中重写 onCreate()、onDestory() 和 onBind() 等方法。onCreate() 方法中注册监听电话状态，onDestory() 方法中取消监听。在注册监听电话状态的内部类 PhoneStateListener 中分别实现电话空闲状态、响铃状态及通话状态下的功能。当监听到通话状态时，创建一个录音机实例（录音机的相关知识在后面章节会详细介绍），并开启录音机开始录音，将录音的音频文件存放在 SD 卡中；当监听到电话空闲状态时，停止录音并释放录音资源。详细代码如下。

```
1    public class PhoneService extends Service {
2        private TelephonyManager tm; //电话管理器
3        private MyListener listener; //监听器对象
4        private MediaRecorder mRecorder; //声明录音机
5        @Override
6        public IBinder onBind(Intent intent) {
7            return null;
8        }
9        @Override
10       public void onCreate() {//服务开启时调用该方法
11           super.onCreate();
12           tm = (TelephonyManager) getSystemService(TELEPHONY_SERVICE);
13           listener = new MyListener();
14           tm.listen(listener, PhoneStateListener.LISTEN_CALL_STATE);
15       }
16       @Override
17       public void onDestroy() {//服务销毁时调用该方法
18           super.onDestroy();
19           tm.listen(listener, PhoneStateListener.LISTEN_NONE);// 取消电话的监听
20           listener = null;
21       }
22       //内部类用于监听到电话的不同状态要实现的功能
23       private class MyListener extends PhoneStateListener {
24           @Override
25           public void onCallStateChanged(int state, String incomingNumber) {
26               super.onCallStateChanged(state, incomingNumber);
27               try {
28                   switch (state) {
29                       case TelephonyManager.CALL_STATE_IDLE://空闲状态
30                           if (mRecorder != null) {
31                               mRecorder.stop();//停止录音
32                               mRecorder.release();//释放资源
33                               mRecorder = null;
34                           }
35                           break;
36                       case TelephonyManager.CALL_STATE_RINGING://响铃状态
37                           break;
38                       case TelephonyManager.CALL_STATE_OFFHOOK://通话状态
39                           mRecorder = getmRecorder();
```

```
40                  //6. 准备开始录音
41                      mRecorder.prepare();
42                      mRecorder.start();
43                      break;
44                  }
45              } catch (Exception e) {
46                  e.printStackTrace();
47              }
48          }
49      }
50      //产生一个录音机对象
51      private MediaRecorder getmRecorder() {
52          //1. 实例化一个录音机
53          MediaRecorder myRecorder = new MediaRecorder();
54          //2. 指定录音机的声音源
55          myRecorder.setAudioSource(MediaRecorder.AudioSource.MIC);
56          //3. 设置录制的文件输出的格式
57  myRecorder.setOutputFormat(MediaRecorder.OutputFormat.DEFAULT);
58          //4. 指定录音文件的名称
59          File file = new File(Environment.getExternalStorageDirectory(),
System.currentTimeMillis() + ".3gp");
60          //5. 设置音频的编码
61          myRecorder.setOutputFile(file.getAbsolutePath());
62  myRecorder.setAudioEncoder(MediaRecorder.AudioEncoder.DEFAULT);
63          return myRecorder;
64      }
65  }
```

3. 启动和关闭 Service

本案例使用启动式 Service 的启动和关闭 Service，其实现代码如下。

```
1   //启动服务
2   Intent intent = new Intent(Context.this, PhoneService.class);
3   startService(intent);
4   //关闭服务
5   stopService(intent);
```

4. 配置文件

Service 是 Android 的四大组件之一，要在应用中使用该组件就必须在注册文件中声明，声明代码如下。

```
1   <service
2           android:name=".PhoneService"
3           android:enabled="true"
4           android:exported="true">
5   </service>
```

其中，enabled 属性值为 true 表示该服务可以开启，为 false 表示该服务不可以开启；exported 属性值为 true 表示该服务可以由其他进程开启，为 false 表示不可以由其他进程开启。

由于监听电话应用中涉及获取电话状态、开启录音机、写 SD 卡等权限，所以必须在配置文件中添加如下权限。

```
1    <uses-permission android:name ="android.permission.READ_PHONE_STATE"/>
2    <uses-permission android:name ="android.permission.RECORD_AUDIO"/>
3    <uses-permission android:name ="android.permission.WRITE_EXTERNAL_STORAGE" />
```

本案例的详细代码读者可以参见 ALLAPP 代码包中 listenphone 文件夹里的内容。

另外，自 Android 6.0 版本开始，有些权限属于 Protected Permission，这类权限只在 AndroidManifest.xml 中声明是无法真正获取到的。例如，本例中由于会用到存储文件到 SD 卡、使用录音机的麦克风及监听电话状态等权限，如果要让该 App 能正常工作，就需要手动开启 Android 6.0 版本及以上版本的相关访问权限，即在模拟器或真机上通过设置→应用→单击要打开权限的 App 出现图 6.4，在图 6.4 中找到"权限"选项，单击后如图 6.5 所示，本案例中需要将存储空间、电话和麦克风的权限全部打开后才能保证 App 正常运行。

图 6.4 App 应用信息界面

图 6.5 App 应用访问授权界面

6.3　短信拦截器的设计与实现

BroadcastReceiver 是 Android 的四大组件之一，用来接收来自系统或其他 App 发出的广播，通常用作事件驱动的起源。例如，当网络状态发生改变时，系统会产生一条广播，BroadcastReceiver 接收到广播后就会及时做出提示和保存数据的操作；当电池电量发生改变时，系统也会产生一条广播，BroadcastReceiver 接收到广播后就能在低电量时提醒用户及时保存当前操作的进展等。本节通过设计开发一个能够拦截指定号码或敏感内容的短信拦截器来详细介绍 BroadcastReceiver 的原理及使用方法。

6.3.1 预备知识

1. BroadcastReceiver 简介

在 Android 中,通常一旦产生某个事件时就可以发送一个广播,App 使用 BroadcastReceiver 接收广播后就会知道系统产生了什么事件,同时也可以根据相应的事件做出相应的处理。所以,BroadcastReceiver 本质上可以认为就是一个监听器,但是通常的监听器是程序级别的监听器,当程序退出时监听器也会随之关闭;而 BroadcastReceiver 是系统级别的监听器,只要与该监听器匹配的 Intent 被广播出来,BroadcastReceiver 就会被激发并根据相应的操作进行处理。

【BroadcastReceiver】

Android 中的广播有以下两种类型。

(1) 普通广播:也可以称为无序广播,它是完全异步的,可以同一时间被所有与广播 Intent 匹配的接收者同时接收,接收者之间相互不会有影响。普通广播对消息的传递效率较高,但接收者不能将接收消息的处理信息传递给下一个接收者,也不能停止消息的传播。

(2) 有序广播:它按照一定的优先级进行消息的接收,即首先发送到跟广播 Intent 匹配的优先级最高的接收者那里,然后再从前面的接收者那里传播到下一个优先级的接收者那里;优先级高的接收者也可以终止这个广播。在优先级的声明中,其取值在 -1 000 ~ 1 000,数值越大,优先级越高,也就越先接收到广播。优先级也可通过 filter. setPriority(10) 方式设置。另外,有序广播的接收者可以通过 abortBroadcast()的方式取消广播的传播,也可以通过 setResultData()方法和 setResultExtras()方法将处理的结果存入 Broadcast 中,然后传递给下一个接收者。下一个接收者可以通过 getResultData()方法和 getResultExtras()方法接收高优先级的接收者存入的数据。

BroadcastReceiver 并没有提供可视化的界面来显示广播信息。在实际应用开发中可以使用 Notification 和 Notification Manager 来实现广播信息的可视化界面,以便显示广播信息的内容、图标及振动等。

2. BroadcastReceiver 的生命周期

BroadcastReceiver 的生命周期很短,当系统或其他 App 发出广播的时候,Android 的包管理对象就会检查所有已安装的包中的配置文件有没有匹配的 action,如果有并且可以接收,那么就调用这个 BroadcastReceiver 获取 BroadcastReceiver 对象,然后执行该对象的 onReceiver()方法。当从该方法返回后,该 BroadcastReceiver 对象被销毁,即生命周期结束,所以在 BroadcastReceiver 的 onReceive()方法中不能做比较耗时的操作。如果要实现耗时的操作,就需要在 onReceive()方法中开启 Service 对象,让 Service 对象来完成这类耗时操作。当然,onReceive()方法中也不能使用线程来操作,因为当父进程被 kill 后,所有子进程也会被 kill。

3. Android 自定义权限的使用

前面章节中已经介绍过,在 Android 中为了保护用户资源的安全,在使用用户资源时必须在 App 中声明权限信息。用户在安装 App 时系统会从 App 中提取出权限信息,通知用户使用该 App 的哪些功能,由用户判断该 App 是否损害安装此 App 的移动终端的安全。除了系统提供的一些使用权限外,开发者也可以根据需要自定义使用权限,下面通过一个案例介绍自定义权限的使用方法。

假设应用程序 A 中有 A_Activity，应用程序 B 中有 B1_Activity 和 B2_Activity，现在要实现通过 A 中的 A_Activity 直接打开 B 中的 B2_Activity。如果不考虑权限的问题，可以使用如下步骤实现。

- 给 A_Activity 中的 Button 绑定监听事件，其关键代码如下。

```
1    Intent intent = new Intent();
2    intent.setClassName("B2_Activity 的包名","B2_Activity 的包名.B2_Activity");
3    startActivity(intent);
```

- 给 B2_Activity 所在的 App 的配置添加如下代码。

```
1    <activity
2        android:name =" B2_Activity 的包名.B2_Activity "
3        android:exported ="true" >
4    </activity>
```

以上第 3 行代码表示可以让外部程序打开 B2_Activity 而没有访问限制，为了增加访问限制，可以使用自定义权限。即只有在声明了此自定义权限的外部程序后才有资格打开 B2_Activity。实现步骤如下。

- 自定义权限。

在 B 中的 AndroidManifest.xml 中通过 permission 标签进行声明（一般在 uses-sdk 标签后），其关键代码如下。

```
1    <permission android:protectionLevel ="normal" android:name ="nnutc.permission.MyBactivity"/>
```

name 是自定义权限的名称，需要遵循 Android 的权限定义命名方案：*.permission.*；protectionLevel 是定义与权限相关的风险级别。表 6-2 所示为 protectionLevel 的风险级别及功能说明。

表 6-2　protectionLevel 的风险级别及功能说明

风险级别	功能说明
normal	表示权限是低风险的，低风险权限只要申请了就可以使用（在 AndroidManifest.xml 中添加 <uses-permission> 标签），安装时不需要用户确认
dangerous	表示权限是高风险的，安装时需要用户确认才可使用
signature	表示权限是高风险的，只有当申请权限的应用程序的数字签名与声明此权限的应用程序的数字签名相同时（如果是申请系统权限，则需要与系统签名相同），才能将权限授给它
signatureOrSystem	表示将权限授给具有相同数字签名的应用程序或 android 包类。这一保护级别适用于非常特殊的情况，比如多个供应商需要通过系统映像共享功能时

- 在 B 中的 B2_Activity 进行权限限定时，需要在 B 的 AndroidManifest.xml 文件中使用如下代码。

```
1    <activity
2        android:name =" B2_Activity 的包名.B2_Activity"
3        android:exported ="true"
4        android:permission =" nnutc.permission.MyBactivity" >
5    </activity>
```

- A 中的 A_Activity 想要直接打开 B 中的 B2_Activity,就需要在其配置文件中添加如下代码。

```
1    <uses-permission android:name =" nnutc.permission.MyBactivity " > </uses-permission >
```

4. BroadcastReceiver 的使用

(1)创建一个继承于 BroadcastReceiver 的子类。由于 BroadcastReceiver 本质上是一种监听器,所以创建 BroadcastReceiver 的子类时,只需要重写该类的 onReceive()方法,具体代码如下。

```
1    public class MyReceiver extends BroadcastReceiver {
2        @Override
3        public void onReceive(Context context, Intent intent) {
4            //用来实现广播接收者收到广播后执行的代码逻辑
5            …
6        }
7    }
```

(2)注册 BroadcastReceiver(本例中是 MyReceiver)。实现了 MyReceiver 类后,需要指定该 MyReceiver 能匹配的 Intent,即注册 BroadcastReceiver。在 Android 应用开发中,注册 BroadcastReceiver 有以下两种方法。

- 静态注册:就是在配置文件 AndroidManifest.xml 中注册,代码如下。

```
1    <receiver android:name ="com.example.test.MyReceiver">
2        <intent-filter>
3            <action android:name ="android.intent.action.MyReceiver"> </action>
4            <category android:name ="android.intent.category.DEFAULT"> </category>
5        </intent-filter>
6    </receiver>
```

配置了以上信息之后,只要广播地址是 android.intent.action.MyReceive,则 MyReceiver 都能够接收到。静态注册的 BroadcastReceiver 是常驻型的,也就是说当应用关闭后,如果有广播信息传来,MyReceiver 也会被系统调用而自动运行,所以也称常驻型广播。

- 动态注册:就是在实现代码中动态地指定广播地址并注册,通常是在 Activity 或 Service 中注册一个广播,代码如下。

```
1    MyReceiver receiver = new MyReceiver();
2    IntentFilter filter = new IntentFilter();
3    filter.addAction("android.intent.action.MyReceiver");
4    registerReceiver(receiver, filter);
```

registerReceiver()是 android.content.ContextWrapper 类中的方法,由于 Activity 和 Service 都继承于 ContextWrapper,所以可以直接调用。在实际应用中,若 Activity 或 Service 中注册了一个 BroadcastReceiver,则当 Activity 或 Service 被销毁(destroy)时需要主动撤销注册,否则会出现异常。所以,使用动态注册方法时开发者需要记住执行解除注册操作,其代码如下。

```
1    @Override
2    protected void onDestroy() {
3        super.onDestroy();
4        unregisterReceiver(receiver);
5    }
```

动态注册的 BroadcastReceiver 是非常驻型的,即它会跟随 Activity 或 Service 的生命周期。

(3) 发送广播。根据以上方法完成注册后,BroadcastReceiver 就可以正常工作。前面已经介绍了广播有普通广播和有序广播两种类型,所以发送广播相应也有以下两种方式。

- 普通广播发送方式,代码如下。

```
1    Intent intent = new Intent("android.intent.action.MyReceiver");
2    intent.putExtra("msg", "hello receiver.");
3    sendBroadcast(intent);
```

- 有序广播发送方式,代码如下。

```
1    Intent intent = new Intent("android.intent.action.MyReceiver");
2    intent.putExtra("msg", "hello receiver.");
3    sendOrderedBroadcast(intent, null);    //没有添加权限
```

有序广播发送方式中使用 sendOrderedBroadcast( )方法,该方法的结构如下。

```
1    public abstract void sendOrderedBroadcast (Intent intent, String receiverPermission)
```

其中,第 1 个参数是指定要接收广播的接收器;第 2 个参数是指定接收器必须拥有的接收权限,如果设为 null,就表示不需要接收权限,所有匹配的接收器都能接收到,如果不为 null,则表示接收者若要接收此广播,就需要自定义与此值相同的权限,即在配置文件中定义一个权限,代码如下。

```
1    <permission android:protectionLevel ="normal" android:name =
"nnutc.permission.MYPERMISSION" />
```

在发送广播的应用程序配置文件中需要声明使用该自定义权限,代码如下。

```
1    <uses-permission android:name ="nnutc.permission.MYPERMISSION" />
```

经过上述设置后,有序广播发送方式代码可以改为如下所示。

```
1    Intent intent = new Intent("android.intent.action.MyReceiver");
2    intent.putExtra("msg", "hello receiver.");
3    sendOrderedBroadcast(intent,"nnutc.permission.MYPERMISSION");    //添加权限
```

由于该发送方式发出广播后,接收器必须是有序的,也就是有优先级的,所以这种优先级必须在注册时配置,其配置代码如下。

```
1    <receiver android:name ="com.example.test.MyReceiver" >
2        <intent-filter android:priority ="10" >
3            <action android:name ="android.intent.action.MyReceiver" >
</action>
4            <category android:name ="android.intent.category.DEFAULT" >
</category>
```

第6章 服务和消息广播

```
5            </intent-filter>
6        </receiver>
```

继承于 BroadcastReceiver 的接收器代码需要修改为如下所示。

```
1    public class MyReceiver extends BroadcastReceiver {
2        @Override
3        public void onReceive(Context context, Intent intent) {
4            //用来实现广播接收者收到广播后执行的代码逻辑
5            ...
6            String msg = intent.getStringExtra("msg");
7            Bundle bundle = new Bundle();
8            bundle.putString("msg", msg + "@FirstReceiver");
9            setResultExtras(bundle);//将处理后的信息传递给下一个接收者
10       }
11   }
```

从上面的 MyReceiver 接收器中可以看到,最后使用了 setResultExtras() 方法将一个 Bundle 对象设置为结果集对象,这样使用的目的是便于将这个接收者接收到的数据经过处理后继续传递到下一个接收器那里。如果要实现这个功能,在优先级低的接收器代码中可以用 getResultExtras() 获取到最新的经过处理的信息集合,其关键代码如下。

```
1    public class SecondReceiver extends BroadcastReceiver {
2        @Override
3        public void onReceive(Context context, Intent intent) {
4            String msg = getResultExtras(true).getString("msg");//获取前一个
接收器处理过的信息
5            Bundle bundle = new Bundle();
6            bundle.putString("msg", msg + "@SecondReceiver");
7            setResultExtras(bundle);//将处理后的信息传递给下一个接收器
8        }
9    }
```

5. BroadcastReceiver 的开机自启动

有时候需要 App 在打开移动终端时就能自动运行,如某个自动从网上更新内容的后台 Service 或其他 App。也就是说,Android 的移动终端在启动后就能自动运行从网上更新内容的 Service 或其他 App。

Android 启动时,系统会自动发出一个系统广播,内容为 ACTION_BOOT_COMPLETED, 它的字符串常量表示为 android. intent. action. BOOT_COMPLETED。即某个 App 如果要实现自启动,只要在 App 中能够监听到这个消息,就立即启动该 App 即可。其实现步骤如下。

(1) 创建一个继承于 BroadcastReceiver 的类,其关键代码如下。

```
1    public class BootBroadcastReceiver extends BroadcastReceiver {
2        static final String ACTION = "android.intent.action.BOOT_COMPLETED";
3        @Override
4        public void onReceive(Context context, Intent intent) {
5            if (intent.getAction().equals(ACTION)) {
```

```
6                    Intent intent = new Intent(context, MainActivity.class);
7                    intent.addFlags(Intent.FLAG_ACTIVITY_NEW_TASK);
8                    context.startActivity(intent);
9            }
10       }
11   }
```

以上代码表示判断接收到的 Intent 是否符合 BOOT_COMPLETED，如果符合，就自动启动 MainActivity。

(2) 在配置文件中注册 (1) 中创建的类，其代码如下。

```
1   <uses-permission android:name ="android.permission.RECEIVE_BOOT_COMPLETED"/>
2   <receiver android:name =".BootBroadcastReceiver" >
3       <intent-filter >
4           <action android:name ="android.intent.action.BOOT_COMPLETED" />
5       </intent-filter >
6   </receiver >
```

此类实际应用中，大多数 App 需要自动运行的不是有界面的应用程序，而是在后台运行的 Service，那么就需要使用 startService( )方法来启动相应的 Service。

Android 4.3 之后的版本，App 可以直接选择安装在 SD 卡上。由于系统开机间隔一小段时间后才能装载 SD 卡，这样安装在 SD 卡上的应用就可能监听不到这个广播，所以既要监听开机广播，又要监听 SD 卡挂载广播。另外，有些手机并没有 SD 卡，实现时就需要在配置文件时将两个广播监听写到不同的 intent-filter 选项中，具体如下。

```
1   <receiver android:name =".MyBroadcastReceiver" >
2       <intent-filter >
3           <action android:name ="android.intent.action.BOOT_COMPLETED"/>
4       </intent-filter >
5       <intent-filter >
6           <action android:name ="ANDROID.INTENT.ACTION.MEDIA_MOUNTED"/>
7           <action android:name ="ANDROID.INTENT.ACTION.MEDIA_UNMOUNTED"/>
8           <data android:scheme ="file"/>
9       </intent-filter >
10  </receiver >
```

6.3.2　短信拦截器的实现

1. 短信拦截器界面的实现

【短信拦截器的实现】

短信拦截器的界面设计比较简单，本案例中放置了两个 Button 和两个 EditText，两个 Button 分别用于注册 BroadcastReceiver 和注销 BroadcastReceiver；两个 EditText 分别用于添加要拦截的号码和敏感词语。

2. SmsReceiver 类的实现

创建一个继承于 BroadcastReceiver 的类（本案例为 SmsReceiver），并实现 onReceive( )方法，其实现代码如下。

```
1   public class SmsReceiver extends BroadcastReceiver {
2       private String tel = "";
3       private String content = "";
4       //构造方法,用于从主调界面传来电话号码和敏感词语
5       public SmsReceiver(String tel, String content) {
6           this.tel = tel;
7           this.content = content;
8       }
9       @Override
10      public void onReceive(Context context, Intent intent) {
11          StringBuilder number = new StringBuilder();//发送短信号码
12          StringBuilder body = new StringBuilder();//短信内容
13          Bundle bundle = intent.getExtras();
14          if (bundle ! = null) {
15              Object[] _pdus = (Object[]) bundle.get("pdus");
16              SmsMessage[] message = new SmsMessage[_pdus.length];
17              for (int i = 0; i <_pdus.length; i++) {
18                  message[i] = SmsMessage.createFromPdu((byte[]) _pdus[i]);
19              }
20              for (SmsMessage msg : message) {
21                  body.append(msg.getDisplayMessageBody());
22                  number.append(msg.getDisplayOriginatingAddress());
23              }
24              String smsbody = body.toString();
25              String smsnumber = number.toString();
26              if (smsbody.contains(content)) {//拦截带敏感词语的短信
27                  Toast.makeText(context, "对不起,来自" + smsnumber + "的短信有敏感词语!", Toast.LENGTH_LONG).show();
28                  this.abortBroadcast();
29              } else if (smsnumber.equals (tel)) {//拦截号码tel发来的短信
30                  Toast.makeText (context, "对不起,来自" + smsnumber + "的短信被拦截", Toast.LENGTH_ LONG).show();
31                  this.abortBroadcast();
32              } else {
33                  //其他短信的处理模块
34              }
35          }
36      }
37  }
```

上述第5～8行代码创建了一个BroadcastReceiver的构造方法,用于从主调模块中传递要拦截的电话号码和敏感词语短信,即从图6.6所示的EditText传递过来的值。onReceive()方法中的代码表示当接收器监听到发送的短信后,使用getDisplayMessageBody()方法获取短信内容,使用getDisplayOriginatingAddress()方法获取短信发送人,然后通过短信内容和发送人判断是否为要拦截的短信。如果是需要拦截的短信,那么就调用abortBroadcast()方法进行拦截;如果不是要拦截的短信,那么就进行相应的短信处理。

图 6.6　短信拦截器界面

图 6.7　短信拦截器效果

3. 注册和使用 BroadcastReceiver

在图 6.6 所示的界面上输入要拦截的电话号码和敏感词语后，通过单击"开启拦截"按钮传递给 BroadcastReceiver。因为短信的广播是有序广播，谁最先收到广播，谁就有权结束广播的传递，所以需要让接收广播的 App（也就是本案例的短信拦截器）权限提到最高，这样该 App 就可以最先收到短信，并结束广播的传递。该功能可以通过采用设置权限最大和动态注册 BroadcastReceiver 两个方法实现，具体实现代码如下。

```
1    String tel =edtNumber.getText().toString();//从 EditText 取得要拦截的电话号码
2    String content = edtContent.getText().toString();//从 EditText 取得要拦截的敏感词语
3    SmsReceiver recevier = new SmsReceiver(tel,content);//实例化 BroadcastReceiver 对象
4    IntentFilter filter = new IntentFilter("android.provider.Telephony.SMS_RECEIVED");
5    filter.setPriority(Integer.MAX_VALUE);;//设置权限最大
6    registerReceiver(recevier, filter);//动态注册 BroadcastReceiver
```

上述第 5 行代码用于设置权限值，实现时就是把注册广播时的优先权设置为最大，其中 Android API 中说明最大权限是 1 000，而在实际中接收的是一个 int 值，而且系统没有判断值的上限，所以本例中直接设置 int 数据类型的最大值，即 Integer.MAX_VALUE。另外还需要在配置文件中注册 BroadcastReceiver，其代码如下。

```
1    <receiver
2              android:name =".SmsReceiver"
3              android:enabled ="true"
4              android:exported ="true" >
5    </receiver>
```

因为本案例实现时使用的是动态注册广播，所以在配置文件中可以不需要使用 <intent-filter> 选项进行静态注册。本案例的全部功能实现代码请读者参见 ALLAPP 代码包中 blockmessage 文件夹里的内容。

6.4 闹钟的设计与实现

闹钟 App 作为常用的基本 App 之一，其重要性不言而喻。在 Android 中如何使用系统提供的 AlarmManager 类开发一个符合应用需求的闹钟是本节介绍的重点。

6.4.1 预备知识

1. AlarmManager 简介

【AlarmManager、PendingIntent】

Android 框架中的底层系统提供了两种类型的时钟：软时钟（Timer）与硬时钟（RTC）。系统在正常运行的情况下，由 Timer 的工作来提供时间服务和闹铃提醒，而在系统进入睡眠状态后，时间服务和闹铃提醒由 RTC 负责。上层应用并不需要关心是由 Timer 提供服务还是由 RTC 提供服务，因为 Android 的 Framework 层把底层细节做了封装并统一提供 API——AlarmManager。Android 中有一个 AlarmManager 的对应服务程序——AlarmManagerService。该服务程序才是真正提供闹铃服务的，它主要维护应用程序注册的各类闹铃并适时地设置即将触发的闹铃给闹铃设备，并且一直监听闹铃设备。一旦有闹铃触发或者是闹铃事件发生，AlarmManagerService 就会遍历闹铃列表找到相应的注册闹铃并发出广播。该服务程序在系统启动时被系统服务程序 system_service 启动并初始化闹铃设备（/dev/alarm）。当然 Java 层的 AlarmManagerService 与 Linux Alarm 驱动程序接口之间还有一层封装，即 Java 本地调用接口 JNI。

AlarmManager 可以实现从指定时间开始。以一个固定的间隔时间执行某项操作，所以经常与 BroadcastReceiver 结合使用实现闹钟等提示功能。实际应用开发中开发者不用关心具体的服务，只须直接通过 AlarmManager 来使用这种服务。AlarmManager 提供的主要方法及功能说明见表 6-3。

表 6-3 AlarmManager 提供的主要方法和功能说明

方 法 名	功 能 说 明
set(int type, long starttime, PendingIntent pintent)	设置一次性闹钟服务
cancel(PendingIntent pintent)	取消 AlarmManager 的闹钟服务
setRepeating(int type, long starttime, long interval, Pending Intent pintent)	间隔固定时间（interval）的重复性闹钟服务
setInexactRepeating(int type, long starttime, long interval, Pending Intent pintent)	间隔不固定时间（interval）的重复性闹钟服务
setExact(int type, long starttime, PendingIntent pintent)	在规定的时间精确执行闹钟服务
setWindow(int type, long windowStartMillis, long windowLengthMillis, PendingIntent pintent)	在给定的时间窗触发闹钟服务
setTimeZone(String timeZone)	设置时区

表6-3中相关方法的参数type用于指明闹钟类型。闹钟的主要类型及功能说明见表6-4。参数starttime用于指明闹钟的第1次执行时间,以毫秒为单位,可以自定义时间,不过一般使用当前时间。这个参数与type参数密切相关,如果type参数对应的闹钟使用的是相对时间(ELAPSED_REALTIME和ELAPSED_REALTIME_WAKEUP),那么starttime就要使用相对时间(相对于系统启动时间来说),如当前时间就设置为SystemClock.elapsedRealtime();如果type参数对应的闹钟使用的是绝对时间(RTC、RTC_WAKEUP或POWER_OFF_WAKEUP),那么starttime就要使用绝对时间,如当前时间就设置为System.currentTimeMillis()。参数intervaltime用于指明闹钟两次间隔时间。参数pintent用于绑定闹钟的执行动作。

表6-4 闹钟的主要类型和功能说明

类 型 名	功 能 说 明	默认值
ELAPSED_REALTIME	闹钟在系统处于睡眠状态时不可用;该状态下闹钟使用相对时间(相对于系统启动开始),可以通过调用SystemClock.elapsedRealtime()获得	3
ELAPSED_REALTIME_WAKEUP	闹钟在系统处于睡眠状态时会唤醒系统并执行提示功能;该状态下闹钟也使用相对时间	2
RTC	闹钟在系统处于睡眠状态时不可用;该状态下闹钟使用绝对时间,即当前时间,可以通过调用System.currentTimeMillis()获得	1
RTC_WAKEUP	闹钟在系统处于睡眠状态时会唤醒系统并执行提示功能;该状态下闹钟使用绝对时间	0
POWER_OFF_WAKEUP	闹钟在系统处于关机状态时也能正常进行提示功能;该状态下闹钟使用绝对时间,不过受Android版本影响	4

2. PendingIntent

PendingIntent继承自Parcelable,可以理解为是Intent的一个封装类,因为该类为final类型,所以没有子类,无法被继承。PendingIntent与Intent的主要区别如下。

- Intent是即时启动的,Intent随Activity消失而消失。PendingIntent用来处理即将发生的事,可以看作对Intent的封装。
- Intent一般用来作为Activity、Service和BroadcastReceiver之间传递数据的桥梁。PendingIntent一般用在Notification上,可以理解为延迟执行的Intent,如在通知Notification中跳转页面,但不是立即跳转。

需要注意的是,如果是通过启动服务来实现闹钟提示,PendingIntent对象的获取就应该采用Pending.getService(Context c, int i, Intent intent, int j)方法;如果是通过广播来实现闹钟提示,PendingIntent对象的获取就应该采用PendingIntent.getBroadcast(Context c, int i, Intent intent, int j)方法;如果是采用Activity的方式来实现闹钟提示,PendingIntent对象的获取就应该采用PendingIntent.getActivity(Context c, int i, Intent intent, int j)方法。如果

错用了这三种方法,虽然不会报错,但是会看不到闹钟提示效果。

开发者可以通过下列三种静态方法获取 PendingIntent 实例。

(1) getActivity(Context, int, Intent, int):通过该方法获得的 PendingIntent 可以直接启动新的 Activity,也就是与调用 Context. startActivity(Intent)一样。需要特别注意的是,如果要让新的 Activity 不再是当前进程存在的 Activity,那么 Intent 中必须使用 Intent. FLAG_ ACTIVITY_ NEW_TASK。代码如下。

```
Intent intent = new Intent(MainActivity.this, ClockActivity.class);
PendingIntentpintent = PendingIntent.getActivity(MainActivity.this , 0 , intent,0);
```

(2) getBroadcast(Context, int, Intent, int):通过该方法获得的 PendingIntent 发起一个广播,也就是与调用 Context. sendBroadcast( )方法一样。当系统通过它发送一个 Intent 时要采用广播形式,并且在该 Intent 中会包含相应的 Intent 接收对象,这个对象可以在创建 PendingIntent 时指定,也可以通过 ACTION 和 CATEGORY 等让系统自动找到该行为处理对象。代码如下。

```
Intent intent = new Intent(MainActivity.this, ClockBroadcast.class);
PendingIntentpintent = PendingIntent.getBroadcast(MainActivity.this,0, intent,0);
```

(3) getService(Context, int, Intent, int):通过该方法获得的 PengdingIntent 可以直接启动新的 Service,也就是与调用 Context. startService( )一样。代码如下。

```
Intent intent =new Intent(MainActivity.this, ClockService.class)
PendingIntent pintent = PendingIntent.getService (MainActivity.this,0, intent,0);
```

以上三个静态方法都包含四个参数,第 1 个参数表示上下文的对象;第 2 个参数表示请求值,请求值不同 Intent 就不同;第 3 个参数表示一个 Intent 对象,包含跳转目标;第 4 个参数有以下四种状态。

- FLAG_CANCEL_CURRENT:如果当前系统中已经存在一个相同的 PendingIntent 对象,那么就先将已有的 PendingIntent 取消,然后重新生成一个 PendingIntent 对象。
- FLAG_NO_CREATE:如果当前系统中不存在相同的 PendingIntent 对象,系统将不会创建该 PendingIntent 对象而是直接返回 null。
- FLAG_ONE_SHOT:该 PendingIntent 只作用一次。在该 PendingIntent 对象通过 send( )方法触发过后,PendingIntent 将自动调用 cancel( )方法进行销毁,如果再调用 send( )方法,系统将会返回一个 SendIntentException。
- FLAG_UPDATE_CURRENT:如果系统中有一个和描述的 PendingIntent 对等的 PendingInent,那么系统将使用该 PendingIntent 对象,但是会使用新的 Intent 来更新之前 PendingIntent 中的 Intent 对象数据。

6.4.2 闹钟的实现

1. 闹钟主界面的实现

闹钟的界面设计比较简单,本案例中放置了两个 Button 和一个 TextView,两个 Button

分别用于设置闹钟时间并启动闹钟和取消闹钟；一个 TextView 用于显示设置的闹钟时间。布局文件代码比较简单，限于篇幅，不再赘述。运行后闹钟主界面效果如图 6.8 所示，时间设置界面效果如图 6.9 所示。

图 6.8　闹钟主界面效果

图 6.9　时间设置界面效果

2. 功能实现

（1）创建一个 Activity（本案例为 ClockActivity）用于当闹钟时间到时实现播放响铃文件和对话提醒功能。

首先在项目的 res 文件夹下创建一个 raw 文件夹用于存放响铃文件（本案例为 pull_event.mp3），然后在 ClockActivity 类的 onCreate()方法中使用如下代码实现功能。

```
1    public class ClockActivity extends AppCompatActivity {
2        private MediaPlayer mediaPlayer;
3        @Override
4        protected void onCreate(Bundle savedInstanceState) {
5            super.onCreate(savedInstanceState);
6            setContentView(R.layout.activity_clock);
7            mediaPlayer = MediaPlayer.create(this,R.raw.pull_event);
8            mediaPlayer.start();
9            //创建一个闹钟提醒的对话框,单击"确定"按钮关闭铃声与页面
10           new AlertDialog.Builder(ClockActivity.this).setTitle("闹钟").setMessage("时间到了! ~")
11                   .setPositiveButton("关闭闹钟", new DialogInterface.OnClickListener() {
12                       @Override
13                       public void onClick(DialogInterface dialog, int which) {
14                           mediaPlayer.stop();
15                           ClockActivity.this.finish();
16                       }
```

```
17                }).show();
18         }
19  }
```

上述第 6 行代码用于设置 ClockActivity 对应的布局文件 activity_clock.xml；第 7～8 行代码用于创建一个 MediaPlayer 对象并实例化后播放；第 10～17 行代码用于创建并显示闹钟提醒对话框。

（2）在项目的 MainActivity 中实现相关功能。

- 定义相关变量

```
private Button btnSet, btnCancel;//设置闹钟、取消闹钟
private TextView tvInfo;//显示闹钟时间信息
private AlarmManager alarmManager;
private PendingIntent pintent;
private Intent intent;
```

- 定义初始化相关对象的方法

```
1   void initView() {
2       btnCancel = (Button) this.findViewById(R.id.btnCancel);
3       btnSet = (Button) this.findViewById(R.id.btnSet);
4       tvInfo = (TextView) this.findViewById(R.id.tvInfo);
5       alarmManager = (AlarmManager) getSystemService(ALARM_SERVICE);
6       //闹钟时间到,启动另一个 Activity
7       intent = new Intent(MainActivity.this, ClockActivity.class);
8       pintent = PendingIntent.getActivity(MainActivity.this, 0, intent, 0);
9   }
```

- 设置闹钟

当单击"设置闹钟"按钮时，弹出时间对话框并进行闹钟的时间设置；如果设置的闹钟时间在当前时间之后，就表示该时间为第 2 天闹钟的时间。

```
1   btnSet.setOnClickListener(new View.OnClickListener() {
2       @Override
3       public void onClick(View v) {
4           Calendar currentTime = Calendar.getInstance();
5           new TimePickerDialog(MainActivity.this, 0,
6               new TimePickerDialog.OnTimeSetListener() {
7                   @Override
8                   public void onTimeSet(TimePicker view, int hourOfDay, int minute) {
9                       Calendar c = Calendar.getInstance();    //日历实例对象
10                      long systemTime = System.currentTimeMillis();//获取当前时间的毫秒值
11                      c.setTimeInMillis(systemTime);//设置日历默认为当前时间
12                      c.set(Calendar.HOUR, hourOfDay);//设置几点
13                      c.set(Calendar.MINUTE, minute);//设置几分
14                      c.set(Calendar.SECOND, 0);//设置几秒
15                      c.set(Calendar.MILLISECOND, 0);//设置几毫秒
```

```
16                          long selectTime = c.getTimeInMillis();//获取新设置的
闹钟时间毫秒值
17                          if (systemTime > selectTime) {
18                              //如果当前时间大于设置的时间,那么从第2天设定的时间开始
19                              c.add(Calendar.DAY_OF_MONTH, 1);
20                          }
21                          //在TextView上显示设置的闹钟时间
22                          tvInfo.setText("你设置的定时闹钟时间:" + hourOfDay +
":" + minute);
23                          //从API 19 版本开始后
24                          if (Build.VERSION.SDK_INT >= Build.VERSION_CODES.KITKAT) {
25                              alarmManager.setExact(AlarmManager.RTC_WAKEUP,
c.getTimeInMillis(), pintent);
26                          } else {   //从API 19 版本开始前
27                              alarmManager.set(AlarmManager.RTC_WAKEUP,
c.getTimeInMillis(), pintent);
28                          }
29                          Toast.makeText(MainActivity.this, "闹钟设置完毕!" +
systemTime + ":" + c.getTimeInMillis(), Toast.LENGTH_SHORT).show();
30                      }
31                  }, currentTime.get(Calendar.HOUR_OF_DAY), currentTime.get
(Calendar.MINUTE), false).show();
32              }
33          });
```

从Android API 19 开始,set()方法设置的闹钟可能会发生deferred(延迟)。基于尽可能减少设备唤醒和电池损耗考虑,Android不推荐设置过多的闹钟。如果实在需要闹钟准时响应,可以采用setExact()方法。基于此原因,上述第24~28行代码判断当前API的版本,如果是API 19之前的版本,使用set()方法;否则使用setExact()方法。如果需要重复使用闹钟功能,可分别将这两个方法修改为setRepeating()方法或setInexactRepeating()方法。例如,闹钟要设置成每天都要执行一次,可以使用如下代码。

```
1   private static final int INTERVAL = 1000 * 60 * 60 * 24;// 24小时
2   alarmManager.setRepeating(AlarmManager.RTC_WAKEUP, c.getTimeInMillis(),
INTERVAL, pintent);
```

● 取消闹钟

取消闹钟的功能实现比较简单,代码如下。

```
1   btnCancel.setOnClickListener(new View.OnClickListener() {
2       @Override
3       public void onClick(View v) {
4           alarmManager.cancel(pintent);
5           Toast.makeText(MainActivity.this, "关闭了闹钟提醒!", Toast.LENGTH_
SHORT).show();
6       }
7   });
```

本案例的全部功能实现代码请读者参见FIRSTAPP代码包中alarmclock文件夹里的内容。

6.5 定时短信发送器的设计与实现

在任何一款手机中短信都是不可或缺的基本应用,且短信的使用频率很高。Android 中发送短信可以直接调用自带的短信程序完成,但应用不够灵活。使用 SmsManager 类可以方便地实现短信群发、短信定时发送等功能。定时短信发送器将定时器与发送短信相结合,在设定的时间到达时实现短信自动发送给收信人。

6.5.1 预备知识

1. SmsManager(短信管理器)

Android 中发送短信有以下两种方法。

第 1 种方法是直接调用 Android 自带的短信 App 发送界面,其主要代码如下。

【SmsManager、Notification】

```
Uri uri = Uri.parse("smsto:5554");//5554 为接收人号码
Intent it = new Intent(Intent.ACTION_SENDTO, uri);
it.putExtra("sms_body", "SMS");//SMS 为短信内容
startActivity(it);
```

第 2 种方法是使用 SmsManager 类,它主要用于管理短信操作,如发送数据、文本和 PDU 短信。SmsManager 的主要方法及功能说明见表 6-5。

表 6-5 SmsManager 的主要方法和功能说明

方 法 名	功 能 说 明
ArrayList <String> divideMessage(String text)	当短信超过 SMS 消息的最大长度时,将短信分割为几块
static SmsManager getDefault()	获取 SmsManager 的默认实例
void SendDataMessage(String destAddress, String scAddress, short destPort, byte[] data, PendingIntent sentIntent, PendingIntent deliveryIntent)	发送一个基于 SMS 的数据到指定的 App 端口
void sendMultipartTextMessage(String destAddress, String scAddress, ArrayList <String> parts, ArrayList <PendingIntent> sentIntents, ArrayList <PendingIntent> deliverIntents)	发送一个基于 SMS 的多部分文本,调用者已经通过调用 divideMessage(String text) 将消息分割成正确的大小
void sendTextMessage(String destAddress, String scAddress, String text, PendingIntent sentIntent, PendingIntent deliveryIntent)	发送一个基于 SMS 的文本

表 6-5 中主要用到九个参数,其功能如下。
- destAddress:短信的目的地址。
- scAddress:服务中心地址或为空(使用当前默认的 SMSC)。

- destPort：短信的目标端口号。
- data：短信的主体，即短信要发送的数据。
- sentIntent：如果不为空，当消息成功发送或失败时，就广播 PendingIntent。
- deliveryIntent：如果不为空，当消息成功传送到接收者时，就广播 PendingIntent。
- sentIntents：与 sentIntent 一样，这里的是一组 PendingIntent。
- parts：有序的 ArrayList <String>，可以重新组合为初始的消息。
- deliverIntents：与 deliveryIntent 一样，这里的是一组 PendingIntent。

2. Notification（状态栏通知）

Notification 是 Android 中用于在状态栏显示通知信息的组件，随着 Android 版本的升级，基于 2.X、4.X、5.X 及以上版本的 Notification 的使用都是不一样的。本节以目前使用较多的 Android 7.1.1 版本的 Notification 为例进行介绍。

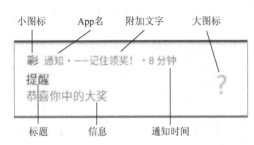

图 6.10 Android 7.1.1 状态栏通知效果

Notification 通常由大图标（Icon）、标题（Title）、信息（Message）、通知时间（Timestamp）、小图标（Secondary Icon）和附加文字（content）等部分组成。但是对于不同版本的 Android，其状态栏通知的显示样式会稍微有点区别。

图 6.10 所示为 Android 7.1.1 状态栏通知效果。

状态栏通知主要涉及 Notification（通知信息类，包含了通知栏的各个属性）和 NotificationManager（状态栏通知管理类，负责发送通知、清除通知等操作）两个类。下面以图 6.10 的实现为例，详细介绍它们的使用步骤。

(1) 获得 NotificationManager 对象。

```
NotificationManager mNManager = (NotificationManager) getSystemService
(NOTIFICATION_SERVICE);
```

(2) 创建一个构造通知栏的 Builder 对象，并进行标题、信息、图标等设置。

```
1    Notification.Builder mBuilder = new Notification.Builder(this);
2    Notification.Builder mBuilder = new Notification.Builder(MainActivity.this);
3    Resources res = MainActivity.this.getResources();
4    Bitmap bmp = BitmapFactory.decodeResource(res, R.mipmap.wenhao);//获取Bitmap对象
5    Intent intent = new Intent(MainActivity.this, SettingsActivity.class);
6     PendingIntent pIntent = PendingIntent.getActivity(MainActivity.this, 1, intent, 0);
7    mBuilder.setTicker("收到信息~")                //收到信息后状态栏显示的
                                                     文字信息
8           .setContentTitle("提醒")               //标题
9           .setContentText("恭喜你中的大奖")       //附加文字
10          .setSubText("——记住领奖！")           //内容下面的一小段文字
11          .setWhen(System.currentTimeMillis())   //设置通知时间
12          .setSmallIcon(R.mipmap.caipiao)        //设置小图标
```

```
13          .setLargeIcon(bmp)                           //设置大图标
14          .setDefaults(Notification.DEFAULT_VIBRATE)   //设置默认振动器
15          .setAutoCancel(true)                         //设置单击后取消
                                                           Notification
16          .setContentIntent(pIntent);                  //设置PendingIntent
17      Notification notify1 = mBuilder.build();
```

上述第7行代码在最新版本中已经无效，为了让开发者开发兼容低版本的App，本例一并将其列出；第14行代码用于向通知添加声音、闪灯和振动效果，可以组合多个属性，具体属性如下。

- Notification. DEFAULT_ VIBRATE：添加默认振动提醒。
- Notification. DEFAULT_ SOUND：添加默认声音提醒。
- Notification. DEFAULT_ LIGHTS：添加默认三色灯提醒。
- Notification. DEFAULT_ ALL：添加默认以上三种全部提醒。

（3）调用NotificationManager的notify()方法发送通知。

```
mNManager.notify(1, notify1);    //指明是普通样式的状态栏通知
```

notify()方法的第1个参数用于指明状态栏通知的类型，具体类型及功能说明见表6-6。

表6-6　状态栏通知的类型及功能说明

类 型 号	功 能 说 明
1	普通通知
2	下载进度通知
3	BigTextStyle通知（单击后可以显示大段文字）
4	InboxStyle通知（单击前如普通通知，单击后展开）
5	BigPictureStyle通知（单击后显示大图）
6	Hangup横幅通知（不显示在通知栏，而是以横幅的模式显示在其他应用上方）
7	MediaStyle（主要是用来关联音频播放服务的，单击后不会自动消失）
8	自定义通知栏布局

例如，下面的代码用于实现图6.11所示状态栏进度条通知的效果。

```
1    NotificationManager mNManager = (NotificationManager)
getSystemService(NOTIFICATION_SERVICE);
2    NotificationCompat.Builder builder = new NotificationCompat.Builder
(MainActivity.this);
3    builder.setSmallIcon(R.mipmap.caipiao);
4    builder.setLargeIcon(BitmapFactory.decodeResource(getResources(),
R.mipmap.wenhao));
5    builder.setAutoCancel(false);//禁止用户单击"删除"按钮删除
6    builder.setOngoing(true); //禁止滑动删除
7    builder.setShowWhen(false);//取消右上角的时间显示
8    builder.setContentTitle("下载中..."+progress +"%");
9    builder.setProgress(100,progress,false);
10   builder.setContentInfo(progress +"%");
```

```
11    builder.setOngoing(true);
12    builder.setShowWhen(false);
13    Intent intent = new Intent(MainActivity.this,DownloadService.class);
14    intent.putExtra("command",1);
15    Notification notify2 = builder.build();
16    mNManager.notify(2,notify2);
```

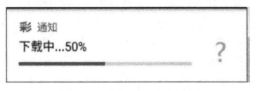

图 6.11　状态栏进度条通知

上述第 9 行代码中 setProgress( )方法的第 3 个 bool 类型的参数表示 progressbar 的 Indeterminate 属性（用于指定是否使用不确定模式）；高版本系统中 progressbar 的进度值可以在 setContentInfo 显示，但是低版本系统中使用这个属性会导致 progressbar 不显示，setContentText 一样。其他类型的状态栏通知的实现与此相仿，限于篇幅，读者可以根据用户的需要查阅相关文档，本书不再赘述。

（4）调用 NotificationManager 的 cancel( )方法取消通知。

```
1    mNManager.cancel(NOTIFYID_1);
```

本案例的全部功能实现代码请读者参见 FIRSTAPP 代码包中 msgnotification 文件夹里的内容。

6.5.2　定时短信发送器的实现

1. 定时短信发送器主界面的设计

布局文件代码比较简单，代码不再列出，请读者参见 ALLAPP 代码包中 timermessage 文件夹里的内容，运行后短信发送器主界面的效果如图 6.12 所示，定时短信发送完的界面效果如图 6.13 所示。

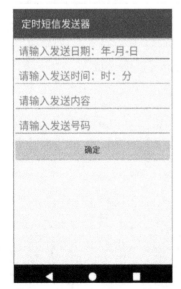

图 6.12　短信发送器主界面的效果

图 6.13　定时短信发送完的界面效果

2. 功能实现

（1）在项目的 MainActivity 中实现相关功能。

- 定义相关变量

```
private EditText edtDate,edtTime,edtContent,edtNum;//日期、时间、短信、号码
private Button btnSend;//设置定时发送
private Calendar calendar=Calendar.getInstance();
```

- 将在图 6.12 上输入日期、时间、短信和号码等内容进行相应处理

```
String date = edtDate.getText().toString();
String dates[] =date.split("-");
int year = Integer.parseInt(dates[0]);//日期——年
int month = Integer.parseInt(dates[1]);//日期——月
int day = Integer.parseInt(dates[2]);//日期——天
String time = edtTime.getText().toString();
String  times[] = time.split(":");
int hour = Integer.parseInt(times[0]);//时间——时
int minute = Integer.parseInt(times[1]);//时间——分
calendar.set(year,month-1,day);//指定定时日期
String num = edtNum.getText().toString();//号码
String content = edtContent.getText().toString();//短信
```

- 使用 AlarmManager 实现一个倒计时的功能，到设定的时间就发送短信。由上节介绍的 AlarmManager 用法可知，AlarmManager 对象需要配合 Intent 对象使用，可以定时开启一个 Activity、广播一个 Broadcast，或者开启一个 Service。设置定时发送短信的代码如下。

```
1    AlarmManager aManager = (AlarmManager)getSystemService(Context.ALARM_SERVICE);
2    Intent intent=new Intent(MainActivity.this,SmsReceiver.class);
3    intent.putExtra("content",content);
4    intent.putExtra("num",num);
5    intent.setAction("ie.nnutc.edu.cn.SmsReceiver");
6    PendingIntent pendingIntent = PendingIntent.getBroadcast(MainActivity.this,0,intent,0);
7    if(Build.VERSION.SDK_INT >= Build.VERSION_CODES.KITKAT){
8         aManager.setExact(AlarmManager.RTC,calendar.getTimeInMillis(),pendingIntent);
9    } else {  //API 19 版本以前
10        aManager.set(AlarmManager.RTC , calendar.getTimeInMillis(),pendingIntent);
11   }
```

上述第 3～4 行代码将电话号码和短信内容通过 Intent 传递给广播接收器，第 8 行和第 10 行代码表示设置的精准 AlarmManager 使用 RTC，在 calendar 指定的时间启动 pendingIntent 广播。

（2）定义一个继承于 BroadcastReceiver 的 SmsReceiver 类用于对发送出来的 Broadcast 进行过滤接收并响应。详细代码如下。

```
1    public class SmsReceiver extends BroadcastReceiver {
2        String num = null;
```

```
3        String content = null;
4        @Override
5        public void onReceive(Context context, Intent intent) {
6            num = intent.getStringExtra("num");
7            content = intent.getStringExtra("content");
8            if (num ! = null) {
9                Toast.makeText(context, "短信发送成功", Toast.LENGTH_LONG).show();
10               sendMsg(num, content);//发送短信
11               //在状态栏显示普通通知
12               NotificationManager mNManager = (NotificationManager) context.getSystemService(NOTIFICATION_SERVICE);
13               NotificationCompat.Builder mBuilder =new NotificationCompat.Builder(context);
14               Resources res = context.getResources();
15               Bitmap bmp = BitmapFactory.decodeResource(res, R.mipmap.msg1);
16               mBuilder.setTicker("收到信息~")
17                       .setContentTitle("提醒")
18                       .setContentText("你的短信已经发出!")
19                       .setWhen(System.currentTimeMillis())
20                       .setSmallIcon(R.mipmap.msg2)
21                       .setLargeIcon(bmp)
22                       .setDefaults(Notification.DEFAULT_VIBRATE)
23                       .setAutoCancel(true)          ;
24               Notification notify = mBuilder.build();
25               mNManager.notify(1, notify);
26           }
27       }
28       private void sendMsg(String number, String message) {
29           SmsManager smsManager = SmsManager.getDefault();
30           smsManager.sendTextMessage(number, null, message, null, null);
31       }
32   }
```

(3) 修改配置文件。

要实现发送短信的功能，必须要用到发送短信的权限，即需要在 AndroidManifest.xml 文件中添加如下内容。

```
<uses-permission android:name ="android.permission.SEND_SMS"/>
```

另外，还需要在 AndroidManifest.xml 文件中注册广播，添加如下内容。

```
<receiver
    android:name=".SmsReceiver"
    android:enabled="true"
    android:exported="true" >
    <intent-filter >
        <action android:name="ie.nnutc.edu.cn.SmsReceiver" />
    </intent-filter>
</receiver>
```

定时短信发送完的效果如图 6.13 所示。实际开发中，短信发送的日期和时间设置也

可以使用 DatePickerDialog 和 TimePickerDialog，前面章节已经对这两个控件的使用进行了详细的讲解，读者可以在本案例的基础上进行相应的修改，以更好地满足用户需求。

本 章 小 结

本章详细介绍了 Android 中 Service 和 BroadcastReceiver 两大组件的工作机制、生命周期，并通过日常应用中典型案例的实现详细讲述了它们的使用方法。通过对本章的学习，读者可以结合 TelephonyManager、AlarmManager、PendingIntent、SmsMessage 和 Notification 等开发出更多有趣且有用的 App。

习　　题

一、选择题

1. 以下 Android 进程中优先级最高的是（　　）。
   A. 前台进程　　　　　　　　B. 可见进程
   C. 服务进程　　　　　　　　D. 后台进程

2. 如果进程服务被一个 Activity 调用，而这个 Activity 正在接收用户的输入，那么该进程服务是（　　）。
   A. 前台进程　　　　　　　　B. 可见进程
   C. 服务进程　　　　　　　　D. 后台进程

3. 在 Service 的全生命周期中，不包含以下哪个回调函数？（　　）
   A. onCreate( )　　　　　　　B. onStart( )
   C. onStop( )　　　　　　　　D. onDestroy( )

4. 下列关于 Service 生命周期的说法正确的是（　　）。
   A. Service 全生命周期包含 onCreate( )、onStart( )、onStop( ) 及 onDestroy( )
   B. 第 1 次启动的时候只会调用 onCreate( ) 方法
   C. 如果 Service 已经启动，将先后调用 onCreate( ) 和 onStart( ) 方法
   D. 如果 Service 已经启动，只会执行 onStart( ) 方法，不再执行 onCreate( ) 方法

5. 下列关于 BroadcastReceiver 的说法不正确的是（　　）。
   A. 它是用来接收广播 Intent 的
   B. 一个广播 Intent 只能被一个订阅了此广播的 BroadcastReceiver 所接收
   C. 对有序广播，系统会根据接收者声明的优先级别按顺序逐个执行接收者
   D. 接收者声明的优先级别在 <intent-filter> 的 Android:priority 属性中声明，数值越大优先级别越高

6. 下列关于 Service 的说法不正确的是（　　）。
   A. Service 没有界面，且长期用于在后台运行
   B. 需要使用 Intent 对象来启动指定的 Service
   C. 如果使用绑定方式，调用者可以获取 Service 对象，从而调用 Service 中的方法
   D. 如果使用启动方法，调用者不可以获取 Service 对象，关闭调用者，Service 也将随

之关闭

7. Android 关于 Service 生命周期的 onCreate( ) 和 onStart( ) 说法正确的是（　　）。

A. 如果 Service 已经启动，将先后调用 onCreate( ) 和 onStart( ) 方法

B. 第 1 次启动的时候会先后调用 onCreate( ) 和 onStart( ) 方法

C. 第 1 次启动的时候只会调用 onCreate( ) 方法

D. 无论 Service 是否已经启动，都先后调用 onCreate( ) 和 onStart( ) 方法

8. SmsManager 类用于管理短信操作，如发送数据、文本和 PDU 等，下列叙述中不正确的是（　　）。

A. SmsManager 类提供了分隔短信的方法，用于实现短信超过 SMS 最大长度时，将短信分隔为几块

B. SmsManager 类提供了发送数据到指定应用程序端口的方法

C. SmsManager 类提供了构造方法创建实例

D. 使用 SmsManager 时需要导入 android. telephony. SmsManager 包

二、填空题

1. 在绑定式启动服务时，如果绑定成功会调用服务中的_____函数。

2. Service 有两种使用方式：启动式和_____，其中_____方式可以访问 Service 内部的方法。

3. BroadcastReceiver 需要使用_____指定接收广播的类型。

4. 在发送 Notification 时，需要使用_____对象中 notify( ) 方法发送通知。

5. Android 的_____是一种类似于 Activity 的组件，但没有用户操作界面，也不能自己启动，其主要作用是提供后台服务调用。

三、判断题

1. 使用 Service 的方法有两种：启动式和绑定式。绑定式可以调用 Service 里的方法，而启动式不能。（　　）

2. Service 的生命周期和 Activity 一样，也包含 onCreate( )、onStart( )、onStop( ) 及 onDestroy( ) 等方法。（　　）

3. BroadcastReceiver 组件不需要在 Android App 的配置文件中进行注册。（　　）

4. BroadcastReceiver 收到广播时会自动执行 onReceive( ) 方法。（　　）

5. 在绑定式使用 Service 时，会自动调用回调函数 onCreate( )、onStart( )、onBind( )、onUnbind( ) 及 onDestroy( ) 等。（　　）

【第6章参考答案】

第 7 章
数据存储与访问

数据存储是开发 App 时需要解决的最基本的问题，数据必须以某种方式保存，并且能够方便有效地使用和更新处理。随着 Android 设备应用范围的扩大，基于 Android 平台的软件的开发效率、性能及数据的存储访问机制受到了普遍关注。本章从 SharedPreferences、文件、SQLite 和 ContentProvider 四个方面深入阐述 Android 的数据存储与访问的机制和原理，并结合具体的案例介绍它们的使用方法。

教学目标

掌握 SharedPreferences 的存储原理和使用方法。
掌握读写 Android 内文件的方法。
掌握读写 Android 设备的 SD 卡中文件的方法。
了解手动创建 SQLite 数据库、创建表和操作数据库的方法。
掌握 SQLiteOpenHelper、SQLiteDatabase、Cursor 和 ContentValues 等创建、打开数据库和操作 SQLite 数据库的方法。
理解 ContentProvider 的原理和用途。
掌握自定义 ContentProvider 方法和 Android 提供的供用户开发使用的 ContentProvider 的用法。

教学要求

知识要点	能力要求	相关知识
概述	了解 Android 中五种数据存储访问机制	网络请求
幸运抽奖器的设计与实现	(1) 理解 Android 平台设备的存储器 (2) 理解 SharedPreferences 的存储访问机制 (3) 掌握 SharedPreferences 方式存储访问文件的方法 (4) 理解文件存储访问机制 (5) 掌握 Android 中 File 类文件的存储访问方法 (6) 掌握存储访问 SD 卡的方法	输入流和输出流 Environment 类

(续)

知识要点	能力要求	相关知识
实验室安全测试系统	（1）了解 SQLite 数据库的结构与原理 （2）掌握手动创建及访问 SQLite 数据库的方法 （3）掌握使用代码创建及访问 SQLite 数据库的方法	关系型数据库
App 间的数据共享	（1）理解 ContentProvider 的原理与机制 （2）掌握 ContentProvider 和 ContentResolver 的使用方法 （3）掌握访问系统提供的 ContentProvider 的方法	自定义权限

7.1 概　　述

在进行各类应用开发时，涉及的数据存储访问方式有三类：文件、数据库和网络。其中文件和数据库存储方式多用于离线应用开发中，文件使用较方便，不需要数据库管理系统的支持就可以进行存储访问；而数据库用起来较为复杂，但它有其强大的优点，如在处理海量数据时性能优越、能方便地进行数据的增删改查、可以跨平台等；网络存储方式则用于比较重要的项目事务，如在线售票、天气预报等实时数据需要通过网络传输到数据处理中心进行存储、处理。

Android 应用开发也会涉及数据存储访问，其存储方式总体也分为上面三类，但从开发者角度来讲，它包含以下五种存储访问机制。

1. SharedPreferences 存储访问机制

SharedPreferences 是一种用来存储简单配置信息的机制，也是 Android 平台上一个轻量级的存储类，只能存储 Long、Float、Integer、String 和 Boolean 五种基本类型。它是通过"key - value 对"的机制将数据存储在 XML 文件中，存储位置默认在/data/data/<包名>/shared_prefs 目录下。

2. 文件存储访问机制

Android 采用 java.io.* 库所提供的 I/O 接口来实现文件读写，同时引入了资源文件，用于存储 App 所需的一些资源，如图片、字符串等。每一种资源文件的语法、格式、保存位置取决于资源类型，在进行开发时需要在 res/目录下的适当子目录下（如 res/layout、res/drawable、res/raw 等目录）创建和存储资源文件，可以通过 R.resource_type.resource_name 语句来直接引用这些资源。

3. SQLite 数据库数据存储访问机制

SQLite 是一个轻量级嵌入式数据库引擎，支持 SQL 语言，并且只占用很少的内存。SQLite 由 SQL 编译器、内核、后端及附件四个部分组成。SQLite 在创建表时，可以把任何数据类型放入任何列中；在插入数据时，如果该类型与关联的列不匹配，则 SQLite 会尝试将该值转换成该列的类型，如果不能转换，则该值将作为其本身具有的类型存储。

对于熟悉 JavaEE 的开发人员，在 Android 开发中使用 SQLite 比较简单。但是，由于 JDBC 消耗资源太多，所以 JDBC 对于智能终端类内存受限的设备来说并不合适。因此，Android 提供了一些新的 API 来使用 SQLite 数据库。

4. ContentProvider 共享数据机制

Android 中的文件数据和数据库数据都是私有的。一般情况下，多个 App 之间不能进行数据交换。为了解决这个问题，Android 提供了 ContentProvider 类，该类实现了一组标准的方法接口，能够让其他应用保存或读取此 ContentProvider 的各种数据。另外，Android 也提供了一些已经在系统中实现的标准 ContentProvider，如联系人信息、图片库等，开发者可以用这些 ContentProvider 来访问设备上存储的信息。

5. 网络数据存储访问机制

在实际开发时，有时也需要将数据以文件的方式上传到服务器或者从服务器读取。Android 的网络存储使用 HTTP 协议，由于 Android 是使用 Java 来开发的，所以网络开发也使用 J2SE 的包。

上述五种方式各有优缺点，开发者在进行应用开发时应根据不同的实际情况来选择。下面结合各种方式的优缺点谈谈最合适的使用情况。SharedPreferences 是用来存储一些 key-value 对的基本数据类型，最适合使用 SharedPreferences 的地方就是保存配置信息。由于 SharedPreferences 使用十分方便，所以能用它就尽量不要用文件或是数据库。对于数据量大的应用，就需要选择文件或数据库存储。智能终端的网络稳定性及所产生的流量等因素对使用网络存储有很大影响，但若是非常重要的实时数据或是需要发送给远端服务器处理的数据，就必须使用网络进行实时发送。关于网络数据存储访问机制会在第 9 章介绍。

7.2 幸运抽奖器的设计与实现

"抽奖"活动在很多场合都有应用，如商家促销、公司年会、节假日文娱活动等，这种抽奖活动大多是通过特殊的抽奖器抽出数字编号（有的是入场编号、有的是手机号码、有的是会员号等）或文本信息（有的是姓名、有的是公司名称等），这些抽出的数字编号或文本信息即为中奖号码。本节将以开发设计幸运抽奖器为案例介绍 Android 中 SharedPreferences 和文件存储访问机制的原理和使用方法。

7.2.1 预备知识

1. Android 平台设备的存储器

【Android设备存储器、SharedPreferences】

在 Android 应用开发中，经常会涉及内存、内部存储器和外部存储器三个名词。内存即 Memory，当系统和 App 运行时必须将程序的核心模块加载到内存中；内部存储器即 InternalStorage，它的空间十分有限，是 Android 本身和系统 App 主要的数据存储所在位置，一旦内部存储空间耗尽，Android 平台设备也就无法使用了，开发者应该尽量避免在编码时使用内部存储空间；外部存储器即 ExternalStorage，通常包含公有目录和私有目录两类子文件夹，和 SD 卡没有必然的联系。

Android 的目录结构中有以下三个重要的文件夹。

● data 文件夹：即内部存储器对应的文件夹，这个文件夹中的 app 文件夹存放着所有安装的 App 的 apk 文件；这个文件夹中的 data 文件夹里都是一些包名，这些包名中通常包含 shared_prefs（SharedPreferences 的数据持久化本地存储位置）、databases（数据库文件）、files（普通数据存储文件）和 cache（缓存文件）四个文件夹，当用户卸载 App 后，这些数据会一并删除。

● storage 文件夹：即外部存储器对应的文件夹，这个文件夹可能会随着不同厂家出厂的 Android 设备而不同。一般来说，在这个文件夹中有一个 sdcard 文件夹，sdcard 文件夹中包含公有目录和私有目录两类子文件夹，其中的公有目录有九大类，如 DCIM、DOWNLOAD 等这种为系统创建的文件夹；私有目录就是 Android 文件夹，这个文件夹中也有一个 data 文件夹，里边有许多包名组成的子文件夹。外部存储空间和 SD 卡没有必然的联系，保存在 App 私有目录下的文件会在 App 被卸载的同时被删除。

● .mnt 文件夹：同 storage 文件夹。

由于内部存储空间有限，在开发中一般都是使用外部存储空间，Google 官方建议开发者将 App 的数据存储在外部存储器的私有目录中 App 对应的包名下（如/storage/emulated/0/Android/data/包名/），这样当用户卸载掉 App 之后，相关的数据会一并删除。如果直接在/storage/sdcard 目录下创建一个 App 的文件夹，那么当删除 App 的时候，这个文件夹就不会被删除。

2. SharedPreferences 存储访问机制

用户在使用 App 时需要保存一些偏好参数，如是否记住账号密码、保存 App 的配置信息（是否打开音效、是否保存背景等）。保存这些参数时的数据量一般较少，它们通常是 boolean、float、int、long 或 String 等基本类型的值。Android 中对于这种类型的数据通常采用轻量级的存储类——SharedPreferences。

SharedPreferences 是 android.content.SharedPreferences 包中的一个接口，主要负责读取 App 的 Preferences 数据。SharedPreferences 的常用方法及功能说明见表 7-1。

表 7-1　SharedPreferences 的常用方法及功能说明

方 法 名	功 能 说 明
boolean contains(String key)	判断 SharedPreferences 是否包含特定 key 的数据
SharedPreferences.Editor edit()	返回一个 Edit 对象用于操作 SharedPreferences
Map <String,?> getAll()	获取 SharedPreferences 数据里全部的 key-value 对
getXXX(String key, XXX defValue)	获取 SharedPreferences 数据指定 key 所对应的 value，如果该 key 不存在，返回默认值 defValue。其中 XXX 可以是 boolean、float、int、long、String 等基本类型的值

SharedPreferences 是一个接口，且在这个接口里没有提供写入数据和读取数据的能力，但是在其内部有一个 Editor 内部接口。Editor 的常用方法及功能说明见表 7-2。

第7章　数据存储与访问

表7-2　Editor 的常用方法及功能说明

方　法　名	功　能　说　明
SharedPreferences. Editor clear( )	清除 SharedPreferences 里所有的数据
boolean commit( )	当 Editor 编辑完成后，调用该方法可以提交修改，而且必须要调用这个方法才能修改数据
SharedPreferences. Editor putXXX (String key,boolean XXX)	向 SharedPreferences 存入指定的 key 对应的数据，其中 XXX 可以是 boolean、float、int、long、String 等基本类型的值
SharedPreferences. Editor remove (String key)	删除 SharedPreferences 里指定 key 对应的数据项

（1）存储数据。

使用 SharedPreferences 存储数据可以通过以下四个步骤实现。

第1步：获取 SharedPreferences 对象。获取该对象有以下三种方式。

● 通过 Activity 类中的 getPreferences（int mode）方法，获取的是本 Activity 私有的 Preferences，保存在系统中的文件以本 Activity 的名称命名。因此通过这种方式获取的 SharedPreferences 对象，一个 Activity 只能对应一个 XML 存储文件，并且它的操作权限仅属于该 Activity。

● 通过 PreferenceManager 类中的 getDefaultSharedPreferences（this）方法，获取的 SharedPreferences 对象保存的 Preferences 属于整个 App，并以 App 的包名与"_preferences"的组合命名保存在系统中。

● 通过 Context 类中的 getSharedPreferences（String name，int mode）方法，因为 Activity 继承了 ContextWrapper，因此这种方式实际上也是通过 Activity 获取 SharedPreferences 对象，但是它的操作权限属于整个 App，每个 Activity 可以对应多个 XML 存储文件，以第1个参数（name）为文件名保存在系统中（名称不需要带后缀.xml，保存文件时系统自动加后缀），第2个参数（mode）表示文件的操作模式。model 的常用方法及功能说明见表7-3。

表7-3　model 的常用方法及功能说明

方　法　名	功　能　说　明
Context. MODE_PRIVATE	为默认操作模式，代表该文件是私有数据，只能 App 本身使用；在该模式下写入的内容会覆盖原文件的内容，如果想把新写入的内容追加到原文件中，可以使用 Context. MODE_APPEND
Context. MODE_APPEND	该模式会检查文件是否存在，如果存在就向文件中追加内容，否则创建新文件
Context. MODE_WORLD_READABLE	表示当前文件可以被其他 App 读取
Context. MODE_WORLD_WRITEABLE	表示当前文件可以被其他 App 写入

Context. MODE_WORLD_READABLE 和 Context. MODE_WORLD_WRITEABLE 的权限在 Android 4.0 版本中已经声明弃用了，也就是说在将来不再建议和支持 App 间通过

SharedPreferences 的方式来共享数据。在 Android 7.0 版本中会直接给出" java.lang.SecurityException：MODE_WORLD_READABLE no longer supported"错误信息，读者在使用时要注意这个问题。

第 2 步：通过 SharedPreferences 对象的 edit() 方法获得 SharedPreferences.Editor 对象。

第 3 步：通过 Editor 对象的 putXXX() 方法，将不同类型的数据以 key-value 对的形式存储。

第 4 步：完成以上步骤后，此时的数据仍然在内存中，需要通过 Editor 对象的 commit() 方法提交数据，也就是将数据保存到 XML 文件中。

下列自定义方法 save() 用于实现将 username 和 password 以 key-value 对形式保存在 myconfig.xml 文件中。

```
1    public void save(String username,String password){
2        SharedPreferences sp = this.getSharedPreferences("myconfig",Context.MODE_PRIVATE);
3        SharedPreferences.Editor editor = sp.edit();
4        editor.putString ("username", username);
5        editor.putString ("password", password);
6        editor.commit();
7    }
```

(2) 读取数据。

从 SharedPreferences 文件中读取数据的步骤比较简单，通常通过下列步骤实现。

第 1 步：获取 SharedPreferences 对象实例，与存储数据一样，限于篇幅不再赘述。

第 2 步：因为 SharedPreferences 文件中存储的数据通常不止一个，在读取数据时也可以根据实现情况采用以下两种方式。

● 读取单一数据，调用 SharedPreferences 的 getXXX() 方法，分别读取 String、int、long 等类型数据。下列自定义方法 read() 用于实现将前面 myconfig.xml 文件中保存的 username 和 password 读出并分别保存在 uname 和 upwd 中。

```
1    public void read(){
2        SharedPreferences sp = getSharedPreferences("myconfig",Context.MODE_PRIVATE);
3        String uname = sp.getString (" username"," admin"); //不能读出，默认 admin
4        String upwd = sp.getString (" password"," admin"); //不能读出，默认 admin
5    }
```

● 读取全部数据，调用 SharedPreferences 的 getAll() 方法，再遍历 map。

```
1    public Map<String,String> reads(){
2        SharedPreferences sp = getSharedPreferences(" myconfig",Context.MODE_PRIVATE);
3        String content ="";
4        Map<String,?> allContent = sp.getAll ();
5        for (Map.Entry<String,?>entry: allContent.entrySet ()) {
6            Content += (entry.getKey () +entry.getValue ());
7        }
8        return content;
9    }
```

下面用一个登录界面示例来说明其实现过程，当用户输入用户名和密码后，单击"安全登录"按钮,显示"登录成功!",并启动另一个 Activity,下一次启动该登录界面时,用户名和密码仍然需要用户输入,效果如图 7.1 所示;当用户输入用户名和密码,并且选择了"保存密码"和"自动登录"复选项时,显示"登录信息已经保存!",效果如图 7.2 所示。如果前一次登录时没有选择"自动登录"复选项,只选择了"保存密码"复选项,那么启动时将直接将用户名和密码显示在输入框中。如果前一次登录时选择了"自动登录"复选项,那么启动时将直接跳转到另一个 Activity。

图 7.1　登录界面　　　　　　　图 7.2　保存登录信息界面

与图 7.1 所示的布局界面对应的代码比较简单,读者可以自行编写完成,功能实现可以按如下步骤完成。

第 1 步:定义 initView()方法,用于初始化布局界面对象。

```
1    void initView() {
2        btnLogin = (Button) this.findViewById(R.id.btnlogin);
3        edtName = (EditText) this.findViewById(R.id.edtname);
4        edtPwd = (EditText) this.findViewById(R.id.edtpwd);
5        chkAuto = (CheckBox) this.findViewById(R.id.chklogin);
6        chkSave = (CheckBox) this.findViewById(R.id.chksave);
7    }
```

第 2 步:定义 save()方法,用于实现数据存储。

```
1    public void save(String username, String password, boolean issave, boolean isauto) {
2        SharedPreferences sp = this.getSharedPreferences("loginconfig", Context.MODE_PRIVATE);   //以 loginconfig.xml 为文件名
3        SharedPreferences.Editor editor = sp.edit();
4        editor.putString("loginname", username);  //用户名
5        editor.putString("loginpwd", password);   //密码
6        editor.putBoolean("loginsave", issave);   //保存密码
7        editor.putBoolean("loginauto", isauto);   //自动登录
8        editor.commit();
9    }
```

上述第 2 行代码表示在程序运行后,如果选择了"保存密码"和"自动登录"复选项,则

会在/data/data/包名/ shared _prefs 文件夹下生成一个 loginconfig.xml 文件，文件内容如下。

```
1    <?xml version = '1.0' encoding = 'utf-8' standalone = 'yes' ? >
2    <map>
3        <string name ="loginname" >admin </string >
4        <string name ="loginpwd" >123456 </string >
5        <boolean name ="loginsave" value ="true" />
6        <boolean name ="loginauto" value ="true" />
7    </map>
```

如果将第 2 行代码修改为如下代码。

```
1    SharedPreferences sp = getPreferences(Context.MODE_PRIVATE);
```

此时会在/data/data/包名/shared_prefs 文件夹下生成一个 MainActivity.xml 文件。

如果将第 2 行代码修改为如下代码。

```
1    SharedPreferences sp = PreferenceManager.getDefaultSharedPreferences(this);
```

此时会在/data/data/包名/shared_prefs 文件夹下生成一个包名_preferences.xml 文件（本案例生成了 cn.edu.nnutc.ie.sharedpreferencesui_preferences.xml 文件）。

第 3 步：定义 read() 方法，用于实现读出数据。

```
1    public void read() {
2        SharedPreferences sp = getSharedPreferences("loginconfig", Context.MODE_PRIVATE);
3        String uname = sp.getString("loginname", "");
4        String upwd = sp.getString("loginpwd", "");
5        boolean issave = sp.getBoolean("loginsave",false);
6        boolean isauto = sp.getBoolean("loginauto",false);
7        if(issave){//如果前次登录选择了"保存密码"复选项,则用户名和密码被读出后,显示在对应位置
8            edtName.setText(uname);
9            edtPwd.setText(upwd);
10       }
11       if(isauto){//如果前次登录选择了"自动登录",则开启程序后直接跳转到 InfoActivity
12           Intent intent =new Intent(MainActivity.this, InfoActivity.class);
13           MainActivity.this.startActivity(intent);
14       }
15   }
```

第 4 步：在 onCreate() 方法中实现初始化对象、读数据及"安全登录"按钮的监听事件。

```
1    initView();
2    read();
3    btnLogin.setOnClickListener(new View.OnClickListener() {
4        @Override
5        public void onClick(View v) {
6            boolean issave = chkSave.isChecked() ? true : false;
7            boolean isauto = chkAuto.isChecked() ? true : false;
8            String username = edtName.getText().toString();
9            String userpwd = edtPwd.getText().toString();
10           if(issave){
11               save(username,userpwd,issave,isauto);
```

```
12                   Toast.makeText(MainActivity.this,"登录信息已经保存!
",Toast.LENGTH_SHORT).show();
13              }else{
14                   Toast.makeText(MainActivity.this,"登录成功!",
Toast.LENGTH_SHORT).show();
15              }
16          }
17      });
```

以上示例的全部功能实现代码请读者参见 FirstAPP 代码包中 sharedpreferencesui 文件夹里的内容。

【文件存储(内、外部存储器数据)】

3. 文件存储访问机制

Java 提供了一套完整的 IO 流体系用来对文件进行操作，包括 FileInputStream、FileOutputStream 等，通过这些 IO 流可以非常方便地访问磁盘上的文件。Android 同样支持以流的方式来访问 Android 平台设备存储器上的文件，包括设备本身的存储设备（InternalStorage）或外接的存储设备（ExternalStorage）。另外，Android 中的 Context 类对象提供了 openFileOutput() 方法及 openFileInput() 方法分别获得输入流和输出流，从而实现文件的读写操作。

（1）Java IO 流介绍。

在传统 Java IO 流实现文件读写操作时，必须创建 File 类对象。创建 File 类对象的方法及功能说明见表 7-4。

表 7-4 创建 File 类对象的方法及功能说明

方 法 名	功 能 说 明
File(File dir, String name)	dir 为 File 对象类型的目录路径，name 为文件名或目录名
File(String path)	path 为新 File 对象的路径
File(String dirPath, String name)	dirPath 为指定的文件路径，name 为文件名或目录名
File(URI uri)	使用 URI 指定路径来创建新的 File 对象

File 类对象包含四类方法，分别见表 7-5 ~ 表 7-8。

表 7-5 判断 File 类对象的方法

方 法 名	功 能 说 明
boolean canExecute()	判断文件或目录是否可执行
boolean canRead()	判断文件或目录是否可读
boolean canWrite()	判断文件或目录是否可写
boolean equals(Object obj)	判断 obj 和调用的对象是否相同
boolean exists()	判断文件或目录是否存在
boolean isAbsolute()	判断当前文件路径是否为绝对路径
boolean isDirectory()	判断 File 对象是否是文件夹
boolean isFile()	判断 File 对象是否是文件
boolean isHidden()	判断是否为操作系统定义的隐藏文件

表 7-6 File 类对象的属性返回方法

方　法　名	功　能　说　明
File getAbsoluteFile( )	返回一个新的文件，该文件的绝对路径是调用的 File 的路径
String getAbsolutePath( )	返回该文件的绝对路径
long getFreeSpace( )	返回所在分区上剩余的字节数量，包括当前 File 的路径
String getName( )	获得文件名
String getParent( )	获得文件或文件夹的父目录名
String getPath( )	获取相对路径
long getTotalSpace( )	返回分区的总字节大小
long getUsableSpace( )	返回分区的可用字节的大小
long lastModified( )	返回最后一次修改该文件的时间，以毫秒计算，从1970年1月1日开始算
long length( )	返回文件的长度，单位为字节
String toString( )	返回一个对象的字符串表示
URI toURI( )	返回一个文件的 URI

表 7-7 File 类对象的属性设置方法

方　法　名	功　能　说　明
boolean renameTo(File newPath)	修改文件夹名和文件名
boolean setLastModified(long time)	设置最后一次修改该文件的时间，以毫秒计算，从1970年1月1日开始算
boolean setReadOnly( )	设置文件只有读权限
boolean setReadable(boolean readable[, boolean ownerOnly])	设置文件的读状态
boolean setWritable([boolean writable,] boolean ownerOnly])	设置文件的写状态

表 7-8 File 类对象的增删查改方法

方　法　名	功　能　说　明
boolean createNewFile( )	创建文件或文件夹
boolean delete( )	删除文件夹或文件
File [ ] listFiles( )	列出文件夹下的所有文件和文件夹名
boolean mkdir( )	创建一个文件夹，当父目录存在才能成功创建
boolean mkdirs( )	创建一个文件夹，当父目录不存在时，则创建父目录

　　java.io 包中有四个基本类：InputStream、OutputStream、Reader 及 Writer，它们分别处理字节流和字符流。Java 中其他的流均是由它们派生出来的。下面对 File 文件流进行详细的介绍。对文件进行读写操作的类有 FileInputStream、FileOutputStream、FileReader、FileWriter。

　　● FileInputStream：用来处理以文件作为数据输入源的数据流，即从文件读数据到内存。与FileInputStream对象相关的方法及功能说明见表 7-9。

表 7-9 与 FileInputStream 对象相关的方法及功能说明

方 法 名	功 能 说 明
FileInputStream(File file)	创建 FileInputStream 对象
FileInputStream(FileDescriptor fd)	创建 FileInputStream 对象
FileInputStream(String path)	创建 FileInputStream 对象
int read()	读取一个字节,返回值为所读的字节
int read(byte[] buffer, int byteOffset, int byteCount)	读取 byteCount 个字节,放置到以下标 byteOffset 开始的字节数组 buffer 中,返回值为实际读取的字节的数量
int available()	返回值为流中尚未读取的字节的数量
long skip(long byteCount)	读指针跳过 byteCount 个字节不读,返回值为实际跳过的字节数量
void close()	流操作完毕后必须关闭,否则可能导致内存泄漏

从文件读数据到内存通常使用如下代码。

```
1       File file = ...;
2       InputStream in = null;
3       try {
4         in = new BufferedInputStream(new FileInputStream(file));
5         ...
6       finally {
7         if (in ! = null) {
8           in.close();
9         }
10      }
11    }
```

● FileOutputStream:用来处理以文件作为数据输出目的的数据流,即从内存区将数据写入文件。与 FileOutputStream 对象相关的方法及功能说明见表 7-10。

表 7-10 与 FileOutputStream 对象相关的方法及功能说明法

方 法 名	功 能 说 明
FileOutputStream(File file)	创建 FileOutputStream 对象
FileOutputStream(File file, boolean append)	创建 FileOutputStream 对象,当第 2 个参数为 true 时,表示在原文件中追加输出的内容
FileOutputStream(FileDescriptor fd)	创建 FileOutputStream 对象
FileOutputStream(String path)	创建 FileOutputStream 对象
FileOutputStream (String path, boolean append)	创建 FileOutputStream 对象,当第 2 个参数为 true 时,表示在原文件中追加输出的内容
void write(int oneByte)	往流中写一个字节 oneByte
void write(byte[] buffer, int byteOffset, int byteCount)	将字节数组 buffer 中从下标 byteOffset 开始,长度为 byteCount 的字节写入输出流中
void close()	流操作完毕后必须关闭,否则可能导致内存泄露

263

从内存将数据写入文件通常使用如下代码。

```
1    File file = ...
2    OutputStream out = null;
3    try {
4        out = new BufferedOutputStream(new FileOutputStream(file));
5        ...
6    finally {
7        if (out ! = null) {
8            out.close();
9        }
10   }
11   }
```

● FileReader：与 FileInputStream 对应，主要用来读取字符文件。与 FileReader 对象相关的方法及功能说明见表 7-11。

表 7-11　与 **FileReader** 对象相关的方法及功能说明

方 法 名	功 能 说 明
FileReader(File file)	创建 FileReader 对象
FileReader(FileDescriptor fd)	创建 FileReader 对象
FileReader(String filename)	创建 FileReader 对象
int read( )	读取单个字符，返回作为整数读取的字符，如果已达到流末尾，则返回 -1
int read (char [ ] charBuffer)	将字符读入 charBuffer 数组，返回读取的字符数，如果已经到达尾部，则返回 -1
void close( )	关闭此流对象，并释放与之关联的所有资源

● FileWriter：与 FileOutputStream 对应，主要用来将字符类型数据写入文件，使用默认字符编码和缓冲器大小。与 FileWriter 对象相关的方法及功能说明见表 7-12。

表 7-12　与 **FileWriter** 对象相关的方法及功能说明

方 法 名	功 能 说 明
FileWriter(File file)	创建 FileWriter 对象
FileWriter(File file, boolean append)	创建 FileWriter 对象，当第 2 个参数为 true 时，表示在原文件中追加输出的内容
FileWriter(FileDescriptor fd)	创建 FileWriter 对象
FileWriter(String filename)	创建 FileWriter 对象
FileWriter(String filename, boolean append)	创建 FileWriter 对象，当第 2 个参数为 true 时，表示在原文件中追加输出的内容
void write(int oneByte)	往流中写一个字节 oneByte

第7章 数据存储与访问

（续）

方 法 名	功 能 说 明
void write(byte [ ] buffer, int byteOffset, int byteCount)	将字节数组 buffer 中从下标 byteOffset 开始，长度为 byteCount 的字节写入输出流中
viod flush( )	刷新该流中的缓冲，将缓冲区中的字符数据保存到目的文件中去
void close( )	关闭此流对象，并释放与之关联的所有资源

通常用 FileReader 和 FileWriter 分别实现读字符文件和写字符文件的代码如下。

```
1    FileReader fr = null;
2    FileWriter fw = null;
3    try {
4        fr = new FileReader("src.txt");
5        fw = new FileWriter("des.txt");
6        char[] buf = new char[1024];
7        int len = 0;
8        while((len = fr.read(buf)) ! = -1){
9            fw.write(buf, 0, len); // 写入字符
10       }
11   } catch (Exception e) {
12       System.out.println(e.toString());
13   } finally {
14       if (fr ! = null) {
15           try {
16               fr.close();// 关闭字符输出流
17           } catch (Exception e2) {
18               throw new RuntimeException("关闭失败!");
19           }
20       }
21       if (fw ! = null) {
22           try {
23               fw.close();// 关闭字符写入流
24           } catch (IOException e) {
25               throw new RuntimeException("关闭失败!");
26           }
27       }
28   }
```

（2）读写内部存储器的文件数据。

Android 的 Context 类提供了如下两个方法来打开本应用程序的数据文件 IO 流。

• FileInputStream openFileInput（String name）：打开应用程序的数据文件下的 name 文件对应的输入流。

• FileOutputStream openFileOutput（String name, int mode）：打开应用程序的数据文件下的 name 文件对应的输出流，其中 mode 参数见表 7 - 3。

Context 类与文件操作有关的方法及功能说明见表 7 - 13。这些方法都是通过上下文对

象 Context 获取的，新创建的文件和目录都属于本应用程序所有。只要应用程序被卸载，这些文件或目录都会被清空。

表7-13 Context 类与文件操作有关的方法及功能说明

方 法 名	功 能 说 明
File getFilesDir( )	返回本应用程序的数据文件夹的绝对路径，在这里获取到的是"/data/data/<包名>/files/"目录，返回 File 对象
File getCacheDir( )	返回本应用程序默认的缓存文件存放路径，用于获取"/data/data/<包名>/cache/"目录，返回 File 对象
File getExternalCacheDir( )	返回本应用程序在外部存储器中的存储目录，用于获取"/cache/"目录，返回 File 对象
String[ ] fileList( )	返回本应用程序的数据文件夹下的全部文件
boolean deleteFile(String)	删除本应用程序的数据文件夹下的指定文件
File getDatabasePath(String name)	返回以 openOrCreateDatabase(String, int, SQLiteDatabase.CursorFactory)方法创建的数据库的绝对路径
File getFileStreamPath(String name)	返回以 openFileOutput(String, int)方法创建的文件的绝对路径

下面以图 7.3 和图 7.4 的运行效果为例介绍文件存储和读写在 Android 平台中的实现过程，即应用程序运行后显示如图 7.3 所示界面，当用户输入文件名和内容后，单击"保存文件"按钮将输入的内容保存在指定的文件中，单击"读出文件"按钮将保存的文件内容读出到显示位置。

图 7.3 文件操作界面　　　　　　图 7.4 文件写入和读出显示效果

与图7.3所示布局界面相应的代码比较简单,读者可以自行编写完成,功能实现可以按如下步骤完成。

第1步:定义initView()方法,用于初始化布局界面对象。

```
1    void initView() {
2      btnRead = (Button) this.findViewById(R.id.btnread);//读出文件按钮
3      btnWrite = (Button) this.findViewById(R.id.btnwrite);//写入文件按钮
4      edtName = (EditText) this.findViewById(R.id.edtname);//输入文件名
5      edtContent = (EditText) this.findViewById(R.id.edtcontent);//输入文件内容
6      tvInfo = (TextView) this.findViewById(R.id.tvinfo);//显示文件内容
7    }
```

第2步:定义writeFiles()方法,用于实现写入文件。

```
1    public void writeFiles(String content, String fileName) {
2        FileOutputStream fos = null; //文件输出流
3        try {// 以追加模式打开文件输出流
4            fos = openFileOutput(fileName, MODE_APPEND);
5            byte[] bytes = content.getBytes();//将字符串转换为字节数组
6            fos.write(bytes);//向文件写入字节数组
7            fos.close();// 关闭文件输出流
8        } catch (IOException e) {
9            e.printStackTrace();
10       }
11   }
```

第3步:定义readFiles()方法,用于实现读出文件。

```
1        public String readFiles(String fileName) {
2            String content = null; // 定义文件内容字符串
3            FileInputStream fileInputStream = null; // 文件输入流
4            try {
5                FileInputStream fis = openFileInput(fileName);
6                int length = fis.available();
7                byte[] buffer = new byte[length];
8                fis.read(buffer);
9                content = new String(buffer);
10               fis.close();
11           } catch (IOException e) {
12               e.printStackTrace();
13           }
14           return content; // 返回文件内容
15       }
```

第4步:在onCreate()方法中实现初始化对象——"保存文件"按钮及"读出文件"按钮的监听事件。

```
1    initView();
2    btnWrite.setOnClickListener(new View.OnClickListener() {
3        @Override
```

```
4           public void onClick(View v) {
5               String filecontent = edtContent.getText().toString();
6               String filename = edtName.getText().toString();
7               writeFiles(filecontent,filename);
8           }
9       });
10      btnRead.setOnClickListener(new View.OnClickListener() {
11          @Override
12          public void onClick(View v) {
13              String filename = edtName.getText().toString();
14             tvInfo.setText( readFiles(filename));
15          }
16      });
```

应用程序在虚拟机或真机上运行后，就会在"/data/data/应用程序的包名"下创建一个 files 文件夹，上面示例的图 7.4 中输入的 "newfile.txt" 文件就保存在这个文件夹下。

（3）读写外部存储器的文件数据。

当应用程序通过 Context 类的 openFileInput( ) 和 OpenFileOutput( ) 来打开文件的输入流和输出流时，所打开的都是应用程序的数据文件夹里的文件。Android 设备的内部存储空间有限，所以能够存储的文件大小也就很有限，而诸如 SD 卡的外部存储设备可大大地扩充 Android 设备的存储能力。Android 应用开发中常使用 Environment 类获取外部存储器目录。在访问外部存储器之前一定要先判断外部存储器是否已经是可使用状态（即已挂载或可使用），并且需要在 AndroidManifest.xml 配置文件中添加外部存储读和写的权限。

Environment 类中提供如表 7-14 所示的外部存储状态标识、如表 7-15 所示的系统其他标准目录的静态常量及如表 7-16 所示的常用方法。

表 7-14 Environment 类提供的外部存储状态标识

状态标识	功能说明
String MEDIA_BAD_REMOVAL	在没有挂载前存储媒体已经被移除
String MEDIA_CHECKING	正在检查存储媒体
String MEDIA_MOUNTED	存储媒体已经挂载，并且挂载点可读写
String MEDIA_MOUNTED_READ_ONLY	存储媒体已经挂载，挂载点只读
String MEDIA_NOFS	存储媒体是空白或是不支持的文件系统
String MEDIA_REMOVED	存储媒体被移除
String MEDIA_SHARED	存储媒体正在通过 USB 共享
String MEDIA_UNMOUNTABLE	存储媒体无法挂载
String MEDIA_UNMOUNTED	存储媒体没有挂载

表 7-15 系统其他标准目录的静态常量

静 态 常 量	功 能 说 明
public static String DIRECTORY_ALARMS	系统提醒铃声存放的标准目录
public static String DIRECTORY_DCIM	相机拍摄照片和视频的标准目录
static String DIRECTORY_DOWNLOADS	下载的标准目录
static String DIRECTORY_MOVIES	电影存放的标准目录
static String DIRECTORY_MUSIC	音乐存放的标准目录
static String DIRECTORY_NOTIFICATIONS	系统通知铃声存放的标准目录
static String DIRECTORY_PICTURES	图片存放的标准目录
static String DIRECTORY_PODCASTS	系统广播存放的标准目录
static String DIRECTORY_RINGTONES	系统铃声存放的标准目录

表 7-16 Environment 类的常用方法

方 法 名	功 能 说 明
File getDataDirectory()	获得 android data 的目录
File getDownloadCacheDirectory()	获得下载缓存目录
File getExternalStorageDirectory()	获得外部存储媒体目录
File getExternalStoragePublicDirectory（String type）	获得特定类型文件的上一级外部存储目录
String getExternalStorageState()	获得当前外部存储媒体的状态
File getRootDirectory()	获得 android 的根目录
static boolean isExternalStorageEmulated()	判断设备的外部存储媒体是否是内存模拟的，是则返回 true
static boolean isExternalStorageEmulated(File path)	判断指定文件目录的外部存储媒体是否是内存模拟的，是则返回 true
static boolean isExternalStorageRemovable()	判断设备的外存（如 SD 卡）是否可拆卸，是则返回 true
static boolean isExternalStorageRemovable(File path)	判断指定目录下的外存（如 SD 卡）是否可拆卸，是则返回 true

为了更好地存取应用程序的大文件数据，应用程序需要读写 SD 卡上的文件。读写 SD 卡上文件的通常步骤如下。

第 1 步：在 AndroidMainfest.xml 文件中添加读写 SD 卡的权限，代码如下。

```
1    <!-- 在 SD 卡中创建与删除文件的权限 -->
2    <uses-permission android:name ="android.permission.MOUNT_UNMOUNT_FILESYSTEMS" />
3    <!-- 向 SD 卡中写入数据的权限 -->
4     < uses- permission android: name =" android. permission. WRITE _ EXTERNAL _ STORAGE" />
```

第 2 步：判断 Android 设备上是否已插入 SD 卡，应用程序是否具有读写 SD 卡的权限，代码如下。

```
1    Environment.getExternalStorageState().equals(Environment.MEDIA_MOUNTED);
```

第 3 步：调用 getExternalStorageDirectory( )来获得 SD 卡的根目录。

```
1    File SDCardRoot = Environment.getExternalStorageDirectory();
```

第 4 步：读写 SD 卡中的文件。

使用前面介绍的 FileInputStream、FileOutputStream、FileReader 或 FileWrite 来读写 SD 卡中的文件，这一步与读写内部存储器中的文件数据的操作方法基本一样。下面将前面的示例功能用读写 SD 卡的方法实现。

- 定义 writeFilesToSDCard ( )方法，用于实现向 SD 卡写入文件。

```
1    public void writeFilesToSDCard (String content, String fileName) {
2        // 判断 SD 卡是否存在及本程序是否具有读写 SD 卡的权限
3        String sdstatus = Environment.getExternalStorageState();
4        if(sdstatus.equals(Environment.MEDIA_MOUNTED)) {
5            // 获得 SD 卡的根目录
6            String sdCardPath = Environment.getExternalStorageDirectory().toString();
7            //指定文件存放位置
8            String tempPath = sdCardPath + File.separator + "savepath";
9            File filepath = new File(tempPath);
10           if(! filepath.exists()) {//如果 savepath 目录不存在,则新建
11               filepath.mkdir();
12           }
13           File newFile = new File(filepath, fileName);
14           FileOutputStream fileOutputStream = null; // 初始化文件输出流
15           try {// 以追加模式打开文件输出流
16               fileOutputStream = new FileOutputStream(newFile, true);
17                fileOutputStream.write(content.getBytes());
18               fileOutputStream.close();// 关闭文件输出流
19           } catch (IOException e) {
20               e.printStackTrace();
21           }
22       }
23   }
```

上述第 8～12 行代码表示如果 SD 卡根目录下没有 savepath 文件夹，那么在根目录下创建 savepath 文件夹，然后将文件创建在 savepath 文件夹中。当然，也可以直接在 SD 卡的根目录下创建文件。

- 定义 readFilesToSDCard ( ) 方法，用于从 SD 卡读出文件。

```
1    public String readFilesFromSDCard (String fileName) {
2        String content = null;// 定义文件内容字符串
3        FileInputStream fileInputStream = null;// 文件输入流
4        String sdstatus = Environment.getExternalStorageState();
5        if (sdstatus.equals(Environment.MEDIA_MOUNTED)) {
6            String sdCardPath =
Environment.getExternalStorageDirectory().toString();
7            String tempPath = sdCardPath + File.separator + "savepath";
8            try {
9                File newFile = new File(tempPath, fileName);
10               // 打开文件输入流
11               fileInputStream = new FileInputStream(newFile);
12               int length = fileInputStream.available();
13               byte[] buffer = new byte[length];
14               fileInputStream.read(buffer);
15               content = new String(buffer);
16               fileInputStream.close();
17           } catch (IOException e) {
18               e.printStackTrace();
19           }
20       }
21       return content;
22   }
```

（4）读资源文件数据。

Android 的资源文件可以分为两种，第 1 种是 res/raw 目录下的资源文件，这类资源文件会在 R.java 里面自动生成该资源文件的 ID；第 2 种是 assets 目录下存放的原生资源文件，它不会被映射到 R.java 文件中，所以不能通过 R.XXX.ID 的方式访问。下面分别介绍这两类资源文件在 Android 开发中的访问方法。

- res 目录下的资源文件。

要在 res 目录下放置数据文件，通常需要在该目录下手动创建 raw 目录，然后把数据文件复制到该目录下。下面代码表示读出 raw 目录下的 test 文件。

```
1    InputStream fis = null;// 文件输入流
2    try {
3        //得到资源中的 raw 数据流
4        fis = getResources().openRawResource(R.raw.test);
5        int length = fis.available();
6        byte[] buffer = new byte[length];
7        fis.read(buffer);
8        content = new String(buffer);
9        fis.close();
10   } catch (IOException e) {
11       e.printStackTrace();
12   }
13   }
```

- assets目录下的资源文件。

使用Android Studio创建项目时，开发环境不会自动创建assets目录。如果开发者需要使用这个文件夹存储原生资源文件，就需要在项目的res/main文件下手动创建assets文件夹，创建完成后将需要使用的原生资源文件复制到该文件夹中即可。下面代码表示定义一个方法读出assets目录下的fileName文件。

```
1   public String readFilesFromAssets (String fileName) {
2       String content = null;// 定义文件内容字符串
3       InputStream fis = null;// 文件输入流
4       try {
5           //得到资源中的asset数据流
6           fis = getResources().getAssets().open(fileName);
7           int length = fis.available();
8           byte[] buffer = new byte[length];
9           fis.read(buffer);
10          content = new String(buffer);
11          fis.close();
12      } catch (IOException e) {
13          e.printStackTrace();
14      }
15      return content;
16  }
```

7.2.2　幸运抽奖器的实现

【抽奖器的实现】

1. 主界面的设计

本案例的主要界面有三个，分别为图7.5所示的抽奖主界面、图7.6所示的奖项设置界面和图7.7所示的中奖结果显示界面。图7.8所示为抽奖界面选项菜单界面。

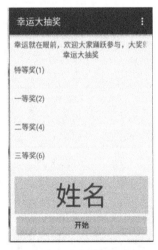

图7.5　抽奖主界面

图7.6　奖项设置界面

第7章 数据存储与访问

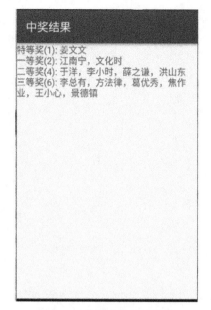

图 7.7 中奖结果显示界面

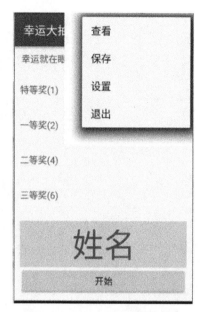

图 7.8 抽奖界面选项菜单界面

抽奖界面最上面一行使用了跑马灯显示效果，然后分别使用了多个 TextView 显示相关信息。单击"开始"按钮时，显示"姓名"的 TextView 上会动态展示参加抽奖人员名单并将按钮上文字变为"停止"；单击"停止"按钮时，将显示"姓名"的 TextView 上的内容填入对应奖项位置。其主要布局代码如下（布局文件名为 main_layout.xml）。

```
1    <?xml version="1.0" encoding="utf-8"?>
2    <RelativeLayout xmlns:android="http://schemas.android.com/apk/res/android"
3        android:layout_width="match_parent"
4        android:layout_height="match_parent"
5        android:padding="10dp">
6        <LinearLayout
7            android:layout_width="match_parent"
8            android:layout_height="wrap_content"
9            android:orientation="vertical">
10           <!-- 跑马灯效果 TextView,此处略 -->
11           <!--幸运大抽奖 TextView,此处略 -->
12           <LinearLayout
13               android:layout_width="match_parent"
14               android:layout_height="wrap_content">
15               <!-- 奖项名 -->
16               <TextView
17                   android:id="@+id/tvj1"
18                   android:layout_width="wrap_content"
19                   android:layout_height="wrap_content"
20                   android:text="一等奖"/>
21               <!-- 对应奖项数 -->
22               <TextView
23                   android:id="@+id/tvj1content"
```

```
24                  android:layout_width="wrap_content"
25                  android:layout_height="wrap_content"
26                  android:lines="2"
27                  android:padding="10dp"
28                  android:text="" />
29          </LinearLayout>
30          <!-- 其他奖项名和奖项数代码类似,此处略 -->
31      <TextView
32          android:id="@+id/tvinfo"
33          android:layout_width="match_parent"
34          android:layout_height="80dp"
35          android:layout_above="@+id/btnStart"
36          android:background="#d9dfd9"
37          android:gravity="center"
38          android:text="姓名"
39          android:textSize="50sp" />
40      <Button
41          android:id="@+id/btnStart"
42          android:layout_width="match_parent"
43          android:layout_height="wrap_content"
44          android:layout_alignParentBottom="true"
45          android:text="开始" />
46  </RelativeLayout>
```

奖项设置界面和显示中奖结果界面的布局代码比较简单,限于篇幅不再赘述,读者可以参见 ALLAPP 代码包中 choujiang 文件夹里的内容。

2. 功能实现

(1) 抽奖界面的相关功能。

抽奖界面一旦运行,就需要读取奖项设置和参加抽奖人员的相关数据,本案例中的这些数据都是在设置界面中实现保存的,奖项设置的数据用 SharedPreferences 以 jiangconfig.xml 为文件名保存在系统默认位置;参加抽奖人员设置的数据用文件方式以"创建日期.txt"为文件名保存在 SD 卡的 jiangpath 目录下。具体实现步骤如下。

● 初始化抽奖 UI 上的组件。

```
1   void initView() {
2       tvj1 = (TextView) this.findViewById(R.id.tvj1);//奖项名1显示位置
3       tvj2 = (TextView) this.findViewById(R.id.tvj2);//奖项名2显示位置
4       tvj3 = (TextView) this.findViewById(R.id.tvj3);//奖项名3显示位置
5       tvj4 = (TextView) this.findViewById(R.id.tvj4);//奖项名4显示位置
6       tvj1content = (TextView) this.findViewById(R.id.tvj1content);//中奖奖项1人员名单
7       tvj2content = (TextView) this.findViewById(R.id.tvj2content);//中奖奖项2人员名单
8       tvj3content = (TextView) this.findViewById(R.id.tvj3content);//中奖奖项3人员名单
9       tvj4content = (TextView) this.findViewById(R.id.tvj4content);//中奖奖项4人员名单
```

```
10      btnStart = (Button) this.findViewById(R.id.btnStart);//抽奖按钮
11      tvInfo = (TextView) this.findViewById(R.id.tvinfo);//动态变化的姓名
12    }
```

- 初始化奖项、奖数和参加抽奖人员名单。

```
1     void initData() {
2         read();//读奖项和奖数设置
3         Calendar calendar = Calendar.getInstance();
4         SimpleDateFormat SDformat = new SimpleDateFormat("yyyy-MM-dd");
5         String filename = SDformat.format(calendar.getTime()) + ".txt";
6         //从SD卡文件中读出参加抽奖人员名单,并分隔
7         String[] temp = readFilesFromSDCard(filename).split(",");
8         for (int i = 0; i <temp.length; i ++) {
9             jiangMember.add(temp[i]);// jiangMember 是 ArrayList 类型
10        }
11    }
12    //从 jiangconfig.xml 文件读出奖项和奖数
13    public void read() {
14        SharedPreferences sp = getSharedPreferences("jiangconfig", Context.MODE_PRIVATE);
15        String n1 = sp.getString("n1", "");
16        String n2 = sp.getString("n2", "");
17        String n3 = sp.getString("n3", "");
18        String n4 = sp.getString("n4", "");
19        int c1 = sp.getInt("c1", 0);
20        int c2 = sp.getInt("c2", 0);
21        int c3 = sp.getInt("c3", 0);
22        int c4 = sp.getInt("c4", 0);
23        if (n1.equals("")) {
24            Toast.makeText(this, "对不起,你本次抽奖参数还没有设置", Toast.LENGTH_SHORT).show();
25        } else {
26            //在对应位置显示奖项名称(数量)
27            tvj1.setText(n1 + "(" + c1 + ")");
28            jiangCount.add(c1);
29            tvj2.setText(n2 + "(" + c2 + ")");
30            jiangCount.add(c2);
31            tvj3.setText(n3 + "(" + c3 + ")");
32            jiangCount.add(c3);
33            tvj4.setText(n4 + "(" + c4 + ")");
34            jiangCount.add(c4);
35        }
36    }
```

上述第 3～10 行代码表示从以当前日期命名的文件（如 2018-01-23.txt）中读出抽奖人员名单，因为该文件在设置界面中以"姓名1,姓名2,……"的格式保存，所以此处调用 readFilesFromSDCard(filename) 方法读出的字符串需要使用 split() 方法进行分隔，并将人员名单保存在 ArrayList(jiangMember) 中。readFilesFromSDCard(filename) 方法的代码与前面介绍的内容完全相同，限于篇幅不再赘述。第 13～36 行代码定义了 read() 方法从

jiangconfig.xml 文件中读出奖项、奖项数的设置内容,并按照"奖项名称(数量)"的格式显示在界面上,效果如图 7.5 所示。

● 定义 Handler 和 Runnable 执行体用于当单击"开始"按钮时取出参加抽奖人员名单并显示在如图 7.5 所示的"姓名"位置处,其代码如下。

```
1   Handler handler = new Handler() {
2       @Override
3       public void handleMessage(Message msg) {
4           super.handleMessage(msg);
5           //产生随机名单的 index
6           useId = (int) (0 + Math.random() * jiangMember.size());
7           tvInfo.setText(jiangMember.get(useId) + "");//将名单显示在 TextView 上
8       }
9   };
10  Runnable runnable = new Runnable() {
11      @Override
12      public void run() {
13          while (flag) {
14              try {
15                  Thread.sleep(100);
16              } catch (InterruptedException e) {
17                  e.printStackTrace();
18              }
19              handler.sendEmptyMessage(0);
20          }
21      }
22  };
23  //开始抽奖
24  btnStart.setOnClickListener(new View.OnClickListener() {
25      @Override
26      public void onClick(View v) {
27          if (flag) {
28              flag = false;
29              btnStart.setText("开始");
30              useCount ++;
31              int n = jiangCount.size();
32              int[] p = new int[n];
33              for (int j = 0; j < p.length; j ++) {
34                  p[j] = (int) jiangCount.get(j);
35              }
36              //分别将中奖名单填写到对应位置
37              if (useCount <= p[0]) {
38                  String j1 = tvj1content.getText().toString();
39                  String t = j1.length() == 0 ? "" : ",";
40                  tvj1content.setText(j1 + t + jiangMember.get(useId));
41                  jiangMember.remove(useId);//中奖的名单退出下一轮抽奖
42              } else if (useCount <= p[0] + p[1]) {
43                  String j2 = tvj2content.getText().toString();
44                  String t = j2.length() == 0 ? "" : ",";
45                  tvj2content.setText(j2 + t + jiangMember.get(useId));
```

```
46              jiangMember.remove(useId);//中奖的名单退出下一轮抽奖
47          } else if (useCount <= p[0] + p[1] + p[2]) {
48              String j3 = tvj3content.getText().toString();
49              String t = j3.length() == 0 ? "" : ",";
50              tvj3content.setText(j3 + t + jiangMember.get(useId));
51              jiangMember.remove(useId);//中奖的名单退出下一轮抽奖
52          } else if (useCount <= p[0] + p[1] + p[2] + p[3]) {
53              String j4 = tvj4content.getText().toString();
54              String t = j4.length() == 0 ? "" : ",";
55              tvj4content.setText(j4 + t + jiangMember.get(useId));
56              jiangMember.remove(useId);//中奖的名单退出下一轮抽奖
57          } else {
58              Toast.makeText(MainActivity.this, "抽奖结束",
Toast.LENGTH_SHORT).show();
59          }
60      } else {
61          flag = true;
62          btnStart.setText("停止");
63          new Thread(runnable).start();//启动线程开始抽奖
64      }
65  }
66  });
```

上述第 6 行代码表示当单击"开始"按钮后，随机产生一个整数表示参加抽奖人员名单在 ArrayList 中的下标，然后将对应的元素显示在"姓名"TextView 上，当单击"停止"按钮后，将这个元素显示到中奖项位置，并将其从 ArrayList 中移除，不再参与下一轮的抽奖。

● 选项菜单及功能，其代码如下。

```
1   @Override
2   public boolean onCreateOptionsMenu(Menu menu) {
3       menu.add(0, 0, 0, "查看");
4       menu.add(0, 1, 0, "保存");
5       menu.add(0, 2, 0, "设置");
6       menu.add(0, 3, 0, "退出");
7       return super.onCreateOptionsMenu(menu);
8   }
9   @Override
10  public boolean onOptionsItemSelected(MenuItem item) {
11      Intent intent = null;
12      switch (item.getItemId()) {
13      case 0://启动查看中奖结果界面
14          intent = new Intent(this, ViewActivity.class);
15          this.startActivity(intent);
16          break;
17      case 1: //将中奖结果保存在以"日期 result.txt"为文件名的文件中
18          String content = tvj1.getText().toString() + ":"
19              + tvj1content.getText().toString() + "\n"
20              + tvj2.getText().toString() + ":"
```

```
21            + tvj2content.getText().toString() + "\n"
22            + tvj3.getText().toString() + ":"
23            + tvj3content.getText().toString() + "\n"
24            + tvj4.getText().toString() + ":"
25            + tvj4content.getText().toString() + "\n";
26        Calendar calendar = Calendar.getInstance();
27        SimpleDateFormat SDformat = new SimpleDateFormat("yyyy-MM-dd");
28        String filename = SDformat.format(calendar.getTime()) + "result.txt";
29        //该方法为写SD卡文件,与前面介绍的方法完全相同,本处略
30        writeFilesToSDCard(content, filename);
31        Toast.makeText(MainActivity.this, "保存成功",Toast.LENGTH_SHORT).show();
32        break;
33    case 2://启动设置界面
34        intent = new Intent(this, SetupActivity.class);
35        this.startActivity(intent);
36        break;
37    case 3://退出应用程序
38        this.finish();
39        break;
40    }
41    return super.onOptionsItemSelected(item);
42  }
```

(2) 奖项设置界面的相关功能。

奖项设置界面如图7.6所示。首先,用户在对应的奖项位置输入奖项名称和对应奖项的数目,单击"保存奖项"按钮,将设置好的值用SharePreferences方法保存在jiangconfig.xml文件中;其次设置参加抽奖人员名单。设置抽奖人员名单有两种方式:一种方式是直接在EditText编辑框中输入"姓名1,姓名2,……"格式的数据,单击"保存人员"按钮保存数据(即选中直接输入单选项);另一种方式是通过文件导入(即选中文件导入单选项)方式将SD卡对应位置的文件读入EditText,此时可以根据用户的需要进行修改后再单击"保存人员"按钮保存数据。具体实现步骤如下。

● 初始化界面对象。

```
1     void initView() {
2         edtName1 = (EditText) this.findViewById(R.id.edtname1);
3         edtName2 = (EditText) this.findViewById(R.id.edtname2);
4         edtName3 = (EditText) this.findViewById(R.id.edtname3);
5         edtName4 = (EditText) this.findViewById(R.id.edtname4);
6         edtCount1 = (EditText) this.findViewById(R.id.edtcount1);
7         edtCount2 = (EditText) this.findViewById(R.id.edtcount2);
8         edtCount3 = (EditText) this.findViewById(R.id.edtcount3);
9         edtCount4 = (EditText) this.findViewById(R.id.edtcount4);
10        btnJiang = (Button) this.findViewById(R.id.btnjiang);
11        btnMember = (Button) this.findViewById(R.id.btnmember);
12        edtMember = (EditText) this.findViewById(R.id.edtmember);
13        radioGroup = (RadioGroup) this.findViewById(R.id.radgroup);
14    }
```

- 将奖项和对应的奖项数用 SharedPreferences 保存。

```
1      public void save(String n1, String n2, String n3, String n4, int c1,
int c2, int c3, int c4) {
2          SharedPreferences sp = this.getSharedPreferences("jiangconfig",
Context.MODE_PRIVATE);
3          SharedPreferences.Editor editor = sp.edit();
4          editor.putString("n1", n1);//奖项1名称
5          editor.putInt("c1", c1);//奖项1数目
6          editor.putString("n2", n2);//奖项2名称
7          editor.putInt("c2", c2);//奖项2数目
8          editor.putString("n3", n3);//奖项3名称
9          editor.putInt("c3", c3);//奖项3数目
10         editor.putString("n4", n4);//奖项4名称
11         editor.putInt("c4", c4);//奖项4数目
12         editor.commit();
13     }
```

- 设置参加抽奖人员名单并保存。

```
1      radioGroup.setOnCheckedChangeListener(new
RadioGroup.OnCheckedChangeListener() {
2          @Override
3          public void onCheckedChanged(RadioGroup group, @IdRes int checkedId) {
4              switch (checkedId){
5              case R.id.radedit://选中直接输入
6                  edtMember.setText("");
7                  break;
8              case R.id.radimport://选中文件导入
9                  Calendar calendar = Calendar.getInstance();
10                 SimpleDateFormat SDformat = new SimpleDateFormat("yyyy-MM-dd");
11                 //以2018-01-23.txt的文件名格式保存文件
12                 String filename = SDformat.format(calendar.getTime()) + ".txt";
13                 edtMember.setText(readFilesFromSDCard(filename));
14                 break;
15             }
16         }
17     });
18     //保存人员
19     btnMember.setOnClickListener(new View.OnClickListener() {
20         @Override
21         public void onClick(View v) {
22             String content = edtMember.getText().toString();
23             Calendar calendar = Calendar.getInstance();
24             SimpleDateFormat SDformat = new SimpleDateFormat("yyyy-MM-dd");
25             String filename = SDformat.format(calendar.getTime()) + ".txt";
26             //以2018-01-23.txt的文件名格式保存文件
27             writeFilesToSDCard(content,filename);
28         }
29     });
```

上述第 13 行代码表示当选中文件导入选项后，调用 readFilesFromSDCard（filename）方法读入文件内容；第 27 行代码表示将 EditText 中的内容调用 writeFilesToSDCard（content，filename）方法保存到 SD 卡中。这两种方法与前面介绍的一样。

中奖结果界面的显示实现起来比较简单，只需要将保存中奖结果的文件读出后显示在对应的 TextView 上就可以了，限于篇幅不再赘述。详细功能代码读者可以参见 ALLAPP 代码包中 choujiang 文件夹里的内容。

7.3 实验室安全测试系统的设计与实现

目前高校的实验室安全已经越来越成为教师、学生关注的重点。对进入实验室的人员进行常规的安全教育是保证实验室安全的前提条件，学生对实验室的安全知识掌握是保证实验安全的根本保证。要了解学生对安全知识的掌握程度就需要有一个科学合理的安全测试平台，在学生进入实验室前对其进行安全测试。这种测试系统需要有大量的数据（题库）支撑，而 SharedPreferences 和文件存储虽然都可以存储信息，然而都有各自的不足：它们一方面不能满足存储大量数据的需要，另一方面对数据操作的灵活度不够。本节以开发设计一个实验室安全测试系统为例，介绍 Android 为开发者提供的一个轻量级数据库（SQLite）的使用方法。

7.3.1 预备知识

1. SQLite 的基本概念

【SQLite基础（创建数据库、表及增删改）】

SQLite 是一个轻量级的关系型数据库。与 C/S 模式的数据库软件不同，SQLite 是进程内的数据库引擎，因此不存在数据库的客户端和服务器。使用时一般只需要附带上 SQLite 的一个动态库，就可以使用它的全部功能。SQLite 运算速度快，占用资源少，很适合在移动设备上使用，不仅支持标准的 SQL 语法，而且还遵循 ACID（数据库事务）原则，使用非常方便。

一般数据库系统采用固定的静态数据类型，而 SQLite 数据库系统采用的是动态数据类型，使用时会根据存入的具体值自动判断其数据类型。通常 SQLite 包含 NULL（空值）、INTEGER（整数）、REAL（浮点数）、TEXT（字符串文本）和 BLOB（二进制对象）五种数据类型，因为其采用动态数据类型，所以对于 varchar、char 等其他数据类型也同样可以保存，如可以在 INTEGER 类型的字段中存放字符串类型。

SQLite 通过文件来保存数据库，一个文件就是一个数据库，一个数据库中可以包含多个表，每个表里又可以有多条记录，每条记录又由多个字段构成，每个字段可以指定其数据类型（主键对应的字段必须指定数据类型）和对应的值。

2. 手动创建数据库

虽然大多数程序都是使用代码创建数据库，并对数据库进行操作，但是手动创建数据库也很重要，它有助于初次接触 Android 应用开发的人员更好地理解数据库的创建和使用。手动创建数据库可以有以下两种途径。

（1）直接在 CMD 命令窗口手动创建。

目前，Android 内置了 SQLite 3 版本的数据库系统，为了便于开发者操作 SQLite 3 的数据库，在系统默认的安装目录中（如 Windows 7 中默认的位置为 C:\Users\用户名\AppData\Local\Android\Sdk）或者 <Android SDK>/platform-tools 目录中有一个 sqlite3.exe 文件，开发者可以将 sqlite3.exe 文件所在目录添加到 Windows 的系统环境变量 Path 中，然后直接在 CMD 窗口中输入 sqlite3 命令，可以看到如图 7.9 所示效果，此时可以使用 SQLite 中的常用命令创建数据库，创建表，进行表的增、删、改、查等操作。这种方式操作的数据库存放在 Windows 的"C:\Users\用户名"文件夹下（以图 7.9 为例）。

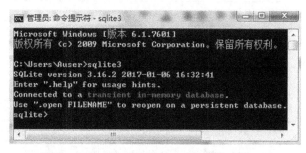

图 7.9　CMD 窗口运行 sqlite3 命令的显示效果

（2）在 Android 中手动创建。

Android 底层是 Linux，开发者可以在 CMD 命令窗口中输入"adb shell"命令，此时系统会进入 Linux 的命令行，在这个命令行下可以直接使用 Linux 命令。如果要进入数据库操作状态，也是直接在命令行上输入"sqlite3"命令即可，显示效果如图 7.10 所示。这种方式操作的数据库存放在 Linux 的"\"文件夹下（以图 7.10 为例），开发者可以根据实际情况使用 Linux 的 cd 命令更改操作目录，这样就可以操作自己指定目录下的数据库文件，一般来说 App 默认数据库的存放位置为/data/data/包名/databases 目录下。

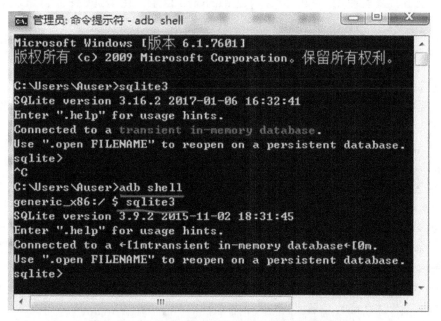

图 7.10　Linux 命令行运行 sqlite3 命令的显示效果

不管开发者使用哪种途径创建 SQLite 数据库，实际操作中都需要使用如表 7-17 所示的常用命令。

表 7-17 SQLite 数据库的常用操作命令

类　别	命　令　名	功　能　说　明
数据库	sqlite3 文件名	打开或创建（没有指定文件名时）数据库文件
查看表	.database	查看数据库信息
	.tables	查看表名称
	.schema	查看创建数据表时的 SQL 命令
	.schema table_name	查看创建表 table_name 时的 SQL 的命令
	.help	查看帮助信息
退出	.exit 或 .quit	退出与 SQLite 数据库交互的界面
操作表	CREATE TABLE 表名(字段名1 字段类型1,……)	创建表
	insert into 表名 values (字段值1,字段值2,……)	向表中插入记录
	drop 表名	删除表
	delete from 表名 where 字段1=值1……	删除符合条件的记录
	update 表名 set 字段名1=值1,字段名2=值2,……	更新表记录内容
	select * from 表名 where 字段1=值1 ……	从表中查询指定条件的记录

需要注意的是，Android 中每个应用程序的数据库文件通常默认保存在/data/data/<package name>/databases 目录下，然而如果是手动建立数据库，在/data/data/<package name>目录中并不存在 databases 目录，而需要自己动手建立。创建时首先在 CMD 窗口中使用 adb shell 命令进入 Linux 命令行，然后使用 Linux 的 cd 命令转到/data/data/<package name>目录下，最后使用 mkdir 命令创建 databases 目录，在 Linux 命令行依次输入如下命令。

```
# cd /data/data/ <package name>
# mkdir databases
```

在创建好 databases 目录后，再使用 cd 命令将操作目录转到 databases 目录下，然后使用 sqlite3 命令打开或创建数据库文件（本文以 school.db 为例），在 Linux 命令行继续输入如下命令。

```
#cd databases
#sqlite3 school.db
sqlite >
```

sqlite3 school.db 命令表示创建一个文件名为 school.db 的数据库文件。如果该数据库文件已经存在，则打开它。在使用 SQL 命令（即表 7-17 中的操作表类命令）时必须使用";"表示命令输入结束。数据库文件打开或创建完成后就可以在该数据库中创建表。下面的命令表示创建一个名为 student 表，其中包含以下三个字段。

- id：类型为 integer 并且设置为主键，自动编号。
- name：类型为 text，不能为空。
- score：类型为 float。

```
create table student(id integer primary key autoincrement, name text not null, score float);
```

可以使用如下命令在 student 表中增加一条新的记录。

```
insert into student(name,score) values('zhangsan',84);
```

其中将 id 字段设置为自动编号，所以在给表增加记录时，不需要给 id 赋值。

可以使用如下命令修改 student 表中 id=1 的记录内容。

```
update student set score=90 where id=1;
```

可以使用如下命名删除 student 表中 id=1 的记录。

```
delete from student where id=1;
```

可以使用如下命令查询 student 表中所有记录内容。

```
select * from student;
```

3. SQLiteDatabase

前面介绍的借助于 sqlite3 工具用 SQL 语句创建数据库、创建表和操作表的方法，通常用于应用程序开发过程中的调试，而实际的软件都是使用代码创建数据库、创建表及对表实现各类操作的。Android 提供了创建和使用 SQLite 数据库的 API，这些 API 被封装在 SQLiteDatabase 类中，该类中提供了下列关于数据库操作的常用方法。

（1）打开或创建数据库的方法。

openOrCreateDatabase(String path, SQLiteDatabae.CursorFactory factory)，该方法的第 1 个参数表示要打开或创建的数据库的所在位置和文件名；第 2 个参数通常为 null。该方法是 SQLiteDatabase 类的静态方法，调用时会检测是否存在 path 参数指定的数据库。如果数据库存在，就打开该数据库；如果数据库不存在，就在指定位置创建该数据库。如果数据库创建成功，就返回一个 SQLiteDatabase 对象，否则会抛出异常 FileNotFoundException。例如：

```
1    File file = MainActivity.this.getDatabasePath("databases").getParentFile();
2    if(file.exists()==false) file.mkdirs();
3    SQLiteDatabase sqLiteDatabase = SQLiteDatabase.openOrCreateDatabase(file.toString()+"/testDB.db",null);
```

上述第 1~2 行代码表示获得当前应用程序的 databases 目录。如果目录不存在，则依次创建完成这些目录后，再在 /data/data/包名/databases 目录下创建 testDB.DB 数据库文件。需要说明的是，在使用这种方法创建数据库文件时必须保证数据库文件存放的位置是存在的，如果不存在，需要使用以上的代码进行创建。当然，上面的代码是在 /data/data/包名/databases 目录下创建数据库文件，其实也可以将数据库文件创建在其他存储位置（如 SD 卡）上，这种方法就和前面文件操作一样了，代码如下。

```
1    String sdpath = Environment.getExternalStorageDirectory().toString();
2    SQLiteDatabase   sqLiteDatabase = SQLiteDatabase.openOrCreateDatabase(sdpath+"/testDB.db",null);
```

（2）创建表的方法。

数据库打开后就可以通过以下两个步骤在打开的数据库中创建表。

第1步：编写创建表的SQL语句，以创建一个Student表结构（表7-18）为例，其代码如下。

表7-18 Student表结构

字段名	字段类型	含义
_id	INTEGER	主键并且自动增加
sname	TEXT	学生姓名
number	TEXT	学生学号
sscroe	FLOAT	学生成绩

```
String createTB ="create table student(_id integer primary key autoincrement, sname text, snumber text, sscore float)";
```

第2步：调用SQLiteDatabase的execSQL(String sql)方法，该方法的SQL参数为要执行的SQL语句，代码如下。

```
sqLiteDatabase.execSQL(createTB);
```

（3）向表中插入记录的方法。

向表中插入记录有两种方法。

① 编写插入记录的SQL语句，然后调用SQLiteDatabase的execSQL(String sql)方法，代码如下。

```
String insertSQL ="insert into student(sname,snumber,sscore) values ('王小明','901005',89.5)";
sqLiteDatabase.execSQL(insertSQL);
```

② 直接调用SQLiteDatabase的insert(String tablename, String nullColumnHack, ContentValues values)方法。该方法的第1个参数表示插入记录的表名；第2个参数用于指定空值字段的名称（通常用null代替）；第3个参数表示要添加的数据，这个数据需要使用ContentValues类型存放，ContentValues类似于MAP。相比MAP，ContentValues提供了存取数据对应的put（String key, XXX value）方法和getAsXXX（String key）方法，key为字段名称，value为字段值，XXX指的是各种常用的数据类型，如String、Integer等。例如，要插入第1种方法对应的记录，需要使用如下代码。

```
1    ContentValues values = new ContentValues();
2    values.put("sname", "王小明");//字段名,字段值
3    values.put("snumber","901005");//字段名,字段值
4    values.put("sscore",89.5); //字段名,字段值
5    //返回新添记录的行号,与主键id无关
6    long rowid = sqLiteDatabase.insert("student", null, values);
```

上述第6行代码的rowid表示执行insert()语句插入新记录后返回的记录号，即如果返回值大于0，说明插入记录成功。

（4）从表中删除记录的方法。

从表中删除记录也有两种方法。

① 编写删除记录的 SQL 语句，然后调用 SQLiteDatabase 的 execSQL（String sql）方法，代码如下。

```
String deleteSQL = "delete from student where _id = 3"; //将_id为3的记录删除
sqLiteDatabase.execSQL (deleteSQL);
```

② 直接调用 SQLiteDatabase 的 delete（String table，String whereClause，String[ ]whereArgs）方法。该方法的第 1 个参数表示删除记录的表名；第 2 个参数表示删除条件；第 3 个参数表示删除条件值数组。例如，要删除第 1 种方法对应的记录，需要使用如下代码。

```
1    String whereClause = "_id=?";//删除条件
2    String[] whereArgs = {"3"};//删除条件参数
3    sqLiteDatabase.delete("student", whereClause, whereArgs); //执行删除
```

如果要删除_id 为 4 或 8 的记录，也可以将上述第 1～2 行代码修改为如下形式。

```
1    String whereClause = "_id=? or _id=?";//删除条件
2    String[] whereArgs = {"4","8"};//删除条件参数
```

（5）修改表中相关记录的内容。

修改表中相关记录的内容也有两种方法。

① 编写删除记录的 SQL 语句，然后调用 SQLiteDatabase 的 execSQL（String sql）方法，代码如下。

```
String updateSQL = "update student set snumber = '9000001' where _id = 5";
sqLiteDatabase.execSQL(updateSQL);
```

② 直接调用 SQLiteDatabase 的 update（String table，ContentValues values，String whereClause，String[ ] whereArgs）方法。该方法的第 1 个参数表示修改记录的表名；第 2 个参数表示要修改的字段及对应的值，需要使用 ContentValues 类型；第 3 个参数表示修改条件；第 4 个参数表示修改值对应的数组。例如，要修改第 1 种方法对应的内容，需要使用如下代码。

```
ContentValues values = new ContentValues();
values.put("snumber", "9000001");//字段名,字段值
String whereClause = "_id=?";//修改条件
String[] whereArgs = {"5"};//修改条件值数组
sqLiteDatabase.update("student", values, whereClause, whereArgs);
```

（6）查询表中相关记录的内容。

查询表中相关记录的内容也有两种方法。

① 编写查询记录的 SQL 语句，然后调用 SQLiteDatabase 的 rawQuery（String sql，String [ ] selectionArgs）方法。该方法第 1 个参数表示要执行的 Select 语句；第 2 个参数表示 Select 语句中的条件值数组；其返回的是 Cursor 类型值，代码如下。

```
1    //查询 student 表中所有记录内容
2    String selectSQL = "select * from student ";
3    //查询 student 表中_id 为 7 的记录内容
4    String selectSQL = "select * from student where _id = 7";
5    Cursor cursor = sqLiteDatabase.rawQuery(selectSQL, null);
```

也可以将上述第 4～5 行代码修改为如下代码，其执行效果一样。

```
1    String selectSQL = "select * from student where _id = ?";
2    String[] selectArgs = {"7"};
3    Cursor cursor = sqLiteDatabase.rawQuery(selectSQL, selectArgs);
```

② 直接调用 SQLiteDatabase 的 query(String table,String[] columns,String selection,String[] selectionArgs,String groupBy,String having,String orderBy,String limit)方法,其对应参数及功能说明见表 7-19。该方法返回值也为 Cursor 类型。例如,要实现第 1 种方法的查询效果,可以使用如下代码。

```
Cursor cursor = sqLiteDatabase.query("student",new String[]{"* "},"_id=?",
new String []{"7"},null,null,null);
```

表 7-19 query( )方法对应参数及功能说明

参 数 名	功 能 说 明
table	表名,相当于 select 语句 from 关键字后面的部分。如果是多表联合查询,可以用逗号将两个表名分开
columns	要查询出来的列名,相当于 select 语句 select 关键字后面的部分
selection	查询条件子句,相当于 select 语句 where 关键字后面的部分,在条件子句中允许使用占位符"?"
selectionArgs	对应于 selection 语句中占位符的值,值在数组中的位置与占位符在语句中的位置必须一致,否则就会有异常
groupBy	相当于 select 语句 group by 关键字后面的部分
having	相当于 select 语句 having 关键字后面的部分
orderBy	相当于 select 语句 order by 关键字后面的部分,如 personid desc,age asc
limit	指定偏移量和获取的记录数,相当于 select 语句 limit 关键字后面的部分

Cursor 是一个游标接口,提供了遍历查询结果的方法。Cursor 常用方法及功能说明见表 7-20。

表 7-20 Cursor 常用方法及功能说明

参 数 名	功 能 说 明
getCount( )	获得总的数据项数
isFirst( )	判断是否为第 1 条记录
isLast( )	判断是否为最后一条记录
moveToFirst( )	移动到第 1 条记录
moveToLast( )	移动到最后一条记录
move(int offset)	移动到指定记录
moveToNext( )	移动到下一条记录
moveToPrevious( )	移动到上一条记录
getColumnIndex(String columnName)	根据列名称获得列索引
getInt(int columnIndex)	获得指定列索引的 int 类型值
getString(int columnIndex)	获得指定列索引的 String 类型值

下面就是用 Cursor 来查询数据库中的数据,并将查询结果显示在 TextView 上,具体代码如下。

```
1    Cursor cursor = sqLiteDatabase.query("student",new String[]{"* "},"_id=?",new String []{"7"},null,null,null);
2        while (cursor.moveToNext()) { //判断游标是否为空
3        {
4            int id = cursor.getInt(0);    //获得_id
5            String sname = cursor.getString(1); //获得姓名
6            String snumber = cursor.getString(2); //获得学号
7            String sscore = cursor.getString(3);//获得成绩
8            tvContent.setText(tvContent.getText().toString() + id + ":" + sname + ":" + snumber + ":" + sscore + "\n");
9        }
```

(7) 删除指定表的方法。

编写删除数据的 SQL 语句,直接调用 SQLiteDatabase 的 execSQL( )方法,其代码如下。

```
String delTBSQL ="DROP TABLE student";    //删除表的 SQL 语句
sqLiteDatabase.execSQL(delTBSQL);//执行 SQL
```

在完成数据库和对应表的所有操作后,还需要调用 close( )方法将数据库关闭。以上示例代码读者可以参见 FirstAPP 代码包中 sqlitedemo 文件夹里的内容。

4. SQLiteOpenHelper

对于涉及数据库的 App,在第 1 次启动时可以使用 openOrCreateDatabase( )方法创建或打开数据库,但是当 App 升级时可能需要修改数据库中的表结构,也就是需要对数据库进行更新。为此,Android 提供了一个名为 SQLiteOpenHelper 的抽象类,当然必须继承它才能使用,它是通过对数据库版本进行管理来实现前面所述需求的。

【表的查操作、SQLite OpenHelper】

SQLiteOpenHelper 类提供了两个重要的方法,分别是 onCreate(SQLiteDatabase db) 方法和 onUpgrade(SQLiteDatabase db, int oldVersion, int newVersion) 方法。前者用于初次使用软件时生成数据库表,后者用于升级软件时更新数据库表结构。

当调用 SQLiteOpenHelper 的 getWritableDatabase( )方法或者 getReadableDatabase( )方法获取用于操作数据库的 SQLiteDatabase 实例的时候,如果数据库不存在,Android 会自动生成一个数据库,接着调用 onCreate( )方法。onCreate( )方法在初次生成数据库时才会被调用,在 onCreate( )方法里可以生成数据库表结构及添加一些应用程序使用到的初始化数据。onUpgrade( )方法在数据库的版本发生变化时会被调用,一般在软件升级时才需改变版本号,而数据库的版本是由开发者控制的。例如,现在的数据库版本是 1,由于业务变更修改了这个数据库中某个表的结构,这样就需要升级软件,而升级软件时希望更新用户 Android 设备里的数据库表结构。为了实现这一目的,可以把原来的数据库版本设置为 2(其他 int 类型值也可以),并且在 onUpgrade( )方法里用功能代码实现表结构的更新。当软件的版本升级次数比较多时,在 onUpgrade( )方法里可以根据原版号和目标版本号进行判断,然后做出相应的表结构及数据更新。

getWritableDatabase( )方法和 getReadableDatabase( )方法都返回一个可读写的数据库对

象，即 SQLiteDatabase 实例，但是当存储空间已满时，getWritableDatabase( )会报异常，而 getReadableDatabase( )不会报错，它此时不会返回可读写的数据库对象，而是仅仅返回一个可读的数据库对象。

下面以在 SD 卡的 databases 目录下创建 school.db 为例（数据库包含的 student 表结构同表 7 - 18），介绍使用 SQLiteOpenHelper 创建数据库和表的操作方法。

（1）自定义一个类继承 SQLiteOpenHelper 类（本示例以 DBHelper.java 为文件名）。

```
1   public class DBHelper extends SQLiteOpenHelper {
2       //构造方法
3       public DBHelper(Context context, String name,
SQLiteDatabase.CursorFactory factory, int version) {
4           super(context, name, factory, version);
5       }
6       @Override
7       public void onCreate(SQLiteDatabase sqLiteDatabase) {
8           String createTB = "create table student(_id integer primary key autoincrement,sname text,snumber text,sscore float)";
9           sqLiteDatabase.execSQL(createTB);
10      }
11      @Override
12      public void onUpgrade(SQLiteDatabase db, int oldVersion, int newVersion) {
13          //往表中增加一列电话号码
14          db.execSQL("ALTER TABLE student ADD sphone VARCHAR(12)");
15      }
16      @Override
17      public void onOpen(SQLiteDatabase db) {
18          super.onOpen(db);
19      }
20  }
```

上述代码中四个方法的执行情况说明如下。

• DBHelper( )：构造方法，第 1 个参数表示上下文；第 2 个参数表示数据库文件标识名；第 3 个参数指定在执行查询时获得一个游标实例的工厂类，设置为 null 代表使用系统默认的工厂类；第 4 个参数表示数据库的版本号。通常在 App 的 Activity 的 onCreate( )方法中调用 DBHelper( )，表示应用程序启动时，首先创建或打开数据库。

• onCreate( )：参数表示要操作的 SQLiteDatabase 对象。在数据库第 1 次生成的时候会调用这个方法，也就是说只有在创建数据库的时候才会调用。一般在这个方法里边生成数据库表。

• onUpgrade( )：第 1 个参数表示要操作的 SQLiteDatabase 对象；第 2 个参数表示数据库的旧版本号；第 3 个参数表示数据库的新版本号。当数据库需要升级的时候，Android 会主动调用这个方法。一般在这个方法里删除旧的数据表，并建立新的数据表，或者修改表的结构等。

• onOpen( )：打开数据库时的回调函数，一般不使用。

（2）在主调模块中用如下代码实现数据库的创建、打开和版本更新等功能。

```
1   String sdpath = Environment.getExternalStorageDirectory().toString();
2   File file = new File(sdpath + File.separator + "databases");
```

```
3      if (file.exists() == false) file.mkdirs();
4      String filename = file.toString() + File.separator + "school.db";
5      int versionId = Integer.parseInt(edtVersion.getText().toString());
6      dbHelper = new DBHelper(MainActivity.this, filename, null, versionId);
7      sqLiteDatabase = dbHelper.getReadableDatabase();
```

上述第 1～4 行代码实现数据库文件标识名功能；第 5 行代码在 EditText 中输入数据库版本号，如果本次输入的版本号与前一次不一样，在实例化 DBHelper 时就会调用 onUpgrade()方法修改 student 的表结构；第 6～7 行代码实现实例化 DBHelper 对象并打开对应的数据库。

到此，使用 SQLiteOpenHelper 类创建、打开和修改数据库已经完成，后面就可以根据用户的需要使用前面介绍的 SQLiteDatabase 来对表进行操作了。

7.3.2 实验室安全测试系统的实现

实验室安全测试系统一共分为四个模块，即登录模块、学习模块、考试模块和回看模块。登录模块的功能比较简单，输入正确的学号和登录密码，并选择学习模块、考试模块或回看模块其中之一就可以完成相关操作。学习模块的功能是显示题库中的所有题目和答案，学生可以进入这个模块学习安全知识。考试模块的功能是从题库中随机抽取 50 道题要求学生在 30min 内完成测试，并保存测试时间、答题结果及考试成绩。回看模块的功能是学生可以根据考试结果查看错误题目和标准答案。

1. 主界面的设计

本案例的主要界面有四个，分别为图 7.11 所示的登录界面、图 7.12 所示的学习界面、图 7.13 所示的考试界面及图 7.14 所示的回看界面。

【安全测试系统的实现1】

图 7.11 所示的登录界面的最上面和最下面分别使用了 ImageView 用于展示图片，中间部分使用了两个 EditText 用于输入学生的学号和密码，使用了一个 Spinner 用于选择操作模式，使用了两个 Button 用于登录和退出。布局文件读者可以自行设计或者参见 ALLAPP 代码包中 securitytest 文件夹里的内容（main_layout.xml）。

图 7.12 所示的学习界面用六个 TextView 分别显示题目内容、选项 A 内容、选项 B 内容、选项 C 内容、选项 D 内容和答案内容，用三个 Button 分别实现向前翻题、向后翻题和返回到登录界面。布局文件读者可以自行设计或者参见 ALLAPP 代码包中 securitytest 文件夹里的内容（study_layout.xml）。

图 7.13 所示的考试界面在学习界面的基础上做了一些修改，增加了开始考试、考试计时和提交试卷等功能，它们分别用三个 TextView 组件实现。布

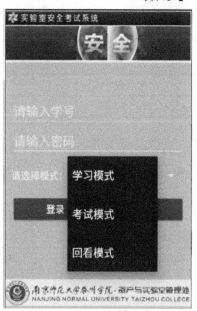

图 7.11 登录界面

局文件读者可以自行设计或者参见 ALLAPP 代码包中 securitytest 文件夹里的内容（test_layout.xml）。

图 7.12 学习界面

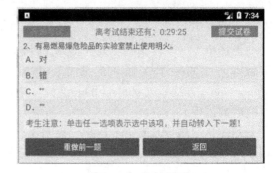

图 7.13 考试界面

图 7.14 回看界面

图 7.14 所示的回看界面在考试界面的基础上做了一些修改，增加了标准答案、考生答案的显示功能，向前查看、向后查看的功能，它们分别用两个 TextView 组件和两个 Button 组件实现。布局文件读者可以自行设计或者参见 ALLAPP代码包中 securitytest 文件夹里的内容（review_layout.xml）。

2. 功能实现

（1）数据库及表的设计。

本案例以在 SD 卡根文件夹的数据库 school.db 为例。其中包含两个表，分别是 stud 表和 tiku 表，stud 表用于保存学生信息，其表结构见表 7-21。tiku 表用于保存题目信息，其表结构见表 7-22。

表 7-21 stud 表结构

字 段 名	字 段 类 型	含 义
stuno	INTEGER	主键并且自动增加
stunumber	TEXT	学号
stupwd	TEXT	密码
stuname	TEXT	姓名
studep	TEXT	所在学院
testids	TEXT	本次测试抽取的题目编号
answer	TEXT	本次测试的标准答案
stuanswer	TEXT	学生测试的答案
stutime	TEXT	学生测试的时间
stuscore	TEXT	学生测试的成绩

第7章 数据存储与访问

表7-22 tiku 表结构

字 段 名	字 段 类 型	含 义
tino	INTEGER	主键并且自动增加
ticontent	TEXT	题目内容
ti1	TEXT	A 选项
ti2	TEXT	B 选项
ti3	TEXT	C 选项
ti4	TEXT	D 选项
tianswer	TEXT	标准答案
ticlassid	INTEGER	题目类别

(2) 登录界面的相关功能。

登录界面运行后,加载 Spinner 组件中的内容(学习模式、考试模式和回看模式),并实现其单击事件,即在 MainActivity 的 onCreate( )方法中加入如下代码。

```
1    private String[ ] modes = new String[ ]{"学习模式","考试模式","回看模式"};
2    private int modeId = 0;//0-学习模式,1-考试模式,2-回看模式
3    //在 Spinner 加载操作模式
4    ArrayAdapter<String> modeArray = new ArrayAdapter<String>(this,
android.R.layout.simple_dropdown_item_1line, modes);
5    spinnerMode.setAdapter(modeArray);
6    //设置 Spinner 选中监听事件
7    spinnerMode.setOnItemSelectedListener(new
AdapterView.OnItemSelectedListener() {
8        @Override
9        public void onItemSelected(AdapterView<?> parent, View view, int position, long id) {
10           modeId = position;//操作模式
11       }
12       @Override
13       public void onNothingSelected(AdapterView<?> parent) {
14       }
15    });
```

接着,编写一个 login(String username, String userpwd) 用于判断在 EditText 中输入的用户名和密码是否正确,此时需要对数据库进行操作,本案例中的题库和学生信息默认存放在 SD 卡的根文件夹下。具体代码如下。

```
1    String login(String username, String usepwd) {
2        String result = "";
3        String sdpath = Environment.getExternalStorageDirectory().toString();
4        String filename = sdpath + File.separator + "school.db";
5        SQLiteDatabase sqLiteDatabase =
SQLiteDatabase.openOrCreateDatabase(filename, null);
6        Cursor cursor = sqLiteDatabase.query("stud", new String[]{"stunumber", "stupwd"}
```

```
7        , "stunumber = ? and stupwd = ?" , new String[]{username, usepwd}
8        , null, null, null);
9        if(cursor.getCount()>0){//输入的学号和密码正确
10            result = "";
11       }else {
12            result = "对不起,你的用户名和密码有误,请与班主任联系!";
13       }
14       return result;
15   }
```

上述第 5 行代码使用 openOrCreateDatabase( ) 方法打开存储在 SD 卡根文件夹下的 school.db 文件,第 6 行代码调用 query( ) 方法在 school.db 数据库中的 stud 表中查询字段 stunumber 和 stupwd 分别与 login( ) 方法中传来的 username(用户名)和 userpwd(密码)参数值相同的记录,并返回给 Cursor 对象,第 9~13 行代码表示如果 Cursor 集合中没有查找到相应的记录,则返回 "对不起,你的用户名和密码有误,请与班主任联系!" 的内容。

最后,实现"登录"按钮监听事件,在该事件中需要获得两个 EditText 中输入的学号和密码,并调用自定义的 login( ) 方法对学号和密码的正确性进行判断。如果密码不正确,用 Toast 给出提示;如果密码正确,则根据在 Spinner 中选择的操作模式转到另一个 Activity。其详细代码如下。

```
1    btnLogin.setOnClickListener(new View.OnClickListener() {
2        @Override
3        public void onClick(View v) {
4            String username = edtLoginName.getText().toString();
5            String usepwd = edtLoginPwd.getText().toString();
6            //调用login方法判断学号和密码的正确性,并返回结果
7            String result = login(username, usepwd);
8            Intent intent = new Intent();
9            if (result.length() == 0) {//学号和密码正确
10               switch (modeId) {
11                   case 0://学习模式
12                       intent = new Intent(MainActivity.this, StudyActivity.class);
13                       //将登录者学号通过ntent传递给下一个Activity
14                       intent.putExtra("username", username);
15                       MainActivity.this.startActivity(intent);
16                       break;
17                   case 1://考试模式
18                       intent = new Intent(MainActivity.this, TestActivity.class);
19                       intent.putExtra("username", username);
20                       MainActivity.this.startActivity(intent);
21                       break;
22                   case 2://回看模式
23                       intent = new Intent(MainActivity.this, ReviewActivity.class);
24                       intent.putExtra("username", username);
25                       MainActivity.this.startActivity(intent);
26                       break;
27                   }
```

```
28              } else {
29                  Toast.makeText(MainActivity.this, result, Toast.LENGTH_SHORT).show();
30              }
31          }
32      });
```

上述第 4 行、第 5 行代码分别表示从输入学号和输入密码的 EditText 中取出输入的内容，然后作为实参传递给 login( ) 方法；第 14 行、第 19 行和第 24 行代码将学号通过 Intent 传递给下一个 Intent，以便保存学生安全测试的题目、分数及错题情况等。

【安全测试系统的实现2】

（3）学习界面的相关功能。

首先，进入学习界面，操作界面自动改变为横屏模式。要实现这一功能，只需要修改配置文件 AndroidManifest.xml，文件只对应 Activity 的配置信息，代码如下。

```
<activity android:name =".StudyActivity" android:screenOrientation ="landscape" />
```

其次，需要再打开数据库，将数据库中的题库表 tiku 中的内容全部查询出来保存到 Cursor 中，以便在学习界面上根据用户单击的"向后学习"按钮或"向前学习"按钮在对应的 TextView 上显示题目内容、选项 A 内容、选项 B 内容、选项 C 内容、选项 D 内容和答案内容。本案例实现时，直接在 StudyActivity 的 onCreate( ) 方法查询出全部的题目的 Cursor，其代码如下。

```
1   String sdpath = Environment.getExternalStorageDirectory().toString();
2   String filename = sdpath + File.separator + "school.db";
3   sqLiteDatabase = SQLiteDatabase.openOrCreateDatabase(filename, null);
4   cursor = sqLiteDatabase.query("tiku", new String[]{"tino", "ticontent", "ti1", "ti2", "ti3", "ti4", "tianswer", "ticlassid"}, null, null, null, null, null);
```

然后，编写显示题目内容、选项 A 内容、选项 B 内容、选项 C 内容、选项 D 内容和答案内容的方法 showTimu（Cursor cursor, boolean flag），其中第 1 个参数 cursor 就是要显示的数据游标集，第 2 个参数 flag 是显示下一题或上一题的标志（若 flag 为 true，则显示下一题，否则显示上一题），详细代码如下。

```
1   void showTimu(Cursor cursor, boolean flag) {
2       if (flag) {//显示下一题
3           if (cursor.moveToNext()) { //判断游标是否为空
4               int id = cursor.getInt(0);   //获得题号
5               String ticontent = cursor.getString(1); //获得题目内容
6               String tva = cursor.getString(2); //获得选项 A
7               String tvb = cursor.getString(3); //获得选项 B
8               String tvc = cursor.getString(4); //获得选项 C
9               String tvd = cursor.getString(5); //获得选项 D
10              String tvanswer = cursor.getString(6); //获得答案
11              tvTimu.setText(id + "、" + ticontent);
12              tv1.setText("A. " + tva);
13              tv2.setText("B. " + tvb);
14              tv3.setText("C. " + tvc);
15              tv4.setText("D. " + tvd);
```

```
16                    tvAnaswer.setText("标准答案:" + tvanswer);
17                }
18            } else {//显示上一题
19                if (cursor.moveToPrevious()) { //判断游标是否为空
20                    //此处内容同第 4～16 行代码,限于篇幅此处略
21                }
22            }
23      }
```

最后,编写"向前学习"按钮和"向后学习"按钮的单击监听事件,该事件中直接调用 showTimu()方法即可,读者可以参见 ALLAPP 代码包中 securitytest 文件夹里的内容(StudyActivity.java)。

(4)考试界面的相关功能。

考试界面显示题目的功能与学习模式界面类似,但选取题目、学生作答、计时及提交答案并评分需要另行设计并实现。具体步骤如下。

① 定义变量。

```
1       //开始考试,考试计时,当前题目内容,当前题目的选项 A、选项 B、选项 C、选项 D,提交试卷
2       private TextView tvStart, tvTime, tvTimut, tv1t, tv2t, tv3t, tv4t, tvSubmit;
3       private Button btnPret, btnReturnt;//重做前一题、返回按钮
4       private SQLiteDatabase sqLiteDatabase;//数据库
5       private Cursor cursor;//题库中所有题目游标
6       private ArrayList indexAL = null;//用于存放抽取的题目索引号
7       private ArrayList answerAL = null;//用于存放标准答案
8       private ArrayList userAL = null;//用于存放学生答案
9       private int showIndex = 0;//用于保存当前显示的题目索引号对应的下标
10      private Intent intent = null;
11      private String stunumber ="";//存放登录学生的学号
```

② 开始考试监听事件。

```
1       //单击开始考试后,计时,并将开始考试禁用
2       tvStart.setOnClickListener(new View.OnClickListener() {
3           @Override
4           public void onClick(View v) {
5               indexAL = new ArrayList<>();
6               answerAL = new ArrayList();
7               userAL = new ArrayList();
8               new myTime(30 * 60 * 1000, 1000).start();//计时
9               tv1t.setEnabled(true);//选项 A 可用
10              tv2t.setEnabled(true);//选项 B 可用
11              tv3t.setEnabled(true);//选项 C 可用
12              tv4t.setEnabled(true);//选项 D 可用
13              tvStart.setEnabled(false);//开始考试禁用
14              getData();//打开数据库并随机抽取 50 道试题
15              showTimu((int) indexAL.get(showIndex));//默认显示抽取题目记录号的第 1 个
16          }
17      });
```

上述第 8 行代码使用了自定义继承于 CountDownTimer 的计时工具实现计时功能。本案

例为计时30min，30min 计时完毕，直接将 A、B、C、D 选项禁用，并提醒学生单击提交试卷，详细代码参见下文的计时器代码块。第 9～12 行代码表示一旦计时开始考试，A、B、C、D 选项设置为可用状态。第 13 行代码将开始考试设置为禁用。第 14 行代码调用自定义方法 getData( ) 打开数据库并抽取 50 道试题，详细代码参见下文获取数据代码块。第 15 行代码调用自定义方法 showTimu( ) 将抽取的第 1 个试题显示在相应位置，并将标准答案保存在 answerAL 中，详细代码参见下文显示题目代码块。

```
1        //计时器代码块
2        private class myTime extends CountDownTimer {
3            public myTime(long millisInFuture, long countDownInterval) {
4                super(millisInFuture, countDownInterval);
5            }
6            @Override
7            public void onTick(long millisUntilFinished) {
8                int hour = (int) millisUntilFinished / 1000 / 3600;//时
9                int minute = (int) millisUntilFinished / 1000 % 3600 / 60;//分
10               int second = (int) millisUntilFinished / 1000 % 3600 % 60;//秒
11               tvTime.setText("离考试结束还有:" + hour + ":" + minute + ":" + second);
12           }
13           @Override
14           public void onFinish() {
15               tv1t.setEnabled(false);
16               tv2t.setEnabled(false);
17               tv3t.setEnabled(false);
18               tv4t.setEnabled(false);
19               Toast.makeText(TestActivity.this, "考试时间到,请提交试卷!", Toast.LENGTH_SHORT).show();
20           }
21       }
22       //获取数据代码块
23       void getData() {
24           String sdpath = Environment.getExternalStorageDirectory().toString();
25           String filename = sdpath + File.separator + "school.db";
26           sqLiteDatabase = SQLiteDatabase.openOrCreateDatabase(filename, null);
27           //从题库中获取所有题目的 Cursor
28           cursor = sqLiteDatabase.query("tiku", new String[]{"tino", "ticontent", "ti1", "ti2", "ti3", "ti4", "tianswer", "ticlassid"}, null, null, null, null, null);
29           //生成 50 个不同记录号放入 ArrayList 中
30           for (int i = 0; i <50; i ++) {
31               int index = (int) (0 + Math.random() * cursor.getCount());
32               while (indexAL.contains(index)) {
33                   index = (int) (0 + Math.random() * cursor.getCount());
34               }
35               indexAL.add(index);
36               answerAL.add("");
37               userAL.add("");
38
```

```
39          }
40          //显示题目代码块
41          void showTimu(int index) {
42              cursor.moveToPosition(index);//将游标移动到指定位置记录
43              int id = cursor.getInt(0);      //获得题号
44              String ticontent = cursor.getString(1);//获得题目内容
45              String tva = cursor.getString(2);//获得选项A
46              String tvb = cursor.getString(3);//获得选项B
47              String tvc = cursor.getString(4);//获得选项C
48              String tvd = cursor.getString(5);//获得选项D
49              String tvanswer = cursor.getString(6);//获得答案
50              answerAL.set(showIndex, tvanswer);//将标准答案添加到ArrayList
51              tvTimut.setText((showIndex + 1) + "、" + ticontent);//题号 = +1
52              tv1t.setText("A. " + tva);
53              tv2t.setText("B. " + tvb);
54              tv3t.setText("C. " + tvc);
55              tv4t.setText("D. " + tvd);
56          }
```

③ 选项单击监听事件。

当单击某个选项时，将当前单击的选项A或B或C或D更新到存放学生答案的ArrayList中，然后判断当前显示题目索引号（从0开始计数），如果题目索引号小于49，则索引号+1，并调用showTimu()方法显示下一个题目内容。四个选项的单击事件一样，此处仅列出单击A选项的监听事件，代码如下。

```
1           //单击A选项监听事件
2           tv1t.setOnClickListener(new View.OnClickListener() {
3               @Override
4               public void onClick(View v) {
5                   userAL.set(showIndex, "A");//将用户答案添加到ArrayList
6                   if (showIndex <49) {
7                       showIndex ++;
8                       showTimu((int) indexAL.get(showIndex));
9                   } else {
10                      Toast.makeText(TestActivity.this, "已到最后一题!", Toast.LENGTH_SHORT).show();
11                  }
12              }
13          });
```

另外，由于学生做完一个题目后，可能还需要往前翻看，甚至修改答案，所以给重做前一题绑定的监听事件如下。

```
1           btnPret.setOnClickListener(new View.OnClickListener() {
2               @Override
3               public void onClick(View v) {
4                   if (showIndex > 0) {
5                       showIndex--;
6                       showTimu((int) indexAL.get(showIndex));
7                   } else {
```

```
8                      Toast.makeText(TestActivity.this,"已到第1题!",
Toast.LENGTH_SHORT).show();
9                  }
10             }
11         });
```

④ 提交试卷监听事件。

当学生单击提交试卷后,将选项全部禁用,并将学生作答的答案与标准答案比较,如果有一项相同,则计2分,并调用自定义saveScore()方法保存考试结果。

```
1    tvSubmit.setOnClickListener(new View.OnClickListener() {
2        @Override
3        public void onClick(View v) {
4            tv1t.setEnabled(false);//一旦交卷,选项A禁止使用
5            tv2t.setEnabled(false);//一旦交卷,选项B禁止使用
6            tv3t.setEnabled(false);//一旦交卷,选项C禁止使用
7            tv4t.setEnabled(false);//一旦交卷,选项D禁止使用
8            //交卷评分
9            int score = 0;
10           for (int i = 0; i <answerAL.size(); i ++) {
11               //在学生已经作答,并且学生答案与标准答案相同
12               if (userAL.get(i).toString().length() ! = 0 &&
answerAL.get(i).equals(userAL.get(i)))
13                   score = score + 2;
14           }
15           Toast.makeText(TestActivity.this, "你的得分为:" + score,
Toast.LENGTH_SHORT).show();
16           saveScore(stunumber,score);
17       }
18   });
```

⑤ 保存考试结果。

根据登录界面的Intent传递来的学号更新school.db数据库中的stud表,通过ContentValues将题目序号、标准答案、学生答案、当前时间、考试分数等更新到对应学号的记录中,以便学生进入回看界面查看错题情况。

```
1    void saveScore(String studnumber,int score) {
2        ContentValues values = new ContentValues();
3        values.put("testids", indexAL.toString());//学生测试的题目序号
4        values.put("answers", answerAL.toString());//标准答案
5        values.put("stuanswers", userAL.toString());//学生答案
6        Calendar calendar = Calendar.getInstance();
7        values.put("stutime", calendar.getTime().toString());//当前时间
8        values.put("stuscore",score + "");//分数
9        String whereClause = "stunumber = ?";//更新条件以登录的学号为依据
10       String[] whereArgs = {studnumber};//登录界面通过Intent传递来的学号
11       sqLiteDatabase.update("stud", values, whereClause, whereArgs);
12   }
```

(5) 回看界面的相关功能。

回看界面启动后,首先根据登录界面的Intent传递来的学号到School.db数据库的stud

表中查询该学生的考试情况信息，也就是该学生的试卷题目索引号、标准答案和学生答案；同时将 tiku 表中的所有题目信息保存在 Cursor 中，这些功能直接写在 onCreate( )方法中；然后单击向前查看和向后查看按钮，通过自定义方法根据试卷题目索引号从 Cursor 中把题目、选项、标准答案和学生答案显示到对应的组件上。具体步骤如下。

① 定义变量。

```
1    private TextView tvTimur, tv1r, tv2r, tv3r, tv4r, tvAnaswerr, tvUseranswer;
2    private Button btnPrer, btnNextr, btnReturnr;
3    private SQLiteDatabase sqLiteDatabase;
4    //保存题库中所有题目信息、保存登录学生的已作答信息
5    private Cursor cursorTiku, cursorStud;
6    private Intent intent = null;
7    private String[] answers = new String[50];//存放学生试题的标准答案
8    private String[] useranswers = new String[50];//存放学生的答案
9    private String[] tiids = new String[50];//存放学生试题的索引号
10   private int showIndex = 0;//记录当前题号
```

② 打开数据库并查询题库、登录学生信息。

```
1    String sdpath = Environment.getExternalStorageDirectory().toString();
2    String filename = sdpath + File.separator + "school.db";
3    sqLiteDatabase = SQLiteDatabase.openOrCreateDatabase(filename, null);
4    cursorTiku = sqLiteDatabase.query("tiku", new String[]{"tino", "ticontent", "ti1", "ti2", "ti3", "ti4", "tianswer", "ticlassid"}
5         , null, null, null, null, null);
6    intent = this.getIntent();
7    String stunumber = intent.getStringExtra("username");
8    String whereClause = "stunumber=?";//查询条件以登录的学号为依据
9    String[] whereArgs = {stunumber};//登录界面通过 Intent 传递来的学号
10   cursorStud = sqLiteDatabase.query("stud", new String[]{"testids", "answers", "stuanswers"}, whereClause, whereArgs, null, null, null);
11   if (cursorStud.moveToNext()) {
12       String ids = cursorStud.getString(0);
13       int end1 = ids.indexOf("]");
14       tiids = ids.substring(1, end1).split(",");
15       Toast.makeText(ReviewActivity.this, tiids[49], Toast.LENGTH_LONG).show();
16       String aws = cursorStud.getString(1);
17       int end2 = aws.indexOf("]");
18       answers = aws.substring(1, end2).split(",");
19       String saws = cursorStud.getString(2);
20       int end3 = aws.indexOf("]");
21       useranswers = saws.substring(1, end3).split(",");
22   }
```

上述第 13～14 行代码表示将 stud 表中 testids（保存有学生试题的索引号）存放的字符中的首尾的 [和] 符号去掉，然后保存在 tiids 数组中。

③ 查看下一题监听事件和查看上一题监听事件。

```
1       //这两个监听事件类似,此处只列出查看下一题监听事件
2       btnNextr.setOnClickListener(new View.OnClickListener() {
3              @Override
4              public void onClick(View v) {
5                   if (showIndex <48) {
6                       showIndex ++;
7                       showTimu(Integer.parseInt(tiids[showIndex].trim()));
8                   }else{
9                       Toast.makeText(ReviewActivity.this,"已到最后一题!",
Toast.LENGTH_LONG).show();
10                  }
11              }
12      });
```

④ 显示题目、选项、标准答案和学生答案的自定义方法。

```
1       //显示题目代码块
2       void showTimu(int index) {
3           cursorTiku.moveToPosition(index);//将游标移动到指定位置记录
4           int id = cursorTiku.getInt(0);    //获得题号
5           String ticontent = cursorTiku.getString(1);//获得题目内容
6           String tva = cursorTiku.getString(2);//获得选项A
7           String tvb = cursorTiku.getString(3);//获得选项B
8           String tvc = cursorTiku.getString(4);//获得选项C
9           String tvd = cursorTiku.getString(5);//获得选项D
10          String tvanswer = cursorTiku.getString(6);//获得答案
11          tvTimur.setText((showIndex + 1) + "、" + ticontent);//题号 = +1
12          tv1r.setText("A. " + tva);
13          tv2r.setText("B. " + tvb);
14          tv3r.setText("C. " + tvc);
15          tv4r.setText("D. " + tvd);
16          tvAnaswerr.setText("标准答案:" + answers[showIndex]);
17          tvUseranswer.setText("考生答案:" + useranswers[showIndex]);
18      }
```

至此,实验室安全测试系统全部设计和开发完毕,读者可以将本案例的设计思路应用到其他考试系统中。

7.4 应用程序间的数据共享

一个软件系统的架构通常如图 7.15 所示,为了降低业务层对底层数据层的依赖,在软件开发时一般需要增加一个数据访问层来解耦。从另外一个角度看,业务层的不同业务就相当于不同的应用程序,即不同的应用程序通过数据访问层访问共享的数据,这些共享的数据可以是应用程序自身的数据,也可以是其他应用程序的数据。Android 中的四大组件之一的 ContentProvier(内容提供者)充当了软件系统架构中数据访问层的角色,所以 ContentProvider 是 Android 为不同应用程序间共享数据提供的一个机制。在实际

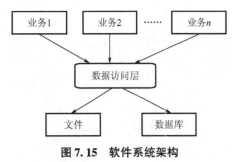

图 7.15 软件系统架构

【ContentProvider、Uri、ContentResolver】

应用开发中，ContentProvider 有以下两种使用情况。

（1）应用程序需要读取或修改其他应用程序中的数据，为此，这些数据就需要通过 ContentProvider 机制暴露出来，如 Android 中内置的通讯录、短信等信息。

（2）应用程序本身需要暴露自己的数据给其他应用程序读取或修改。

7.4.1 预备知识

1. ContentProvider

ContentProvider 为不同的软件之间数据共享提供统一的接口，它以类似数据库中表的方式将数据暴露，即 ContentProvider 就像是向外界提供数据的"数据库"，所以外界在获取其提供的数据时，也就应该使用与获取实际数据库中数据类似的方法，只不过 ContentProvider 采用 Uri 来表示外界需要访问的"数据库"。至于如何从 Uri 识别出外界需要的是哪个"数据库"，这个是 Android 底层需要完成的工作。也就是说，如果开发者想让其他应用程序使用自己应用程序内的数据，就可以使用 ContentProvider 定义一个对外开放的接口，让其他应用程序可以使用自己应用程序中的数据。

当然，开发者开发的应用程序需要给其他应用程序共享数据信息的需求一般是比较少见的，但在 Android 中，很多系统内置程序都向外提供了访问接口用于共享数据，如联系人信息、短信信息、图片库、音频库、日历等，Google 工程师已经将其封装好并向开发者提供了 Uri，开发者可以根据需要使用 Uri 去直接访问这些信息。

开发者使用 ContentProvider 向外暴露数据时需要创建一个继承于 ContentProvider 的子类，并在该类中重写如下方法。

- public boolean onCreate()：在创建 ContentProvider 时调用。
- public Cursor query(Uri uri, String[] projection, String selection, String[] selectionArgs, String sortOrder)：用于查询指定 Uri 的 ContentProvider，返回一个 Cursor。
- public Uri insert(Uri uri, ContentValues values)：用于添加数据到指定 Uri 的 ContentProvider 中。
- public int update(Uri uri, ContentValues values, String selection, String[] selectionArgs)：用于更新指定 Uri 的 ContentProvider 中的数据。
- public int delete(Uri uri, String selection, String[] selectionArgs)：用于从指定 Uri 的 ContentProvider 中删除数据。
- public String getType(Uri)：用于返回指定的 Uri 中的数据的 MIME 类型。

2. Uri

Uri 用来定位任何远程或本地的可用资源，也就是要操作的数据。在 Android 中，所有的可用资源（如联系人、图像、视频、短信等）都可以用 Uri 表示。Uri 的结构通常由 scheme、authority、path、query 和 fragment 组成，其中 authority 又分为 host 和 port。它的格式根据划分的详细程度可以分为以下三种。
- [scheme:][scheme-specific-part][#fragment]
- [scheme:][//authority][path][?query][#fragment]
- [scheme:][//host:port][path][?query][#fragment]

其中：scheme 表示定位资源的方式，ContentProvider 中用 content 表示；path 表示要访问数据的路径；query 表示访问对应参数指定的数据，可以带参数值，也可以不带，如果带参数值就需要使用"="连接。该参数可以有多个，如果带多个参数值，就需要使用"&"连接。

例如，http://www.nnutc.com:8080/web/index.htm?id=10&name=us#harvic 中 Uri 值对应部分分别为：scheme 对应 http；fragment 对应 harvic；authority 对应 www.nnutc.com:8080（其中 host 对应 www.nnutc.com:8080，port 对应 8080）；query 对应 id=10&name=us（其中 id 和 name 是两个参数，值分别为 10 和 use）。

在 Android 应用中，Uri 组成部分中的 scheme、authority 和 path 是不能省略的，其他部分可以省略。所以对 ContentProvider 中的什么数据进行操作，是通过指定的 Uri 决定的，一个操作 ContentProvider 的 Uri 由以下几个部分组成。

content://authority 或主机名/路径/id

其中：content://是固定的；authority（或主机名）用于唯一标识 ContentProvider，外部调用者可以根据这个标识找到它；路径可以用来表示要操作的数据，它的构建根据需要而定。例如：

要操作 person 表中 id 为 10 的记录，可以构建这样的路径：/person/10；

要操作 person 表中 id 为 10 的记录的 name 字段，可以构建这样的路径：person/10/name；

要操作 person 表中的所有记录，可以构建这样的路径：/person；

要操作 xxx 表中的记录，可以构建这样的路径：/xxx。

当然，要操作的数据不一定来自数据库，也可以是来自文件等其他存储方式。例如，要操作 XML 文件中 person 节点下的 name 节点，可以构建这样的路径：/person/name。

如果要把一个字符串转换成 Uri，可以使用 Uri 类中的 parse() 方法。例如，Uri uri = Uri.parse("content://com.xxx.provider.myprovider/person")。

3. UriMatcher 和 ContentUris

由于 Uri 代表了要操作的数据，所以需要对 Uri 进行解析，并从中获取要操作的数据。Android 提供了两个用于操作 Uri 的工具类，即 UriMatcher 和 ContentUris。

（1）UriMatcher。

UriMatcher 类用于匹配 Uri，该类提供了如下三个方法。

- public UriMatcher(int code)：创建一个 UriMatcher 对象。
- public void addURI(String authority, String path, int code)：为 ContentProvider 添加一个用于匹配的 Uri，当匹配成功时返回 code。其中：authority 为 AndroidManifest.xml 中注册的 ContentProvider 的 authority 属性；path 为一个路径，可以设置通配符；"#"表示任意数字；"*"表示任意字符；code 为自定义的一个 Uri 代码。
- public int match(Uri uri)：对 Uri 进行匹配验证，如果匹配成功，返回 addURI() 传递

的 code 参数。Uri 中可以包含"*"和"#"通配符。例如,传过来的 Uri 值为"content://cn.edu.nnutc.ie/student/#",其中"#"可以表示任意数字,所以可以匹配"content://com.example.test/student/10" Uri 值。

使用时,首先将需要匹配的 Uri 路径全部注册。然后,使用 match() 方法对输入的 Uri 进行匹配。如果匹配成功,返回对应的匹配码,匹配码就是调用 addURI() 方法传入的第 3 个参数值,如果匹配不成功,一般返回 UriMatcher.NO_MATCH。最后,根据匹配码的值决定实现需要的操作。

(2) ContentUris。

ContentUris 类用于获取 Uri 路径后面的 id 部分,它有以下两个很有用的方法。

• public static Uri withAppendedId(Uri contentUri, long id):将 id 和 contentUri 连接成一个新的 Uri。

• public static long parseId(Uri contentUri):将 contentUri 最后的 id 解析出来。

例如:

Uri uri = Uri.parse("content://com.xxx.provider.myprovider/person");
Uri resultUri = ContentUris.withAppendedId(uri, 10);

以上代码执行后生成后的 Uri 为"content://com.xxx.provider.myprovider/person/10"。

Uri uri = Uri.parse("content://com.xxx.provider.myprovider/person/10");
long personid = ContentUris.parseId(uri);

以上代码执行后 personid 获取的结果为 10。

4. ContentResolver

ContentProvider(内容解析器)共享数据机制是通过定义一个对外开放的统一接口实现的,其他应用程序并不是直接调用这个外部接口中的方法,而是使用一个 ContentResolver 对象来调用这些方法从而实现对共享数据的访问。当其他应用需要对 ContentProvider 中的数据进行添加、删除、修改和查询等操作时,可以使用 Context 提供的 getContentResolver() 方法获得 ContentResolver 对象,即

ContentResolver cr = getContentResolver();

所以,ContentProvider 是用于向其他应用程序提供数据,而 ContentResolver 是负责获取 ContentProvider 提供的数据,并实现数据的添加、删除、修改和查询等操作。ContentResolver 类也提供了以下与 ContentProvider 类相对应的方法。

• public Uri insert(Uri uri, ContentValues values):用于向 ContentProvider 添加数据。

• public int delete(Uri uri, String selection, String[] selectionArgs):用于从 ContentProvider 中删除数据。

• public int update(Uri uri, ContentValues values, String selection, String[] selectionArgs):用于更新 ContentProvider 中的数据。

• public Cursor query(Uri uri, String[] projection, String selection, String[] selectionArgs, String sortOrder):用于从 ContentProvider 中获取数据。

7.4.2 学生信息共享应用的实现

下面以一个学生信息共享应用案例介绍应用程序间数据共享的实现步骤。本案例涉及

两个应用程序,其中 ProviderApp 用于向外暴露数据源,ProviderUseApp 用于使用 Provider-App 向外暴露无遗的数据。数据源使用 SQLite 数据库(数据库名为 jwxt.db,学生信息表为 student,表结构见表 7-23)。

表 7-23 student 表结构

字 段 名	字 段 类 型	含 义
id	INTEGER	主键并且自动增加
sname	varchar(10)	姓名
ssex	varchar(2)	性别
saddress	TEXT	家庭住址
stel	varchar(15)	联系电话

1. 创建继承于 SQLiteOpenHelper 的数据库帮助类——DBOpenHelper

在 ProviderApp 目录下创建 DBOpenHelper 子类,其关键代码如下。

```
1    public class DBOpenHelper extends SQLiteOpenHelper {
2        private final static String DB_NAME = "jwxt.db";
3        public final static String TABLE_NAME = "student";
4        private final static int DB_VERSION = 1;
5        public DBOpenHelper(Context context) {
6            super(context, DB_NAME, null, DB_VERSION);
7        }
8        @Override
9        public void onCreate(SQLiteDatabase db) {
10           String sql = "create table " + TABLE_NAME + "(id integer primary key autoincrement,sname varchar(10),ssex varchar(2),saddress text,stel varchar(15))";
11           db.execSQL(sql);
12       }
13       @Override
14       public void onUpgrade(SQLiteDatabase db, int oldVersion, int newVersion) {
15       }
16   }
```

上述代码创建了数据库 jwxt.db 和 student 表。

2. 创建继承于 ContentProvider 的子类——StudentProvider

在 ProviderApp 目录下创建 StudentProvider 子类,该子类中有多个方法需要实现重写实现,分别是 onCreate()方法、query()方法、getType()方法、insert()方法、delete()方法和 update()方法,主要用于实现数据的添加、删除、修改和查询等操作。应用程序启动时会调用 onCreate()方法,通常在这个方法里实现一些初始化操作(如创建数据库或打开数据库的操作),其他方法具体实现添加、删除、修改和查询等操作。

(1) 定义变量和 Uri 匹配规则。

• 定义变量

```
1    private static UriMatcher mUriMatcher;
2    public static final String AUTHORITY = "cn.edu.nnutc.ie.StudentProvider";
//授权
3    private Context mContext;//上下文
4    private SQLiteDatabase mDatabase;//数据库
5    private String mTable;//表
```

- 定义 Uri 匹配规则

```
1    static {
2        //不匹配时的返回码为 UriMatcher.NO_MATCH
3        mUriMatcher = new UriMatcher(UriMatcher.NO_MATCH);
4        //匹配表,即 student 表中所有学生记录
5        mUriMatcher.addURI(AUTHORITY, "student", 1);
6        //匹配任何数字,即 student 表中某一学生单条记录
7        mUriMatcher.addURI(AUTHORITY, "student/#", 2);
8        //匹配任何文本
9        mUriMatcher.addURI(AUTHORITY, "student/* /* ", 3);
10       mUriMatcher.addURI(AUTHORITY, "student/* ", 4);
11
12   }
```

此处在静态模块中初始化 UriMatcher,并向其中添加匹配规则。例如,上述第 7 行代码表示如果需要验证的 Uri 为 "content://cn.edu.nnutc.ie.StudentProvider/student",那么匹配结果为 "1";如果需要验证的 Uri 为 "content://cn.edu.nnutc.ie.StudentProvider/student/1",那么匹配结果为 "2";如果需要验证的 Uri 为 "content://cn.edu.nnutc.ie.StudentProvider/student/saddress/江苏",那么匹配结果为 "3";如果需要验证的 Uri 为 "content://cn.edu.nnutc.ie.StudentProvider/student/zhangsan",那么匹配结果为 "4"。

(2) 重写 onCreate() 方法。

应用程序一旦启动,这个方法就会调用执行,所以通常将打开(创建)数据库的功能放在这个方法中,代码如下。

```
1    @Override
2    public boolean onCreate() {
3        mTable = DBOpenHelper.TABLE_NAME;
4        mContext = this.getContext();
5        mDatabase = new DBOpenHelper(mContext).getWritableDatabase();//打开
数据库
6        return true;
7    }
```

(3) 重写 insert()、delete()、update() 和 query() 等方法实现功能。

- 重写 insert() 方法,代码如下。

```
1    @Override
2    public Uri insert(Uri uri, ContentValues values) {
3        int code = mUriMatcher.match(uri);
```

```
4       Uri result = null;
5       switch (code) {
6         case 1://插入记录
7           long index = mDatabase.insert(DBOpenHelper.TABLE_NAME, null, values);
8           if (index > 0)//插入成功
9           {
10            //在前面已有的 Uri 后面追加 id
11            result = ContentUris.withAppendedId(uri, index);
12            //通知数据已经发生改变
13            getContext().getContentResolver().notifyChange(result, null);
14            return result;
15          }
16          break;
17      }
18      return result;
19    }
```

执行上述第 7 行代码时,如果外部应用程序插入记录成功,会返回一个大于 0 的 long 型值,否则表示插入不成功。

- 重写 delete( )方法,代码如下。

```
1     @Override
2     public int delete(Uri uri, String selection, String[] selectionArgs) {
3       int code = mUriMatcher.match(uri);
4       int result = 0;
5       switch (code) {
6         case UriMatcher.NO_MATCH:
7           break;
8         case 1: //删除表中所有记录
9           result = mDatabase.delete(DBOpenHelper.TABLE_NAME, null, null);
10          break;
11        case 2:
12          // content://cn.edu.nnutc.ie.StudentProvider/student/10
13          // 按 id 条件删除,使用 ContentUris 取出传入 Uri 中的 id 值
14          result = mDatabase.delete(DBOpenHelper.TABLE_NAME, "id=?",
new String[]{ContentUris.parseId(uri) + ""});
15          break;
16        case 3:
17          //content://cn.edu.nnutc.ie.StudentProvider/student/zhangsan
18          //uri.getPathSegments()获取到一个 List<String>,针对 17 行的 Uri,
List 中的值分别是 0 索引对应"student"值、1 索引对应"zhangsan"值
19          List<String> list
20          result = mDatabase.delete(DBOpenHelper.TABLE_NAME, "sname=?",
new String[]{uri.getPathSegments().get(1)});
21        }
22        return result;
23    }
```

上述第 14 行代码表示删除表中 id = 10 的记录;第 20 行代码表示删除表中 sname 为 "zhangsan" 的记录。

- 重写 update( )方法，代码如下。

```
1       @Override
2       public int update(Uri uri, ContentValues values, String selection,
String[] selectionArgs) {
3           int code = mUriMatcher.match(uri);
4           int result = 0;
5           switch (code) {
6               case 1://更新表中所有记录
7                   result = mDatabase.update(DBOpenHelper.TABLE_NAME,
values, selection, selectionArgs);
8                   break;
9               case 2://根据 Uri 中指定的 id 更新记录
10                  result = mDatabase.update(DBOpenHelper.TABLE_NAME,
values, "id = ?", new String[]{ContentUris.parseId(uri) + ""});
11                  break;
12              case 3://根据 Uri 中指定字段的值更新记录
13                  result = mDatabase.update(DBOpenHelper.TABLE_NAME, values,
14                      uri.getPathSegments().get(1) + " = ?",
15                      new String[]{uri.getPathSegments().get(2)});
16                  break;
17          }
18          return result;
19      }
```

上述第 13～15 行代码表示根据 Uri 中指定的字段及字段值更新记录。例如，Uri 为 "content://cn.edu.nnutc.ie.StudentProvider/student/ssex/男"，则更新 student 表中 ssex 字段值为"男"的记录；Uri 为 "content://cn.edu.nnutc.ie.StudentProvider/student/saddress/江苏"，则更新 student 表中 saddress 字段值为"江苏"的记录。其中 uri.getPathSegments( ) 方法用于返回 Uri 中的路径，返回值为一个 List<String> 类型的数据。此处的 List 中有三个元素，即 0 索引对应 student，1 索引对应 ssex 或 saddress，2 索引对应"男"或"江苏"。对于不同的应用需求，读者可以根据需要修改这些代码来实现。

- 重写 query( )方法，代码如下。

```
1       @Override
2       public Cursor query(Uri uri, String[] projection, String selection,
String[] selectionArgs,String sortOrder) {
3           int code = mUriMatcher.match(uri);
4           Cursor result = null;
5           switch (code) {
6               case 1://查询所有记录
7                   result = mDatabase.query(DBOpenHelper.TABLE_NAME,
8                       projection, selection, selectionArgs,
9                       null, null, sortOrder);
10                  break;
11              case 2://根据 Uri 中指定 id 查询记录
12                  result = mDatabase.query(DBOpenHelper.TABLE_NAME,
13                      projection, "id = ?",
14                      new String[]{ContentUris.parseId(uri) + ""},
15                      null, null, sortOrder);
```

第7章 数据存储与访问

```
16                    break;
17            case 3://根据 Uri 中指定的字段名查询记录
18                    result = mDatabase.query(DBOpenHelper.TABLE_NAME,
19                        projection, uri.getPathSegments().get(1) + " = ?",
20                        new String[]{uri.getPathSegments().get(2)},
21                        null, null, sortOrder);
22                    break;
23            }
24            return result;
25        }
```

3. 注册 ContentProvider——StudentProvider

ContentProvider 是 Android 的四大组件之一，需要在 ProviderApp 的配置文件中进行注册，即需要在 AndroidManifest.xml 的 <application> 节点中增加如下代码。

```
1   <provider
2       android:name = ".StudentProvider"
3       android:exported = "true"
4       android:authorities = "cn.edu.nnutc.ie.StudentProvider" >
5   </provider>
```

为了保证外部应用程序能够操作 ContentProvider 提供的数据，在注册 ContentProvider 时有以下两点需要读者注意。

（1）authorities 属性值一定要与 UriMatcher 中定义的值完全一样。

（2）exported 属性值一定为 true，表示允许其他应用程序拥有操作权限。

4. 通过 ContentResolver 和 Uri 使用 ContentProvider

创建 ProviderUseApp 并设计如图 7.16 和图 7.17 所示界面，用于在界面上输入要插入的学生信息（姓名、性别、家庭住址和联系电话）、输入要查询的学生的 id 及输入要查询的字段和字段值。

【通过ContentResolver和Uri使用Content Provider】

图 7.16　向 ContentProvider 插入数据　　图 7.17　查询 id 为 1 的学生信息

界面设计比较简单，限于篇幅不再赘述，读者可以参见 FirstAPP 代码包中 provideruseapp 文件夹里的内容（main_layout.xml）。

- 插入学生信息

```
1    btnInsert.setOnClickListener(new View.OnClickListener() {
2        @Override
3        public void onClick(View v) {
4            String sname = edtIsname.getText().toString();//
5            String ssex = edtIssex.getText().toString();
6            String saddress = edtIsaddress.getText().toString();
7            String stel = edtIstel.getText().toString();
8            ContentValues values = new ContentValues();
9            values.put("sname",sname);
10           values.put("saddress",saddress);
11           values.put("ssex",ssex);
12           values.put("stel",stel);
13           Uri uri = Uri.parse("content://cn.edu.nnutc.ie.StudentProvider/student");
14           ContentResolver contentResolver = this.getContentResolver();
15           Uri result = contentResolver.insert(uri,values);
16           if(uri!=null){
17               Toast.makeText(MainActivity.this,
18                   result.toString()+"插入成功!",Toast.LENGTH_LONG).show();
19           }else {
20               Toast.makeText(MainActivity.this,"插入不成功!",
21                   Toast.LENGTH_LONG).show();
22           }
23       }
24   });
```

上述第 4~7 行代码表示从图 7.16 所示界面的姓名、性别、地址和电话的 EditText 中获得数据；第 8~12 行代码表示将数据封装到 ContentValue 中；第 13~15 行代码表示获得 ContentResolver 对象，使用 Uri 调用 ProviderApp 中暴露的 insert() 方法并向其插入数据。运行效果如图 7.16 所示。

- 根据学生的 id 查询学生信息

```
1    btnQID.setOnClickListener(new View.OnClickListener() {
2        @Override
3        public void onClick(View v) {
4            String id = edtMid.getText().toString();
5            Uri uri = Uri.parse("content://cn.edu.nnutc.ie.StudentProvider/student"+"/"+id);
6            ContentResolver contentResolver = this.getContentResolver();
7            Cursor cursor = contentResolver.query(uri,
8                new String[]{"sname", "ssex", "saddress", "stel"}, null, null, null);
9            while (cursor.moveToNext()){
10               String sname = cursor.getString(0);
11               String ssex = cursor.getString(1);
```

```
12          String saddress = cursor.getString(2);
13          String stel = cursor.getString(3);
14              tvInfo.setText(sname + "," + ssex + "," + saddress + "," +
stel + "\n");
15        }
16      }
17    });
```

上述代码表示根据 edtMid 中输入的 id 对应查询学生信息，并将查询结果显示在 tvInfo 上。运行效果如图 7.17 所示。

- 根据学生的 id 修改联系电话

```
1  btnMID.setOnClickListener(new View.OnClickListener() {
2    @Override
3    public void onClick(View v) {
4      String id = edtMid.getText().toString();
5      Uri uri = Uri.parse("content://cn.edu.nnutc.ie.StudentProvider/student"
+"/"+id);
6      ContentResolver contentResolver = this.getContentResolver();
7      ContentValues values = new ContentValues();
8      values.put("stel","12345");
9      contentResolver.update(uri,values,null,null);
10    }
11 });
```

上述代码表示将输入的 id 对应学生的联系电话修改为"12345"。

- 根据指定的字段修改

```
1  btnMFIELD.setOnClickListener(new View.OnClickListener() {
2    @Override
3    public void onClick(View v) {
4      String fieldname = edtMfieldname.getText().toString();
5      String fieldvalue = edtMfieldvalue.getText().toString();
6      Uri uri =
Uri.parse("content://cn.edu.nnutc.ie.StudentProvider/student" + "/" + fieldname
+"/"+fieldvalue);
7      ContentValues values = new ContentValues();
8      values.put("stel","118");
9      int result = contentResolver.update(uri,values,null,null);
10    }
11 });
```

上述代码中，如果在 edtMfieldname 中输入"saddress"、edtMfieldvalue 中输入"江苏"，则表示将 student 表中家庭住址为江苏的学生的联系电话修改为"118"。

至此，ProviderApp 提供学生信息共享数据和 ProviderUserApp 使用共享数据已全部介绍结束，两个应用程序的源代码目录结构分别如图 7.18 和图 7.19 所示。

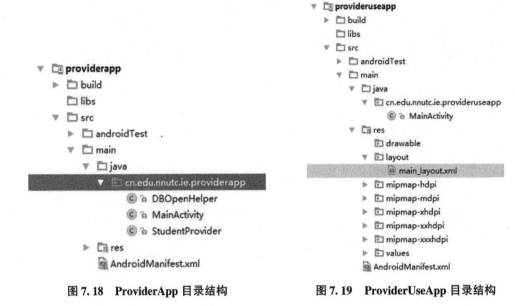

图 7.18　ProviderApp 目录结构　　　　图 7.19　ProviderUseApp 目录结构

7.4.3　使用 Android 提供的 ContentProvider

1. 短信

（1）读短信。

Android 向其他应用程序提供访问短信的 Uri 为" content://sms/"，所以读出 Android 中所有短信的代码如下。

```
1    Uri uri = Uri.parse("content://sms/");
2    ContentResolver resolver = getContentResolver();
3    Cursor cursor = resolver.query(uri, new
String[]{"address","date","type","body"}, null, null, null);
4    while(cursor.moveToNext()){
5        String address = cursor.getString(0);
6        SimpleDateFormat sdf = new SimpleDateFormat("yyyy-MM-dd HH:mm:ss");
7        String date = sdf.format( new
Date(Long.parseLong(cursor.getString(1))));
8        String type = cursor.getString(2).equals("1")?"收到":"发出";
9        String body = cursor.getString(3);
10       String result = "号码:" + address +","+"时间:" + date +","+"类型:
" + type +","+"内容:" + body + "\n";
11       edtRead.setText( edtRead.getText().toString() +result);
12   }
13   cursor.close();
```

上述第 3 行代码中的 address、date、type、body 分别对应短信中的电话号码、短信时间、短信类型（1—收到短信，2—发布的短信）、短信内容；第 11 行代码表示将读出的全部短信内容显示在 edtRead 上。

另外，需要在配置文件中添加读短信权限，即

```
    <uses-permission android:name ="android.permission.READ_SMS" > </uses-per-
mission >
```

(2) 写短信。

```
1    ContentResolver resolver = getContentResolver();
2    Uri uri = Uri.parse("content://sms/");
3    ContentValues conValues = new ContentValues();
4    conValues.put("address", "123456789");
5    conValues.put("type", 1);
6    conValues.put("date", System.currentTimeMillis());
7    conValues.put("body", edtInsert.getText().toString());
8    resolver.insert(uri, conValues);
9    Toast.makeText(MainActivity.this,"短信插入完毕 ~ ",Toast.LENGTH_LONG).show();
```

从上述代码可以看出,这种方法可以将指定号码、内容等的信息插入短信库中,这样给伪造短信号码和短信内容提供了方便。为了避免第三方应用发送虚假短信并保证用户的信息安全,自 Android 5.0 开始已经不能直接使用这个方法实现写短信了。

2. 联系人

在 Android 的通讯录中、联系人信息全部都存放在系统的数据库中,如果需要获得通讯录里联系人的信息就需要访问系统的数据库,通讯录数据库文件 contacts2.db 存放在/data/data/ com.android.providers.contacts 目录下的 databases 文件夹中,如图 7.20 所示。contacts2.db 中包含很多表,其中有四个关键表的表名及相关说明见表 7 - 24。

图 7.20 通讯录数据库

表 7 - 24 contacts2.db 中的表的表名及相关说明

表 名	相 关 说 明
contacts	包含头像的 ID、与联系人通话的次数、最后通话的时间等信息
data	包含联系人的姓名、电话号码、电子邮件、地址等信息
raw_contact	包含联系人姓名、删除标志、最后联系时间等
phone_lookup	包含 Data 表的 ID、raw_contact 表的 ID、电话号码的逆序等

为了让开发者访问 Android 中的通讯录数据,工程师编写了系统的 ContentProvider,开发者只要通过指定的 Uri 就可以对 contacts2.db 中的数据进行访问。访问 contacts2.db 的 URI 为 Phone.CONTENT_URI,Phone 类在 android.provider.ContactsContract.CommonDataKinds 包中,获取联系人的时候需要通过这个 Uri 去访问数据。它所指向的 Uri 就是 "content://com.android.contacts/data/phones"。

(1) 根据电话号码查询联系人姓名。

```
1    String phone = "110";
2    //uri=  content://com.android.contacts/data/phones/filter/#
3    Uri uri =
Uri.parse("content://com.android.contacts/data/phones/filter/" +phone);
```

```
    4    ContentResolver resolver = this.getContext().getContentResolver();
    5    Cursor cursor = resolver.query(uri, new String[]{Data.DISPLAY_NAME},
null,null,null);
    6    if(cursor.moveToFirst()){
    7        edtName.setText("指定电话号码的联系人为:"+cursor.getString(0)));
    8    }
```

上述第 5 行代码表示从 raw_contact 表中返回 display_name 列，该列对应的就是联系人姓名。

（2）查询所有联系人姓名和号码。

```
    1    ContentResolver resolver = getContentResolver();
    2    Uri uri = ContactsContract.CommonDataKinds.Phone.CONTENT_URI;
    3    //查询联系人数据
    4    cursor = resolver.query(uri, null, null, null, null);
    5    while(cursor.moveToNext()){
    6    String cName =
cursor.getString(cursor.getColumnIndex(ContactsContract.CommonDataKinds.Phone.
DISPLAY_NAME));
    7        String cNum =
cursor.getString(cursor.getColumnIndex(ContactsContract.CommonDataKinds.Phone.
NUMBER));
    8        String result = "姓名:" + cName + "," +"号码:" + cNum + "\n";
    9        editAll.setText(editAll.getText().toString() + result);
    10   }
    11   cursor.close();
```

上述第 6 行代码表示获取联系人姓名，第 7 行代码表示获取联系人号码；第 9 行代码表示将查询的所有联系人姓名和号码显示在 editAll 上。

（3）插入联系人信息。

```
    1    Uri uri = Uri.parse("content://com.android.contacts/raw_contacts");
    2    ContentResolver resolver = this.getContext().getContentResolver();
    3    ContentValues values = new ContentValues();
    4    //插入 raw_contacts 表,并获取_id 属性
    5    long contact_id = ContentUris.parseId(resolver.insert(uri, values));
    6    uri = Uri.parse("content://com.android.contacts/data");
    7    //姓名
    8    values.put("raw_contact_id", contact_id);
    9    values.put(Data.MIMETYPE,"vnd.android.cursor.item/name");
    10   values.put("data2", "小新");
    11   values.put("data1", "徐小新");
    12   resolver.insert(uri, values); //插入 data 表
    13   values.clear();
    14   //电话
    15   values.put("raw_contact_id", contact_id);
    16   values.put(Data.MIMETYPE,"vnd.android.cursor.item/phone_v2");
    17   values.put("data2", "2");
    18   values.put("data1", "02587654321");
    19   resolver.insert(uri, values);
```

```
20    values.clear();
21    //邮箱
22    values.put("raw_contact_id", contact_id);
23    values.put(Data.MIMETYPE,"vnd.android.cursor.item/email_v2");
24    values.put("data2", "2");
25    values.put("data1", "tznkf@qq.com");
26    resolver.insert(uri, values);
```

（4）删除指定姓名的联系人信息。

```
1    String name = "徐小新";
2    //根据姓名求 id
3    Uri uri = Uri.parse("content://com.android.contacts/raw_contacts");
4    ContentResolver resolver = this.getContext().getContentResolver();
5    Cursor cursor = resolver.query(uri, new
String[]{Data._ID},"display_name=?", new String[]{name}, null);
6    if(cursor.moveToFirst()){
7        int id = cursor.getInt(0);
8        //根据 id 删除 data 中的相应数据
9        resolver.delete(uri, "display_name=?", new String[]{name});
10       uri = Uri.parse("content://com.android.contacts/data");
11       resolver.delete(uri, "raw_contact_id=?", new String[]{id+""});
12   }
```

（5）更新指定 id 的联系人电话号码。

```
1    int id = 1;
2    String phone = "999999";
3    //对 data 表的所有数据操作
4    Uri uri = Uri.parse("content://com.android.contacts/data");
5    ContentResolver resolver = this.getContext().getContentResolver();
6    ContentValues values = new ContentValues();
7    values.put("data1", phone);
8    resolver.update(uri, values, "mimetype=? and raw_contact_id=?", new
String[]{"vnd.android.cursor.item/phone_v2",id+""})
```

本 章 小 结

本章详细介绍了 Android 中数据存储的技术，包括 SharedPreferences、内部文件存储、SD 卡文件存储、SQLite 数据库存储和 ContentProvider 数据共享机制等，读者通过对本章的学习能够掌握基本的数据存储相关的知识，对编写一些数据密集型的软件很有帮助。

习　　题

一、选择题

1. 如果某个文件或文件夹的权限为 drw-rw----，则下面说法正确的是（　　）。

A. 这是一个普通的文件而不是文件夹

B. 创建者具有读文件、写文件及执行文件的功能

C. 同组人只具有读文件和写文件的功能
D. 其他用户只具有读文件的功能

2. Android 模拟器中的 SD 卡文件夹在（　　）。
   A. /data/data/包名/files/　　　　　　B. /mnt/
   C. /data/data/包名/shared_prefs/　　D. /mnt/包名/

3. 手动创建数据库时，如果创建一个 people.db 的数据库，需要执行命令（　　）。
   A. create people.db　　　　　　　　B. sqlite people.db
   C. sqlite3 people.db　　　　　　　　D. create databse people.db

4. 在手机开发中常用的数据库是（　　）。
   A. sqlLite　　　B. Oracle　　　C. Sql Server　　　D. Db23

5. 在多个应用中读取共享存储数据时，需要用到的 query 方法，是（　　）对象的方法。
   A. ContentResolver　　　　　　　　B. ContentProvider
   C. Cursor　　　　　　　　　　　　D. SQLiteHelper

6. SQLiteDatabase 类中的 query( )方法用于从数据库表中查询数据信息，该方法执行查询后返回值的类型为（　　）。
   A. Table　　　　　　　　　　　　　B. ContentValues
   C. Cursor　　　　　　　　　　　　D. SQLiteHelper

7. 下列关于 ContentProvider 组件说法中不正确的是（　　）。
   A. ContentProvider 是四大组件之一，需要在配置文件中注册
   B. ContentProvider 用于应用程序之间共享数据，其共享的数据可以是数据库、文件或网络数据
   C. 访问 ContentProvider 共享的数据时，需要创建 ContentResolver 对象
   D. ContentProvider 只能共享自定义应用程序的数据，不能共享 Android 内部应用程序的数据

8. Google 为 Andriod 的较大的数据处理提供了 SQLite，它在数据存储、管理、维护等各方面都相当出色，功能也非常的强大。下列关于 SQLite 数据库的描述不正确的是（　　）。
   A. SQLite 是轻量级的，使用 SQLite 只需要带一个动态库，就可以享受它的全部功能，而且那个动态库的尺寸相当小
   B. SQLite 是独立性的，它的核心引擎不需要依赖第三方软件，但用时也需要用户单独安装
   C. SQLite 是隔离性的，数据库中所有的信息（如表、视图、触发器等）都包含在一个文件夹内，方便管理和维护
   D. SQLite 是跨平台的，目前支持大部分操作系统，在 PC 机和手机平台上都可以运行

9. 下列对 SharePreferences 存、取文件的说法中不正确的是（　　）。
   A. 属于移动存储解决方案
   B. sharePreferences 处理的就是 key-value 对
   C. 读取 XML 文件的路径是/sdcard/shared_prefs
   D. 信息的保存格式是 xml

10. 下列关于 Sqlite 数据库的说法不正确的是（　　）。

A. SqliteOpenHelper 类主要是用来创建数据库和更新数据库的

B. SqliteDatabase 类是用来操作数据库的

C. 在每次调用 SqliteDatabase 的 getWritableDatabase( ) 方法时，都会执行 SqliteOpenHelper 的 onCreate 方法

D. 当数据库版本发生变化时，可以自动更新数据库结构

11. 下列关于 ContentValues 类的说法正确的是（　　）。

A. 它和 Hashtable 比较类似，也是负责存储一些键值对，但是它存储的键值对当中的名是 String 类型，而值都是基本类型

B. 它和 Hashtable 比较类似，也是负责存储一些键值对，但是它存储的键值对当中的名是任意类型，而值都是基本类型

C. 它和 Hashtable 比较类似，也是负责存储一些键值对，但是它存储的键值对当中的名可以为空，而值都是 String 类型

D. 它和 Hashtable 比较类似，也是负责存储一些键值对，但是它存储的键值对当中的名是 String 类型，而值也是 String 类型

12. 数据源如果为 SQLite 数据库中查出的信息，最适合的适配器为（　　）。

A. SimpleAdapter　　　　　　　　B. SimpleCursorAdapter

C. ArrayAdapter　　　　　　　　　D. ListAdapter

二、填空题

1. SharedPreferences 支持三种访问模式：_____，MODE_WORLD_READABLE，MODE_WORLD_WRITEABLE。如果需要设置其访问模式为全局可读可写应该设置为_____。

2. 启动 Linux 的命名行界面的方法是在 CMD 中输入_____命令。

3. 使用 openFileOutput( ) 写内部文件时，需要设置文件的访问模式，如果向文件中追加内容，则要将访问模式设置为_____。

4. 如果某个文件或文件夹的权限为 drw-rw--wx，其他用户具备_____权限。

5. Android 提供了读写内部文件的方法：openFileInput( ) 和_____，该方法的返回值类型是_____。

6. Android 提供了一个帮助访问数据库的辅助类_____，该类可以获得一个可读的数据库或可写的数据库。

7. 如果手动创建一个数据库 person.db，需要使用_____命令。

8. Android 应用开发中可以使用 SQLite、ContentProvider、File 和_____四种存储方式。

9. 使用 SharedPreferences 保存数据时，保存文件的位置是_____。

10. 如果一个应用的包名为"com.myapp"，那个该应用程序写的数据库文件在_____目录中。

三、判断题

1. 使用 SharedPreferences 只能写入 name/value 这样的键值对，不可以写入任意内容的文件。　　　　　　　　　　　　　　　　　　　　　　　　　　　　（　　）

2. 使用OpenFileOutput( )函数写文件时，需要指定文件的名字和文件的访问方式，文件的名字可以包含路径，如D：\ file. txt。   （   ）

3. Android程序中的数据库文件保存在/data/data/包名/databases目录中。   （   ）

4. 写SD卡文件时只需要给模拟器设置一个模拟SD卡，不需要加入任何权限设置。
                                                                      （   ）

5. 写Android内部文件时，默认路径就是data/data/<包名>/files，因此不需要加入任何文件的路径。   （   ）

6. 某个文件夹的权限可以为-rw-rw---x。   （   ）

7. Android应用程序创建的数据库文件在/data/data/<包名>/databases目录下，不管是代码创建数据库还手动创建数据库，Android都会自动创建该目录。   （   ）

8. Android不仅有自己的读写文件方法，也可以使用Java中的文件读写方法。（   ）

【第7章参考答案】

第 8 章 多媒体应用开发

在移动终端迅速发展的今天,一个明显的趋势是它们提供的多媒体功能与网络功能不断增强。例如,在手机 App 中,图像、视频、声音、移动上网和微博等早已广泛应用。随着技术的日益更新,越来越多的移动终端设备拥有更为专业的视频播放和网络性能。用户经常使用手机来拍摄和浏览照片,录制声音和观看视频,上网聊天和浏览网络信息等,那么用户实现这些操作的应用程序是怎么开发的呢?本章将详细介绍它们的开发过程和实现方法。

教学目标

理解 Android 中多媒体组件的体系结构和原理。

掌握 Android 中 MediaPlayer、MediaRecorder、VideoView、Camera 等多媒体类的常用方法。

掌握使用 Android 中的多媒体类开发多媒体应用软件的方法。

掌握 Intent 启动 Activity 获取返回值的方法。

教学要求

知识要点	能力要求	相关知识
概述	(1) 了解 Android 多媒体框架的核心 (2) 熟悉 OpenCore 中包含的常用多媒体类	Android 系统框架
音乐播放器的设计与实现	(1) 掌握 MediaPlayer 类、AudioManager 类的使用方法 (2) 掌握 SeekBar 的使用方法和应用场景 (3) 掌握读取 SD 卡中的指定类型文件的方法	ListView
视频播放器的设计与实现	掌握使用 MediaPlayer、VideoView 实现视频播放的方法	Gallery
录音机的设计与实现	掌握使用 MediaRecorder 实现声音录制的方法	文件操作
照相机的设计与实现	(1) 掌握 Camera 类的常用方法 (2) 掌握系统相机的调用和自定义相机的开发方法	Intent

8.1 概　　述

Android 软件开发包提供了一系列的方法来处理音频及视频媒体，包括对于多种媒体类型和格式的支持。Android 的多媒体框架核心（Multimedia Framework）是 OpenCore，也称 PacketVideo，它兼顾了跨平台的移植性，并且也有较好的稳定性。与其他 Android 程序库相比，OpenCore 的代码非常庞大，需要耗费相当多的时间对其进行维护。

OpenCore 是一个基于 C++的实现，定义了全功能的操作系统移植层，各种基本的功能均被封装成类的形式，各层次之间的接口多使用继承等方式。程序员可以通过 OpenCore 方便迅速地开发出想要的多媒体应用程序，如录音、播放、回放、视频会议、流媒体播放等。

（1）OpenCore 是一个多媒体的框架，从宏观上来看，它主要包含了以下两大方面的内容。

● PVPlayer：提供媒体播放器的功能，实现各种音频（Audio）流和视频（Video）流的回放（Playback）。

● PVAuthor：提供媒体流记录的功能，实现各种音频（Audio）流、视频（Video）流及静态图像的捕获。

PVPlayer 和 PVAuthor 以 SDK 的形式提供给开发者，开发者可以在这个 SDK 之上构建多种应用程序和服务。

（2）OpenCore 主要提供了如下几个多媒体类。

● MediaPlayer 类：可以用于播放音频、视频和流媒体，包含了 Audio 和 Video 的播放功能。在 Android 的界面上，音频和视频的播放都是调用 MediaPlayer 实现的。MdeiaPlayer 类可以获得媒体文件和各种属性当前的播放状态，并可以开始和停止文件的播放。

● MediaRecorder 类：用来进行媒体采样，包括音频和视频。MediaRecorder 作为状态机运行，需要设置不同的参数，如源格式和源设备，设置后可以执行任意长度的录制，直到用户停止。

● VideoView 类：主要用来显示一个视频文件，是 SurfaceView 类的一个子类，且实现了 MediaControl 接口。

● Camera 类：用来处理系统中与相机相关的事件，Camera 类是一种专门用来连接和断开相机服务的类。

8.2 音乐播放器的设计与实现

【MediaPlayer、AudioManager、SeekBar】

音乐播放器对于所有读者来说并不陌生，使用者通常可以使用音乐播放器实现音乐文件的播放、暂停、下一首、上一首、播放列表、进度条和音量控制等功能。本节将通过普通音乐播放器的实现介绍 Android 中与声音有关的组件的使用方法。

8.2.1 预备知识

1. MediaPlayer

Android SDK 提供了 MediaPlayer 类，以便在 Android 中实现多媒体服务，如音频、视

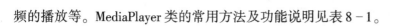

频的播放等。MediaPlayer 类的常用方法及功能说明见表 8-1。

表 8-1 MediaPlayer 类的常用方法及功能说明

方 法 名	功 能 说 明
MediaPlayer( )	构造方法
create(Context context, Uri uri)	通过 Uri 创建一个多媒体播放器
create(Context context, int resid)	通过资源 ID 创建一个多媒体播放器
create(Context context, Uri uri, SurfaceHolder holder)	通过 Uri 和指定 SurfaceHolder(抽象类)创建一个多媒体播放器
getCurrentPosition( )	返回当前播放位置
getDuration( )	返回文件播放总时间
getVideoHeight( )	返回视频的高度
getVideoWidth( )	返回视频的宽度
isLooping( )	是否循环播放
isPlaying( )	是否正在播放
pause( )	暂停
prepare( )	准备同步
prepareAsync( )	准备异步
release( )	释放 MediaPlayer 对象
reset( )	重置 MediaPlayer 对象
seekTo(int msec)	指定播放的位置(以毫秒为单位的时间)
setAudioStreamType(int streamtype)	指定流媒体的类型
setDataSource(String path)	设置多媒体数据来源(根据路径)
setDataSource(FileDescriptor fd, long offset, long length)	设置多媒体数据来源(根据 FileDescriptor)
setDataSource(FileDescriptor fd)	设置多媒体数据来源(根据 FileDescriptor)
setDataSource(Context context, Uri uri)	设置多媒体数据来源(根据 Uri)
setDisplay(SurfaceHolder sh)	设置用 SurfaceHolder 来显示多媒体
setLooping(boolean looping)	设置是否循环播放
setScreenOnWhilePlaying(boolean screenOn)	设置是否使用 SurfaceHolder 显示
setVolume(float leftVolume, float rightVolume)	设置音量
start( )	开始播放
stop( )	停止播放

MediaPlayer 类的常用事件及功能说明见表 8-2。

表 8-2 MediaPlayer 类的常用事件及功能说明

事 件 名	功 能 说 明
setOnBufferingUpdateListener(MediaPlayer. OnBufferingUpdateListener listener)	缓冲监听
setOnPreparedListener(MediaPlayer. OnPreparedListener listener)	就绪监听
setOnCompletionListener(MediaPlayer. OnCompletionListener listener)	播放结束监听
setOnErrorListener(MediaPlayer. OnErrorListener listener)	出错错误监听
setOnVideoSizeChangedListener(MediaPlayer. OnVideoSizeChangedListener listener)	视频尺寸监听

实际应用开发时，可以使用如下步骤实现。

（1）可以直接用 new 方法或者调用 create 方法创建 MediaPlayer 实例，代码如下。

MediaPlayer mp = new MediaPlayer();

MediaPlayer mp = MediaPlayer.create(this, R.raw.test);   //无需再调用 setDataSource

另外，create()方法还有这样的形式：create (Context context, Uri uri, SurfaceHolder holder)，即可以通过 Uri 和指定 SurfaceHolder 创建一个多媒体播放器，这将在下一节介绍。

（2）设置播放文件。

- raw 下的资源

MediaPlayer.create(this, R.raw.test);

- 本地文件路径

mp.setDataSource("/sdcard/test.mp3");

- 网络 URL 文件

mp.setDataSource("http://www.xxx.com/music/test.mp3");

另外 setDataSource()方法有多个，里面有这样一个类型的参数：FileDescriptor，在使用这个 API 的时候，需要把文件放到与 res 文件夹平级的 assets 文件夹里，然后使用下述代码设置 DataSource。

```
AssetFileDescriptor fileDescriptor = getAssets().openFd("rain.mp3");
mp.setDataSource(fileDescriptor.getFileDescriptor(),fileDescriptor.getStartOffset(), fileDescriptor.getLength());
```

（3）准备及播放。

```
mp.prepare();
mp.start();
```

如果播放的是 res/raw 目录下的音频文件，创建 MediaPlayer 调用的是 create()方法，则第 1 次启动播放前不需要再调用 prepare()方法；如果是使用 MediaPlayer()构造方法构造的 MediaPlayer 对象，则需要调用一次 prepare()方法。

2. AudioManager

（1）AudioManager 类位于 Android.Media 包下，提供了音量控制与铃声模式相关操作。该类的常用方法及功能说明如下。

① adjustVolume(int direction, int flags)：控制手机音量，调大或者调小一个单位，根据第 1 个参数进行判断。AudioManager.ADJUST_LOWER 可，调小一个单位；AudioManager.ADJUST_RAISE，可调大一个单位。

② adjustStreamVolume(int streamType, int direction, int flags)：同上，不过可以选择调节的声音类型。

第 1 个参数（streamType）用于指定声音类型，常见的声音类型如下。

- STREAM_ALARM（手机闹铃）
- STREAM_MUSIC（手机音乐）
- STREAM_RING（电话铃声）
- STREAM_SYSTEAM（手机系统）

- STREAM_DTMF（音调）
- STREAM_NOTIFICATION（系统提示）
- STREAM_VOICE_CALL（语音电话）

第2个参数（direction）和第1个参数含义一样；第3个参数（flags）表示可选的标志位。例如：

- AudioManager.FLAG_SHOW_UI（显示进度条）
- AudioManager.PLAY_SOUND（播放声音）

③ setStreamVolume(int streamType, int index, intflags)：直接设置音量大小。
④ getMode()：返回当前的音频模式。
⑤ setMode()：设置声音模式，有下述几种模式。

- MODE_NORMAL（普通）
- MODE_RINGTONE（铃声）
- MODE_IN_CALL（打电话）
- MODE_IN_COMMUNICATION（通话）

⑥ getRingerMode()：返回当前的铃声模式。
⑦ setRingerMode(int streamType)：设置铃声模式，有下述几种模式。

- RINGER_MODE_NORMAL（普通）
- RINGER_MODE_SILENT（静音）
- RINGER_MODE_VIBRATE（振动）

⑧ getStreamVolume(int streamType)：获得手机的当前音量，最大值为7，最小值为0，当设置为0的时候，会自动调整为振动模式。
⑨ getStreamMaxVolume(int streamType)：获得手机某个声音类型的最大音量值。
⑩ setStreamMute(int streamType, boolean state)：将手机某个声音类型设置为静音。
⑪ setSpeakerphoneOn(boolean on)：设置是否打开扩音器。
⑫ setMicrophoneMute(boolean on)：设置是否让麦克风静音。
⑬ isMicrophoneMute()：判断麦克风是否静音或是否打开。
⑭ isMusicActive()：判断是否有音乐处于活跃状态。
⑮ isWiredHeadsetOn()：判断是否插入了耳机。

（2）实际应用开发时，可以使用如下步骤实现。
① 获得 AudioManager 对象实例。

```
AudioManageraManager = (AudioManager)context.getSystemService(Context.AUDIO
_SERVICE);
```

② 常用示例。

- 指定调节音乐的音频，调高音量，而且显示音量图形示意，代码如下。

```
aManager.adjustStreamVolume(AudioManager.STREAM_MUSIC,
                AudioManager.ADJUST_RAISE,
                AudioManager.FLAG_SHOW_UI);
```

- 指定调节音乐的音频，调低音量，只有声音，不显示图形条，代码如下。

```
aManager.adjustStreamVolume(AudioManager.STREAM_MUSIC,
                            AudioManager.ADJUST_LOWER,
                            AudioManager.FLAG_PLAY_SOUND);
```

- 取消静音，代码如下。

```
//API 23 前版本用
aManager.setStreamMute(AudioManager.STREAM_MUSIC, true);
//API 23 后版本用
Manager.adjustStreamVolume (AudioManager.STREAM_MUSIC,
                           AudioManager.ADJUST_MUTE,
                           AudioManager.FLAG_SHOW_UI);
```

- 设置静音，代码如下。

```
//API 23 前版本用
aManager.setStreamMute(AudioManager.STREAM_MUSIC, false);
//API 23 后版本用
aManager.adjustStreamVolume (AudioManager.STREAM_MUSIC,
                            AudioManager.ADJUST_UNMUTE,
                            AudioManager.FLAG_SHOW_UI);
```

3. SeekBar

SeekBar（拖动条）是 ProgressBar 的子类，除了可以使用 ProgressBar 类包含的属性和方法外，SeekBar 还包含一个可以定制滑块的属性 android：thumb 和一个重要的监听事件 SeekBar. OnSeekBarChangeListener，该事件中有以下三个需要重写的方法。

- onProgressChanged：进度发生改变时会触发。
- onStartTrackingTouch：按住（拖动开始）SeekBar 时会触发。
- onStopTrackingTouch：放开（拖动停止）SeekBar 时触发。

例如，如果要实现如图 8.1 所示效果，可以先在布局文件中使用如下关键代码。

```
1    <SeekBar
2        android:id="@+id/sb1"
3        android:layout_width=" match_parent"
4        android: layout_height =" wrap_content"
5        android: thumb =" @mipmap/search_icon" />
6    <TextView
7        android: id=" @+id/tvinfo"
8        android: layout_width=" match_parent"
9        android: layout_height =" wrap_content"
10       android: gravity =" center_horizontal" />
```

然后在功能模块中使用如下关键代码。

```
1    mContext = this;
2    tvInfo = (TextView) this.findViewById(R. id. tvinfo);
3    sb1 = (SeekBar) this.findViewById(R. id. sb1);
4    sb1.setOnSeekBarChangeListener(new SeekBar.OnSeekBarChangeListener() {
5        @Override
6        public void onProgressChanged(SeekBar seekBar, int progress, boolean fromUser) {
```

```
7              tvInfo.setText("当前进度值:" + progress + " / 100 ");
8          }
9          @Override
10         public void onStartTrackingTouch(SeekBar seekBar) {
11             Toast.makeText(mContext, "触碰 SeekBar", Toast.LENGTH_SHORT).show();
12         }
13         @Override
14         public void onStopTrackingTouch (SeekBar seekBar) {
15             Toast.makeText (mContext, " 放开 SeekBar", Toast.LENGTH_SHORT).show();
16         }
17     });
```

以上代码运行后可以看到 SeekBar 的滑块变成了开发者指定的图片，即在布局文件中将 thumb 属性值设置为保存在 mipmap 文件夹下的 search_icon.png。

如果要实现如图 8.2 所示的定制效果，即默认滑块图标为 sb_normal.jpg，按钮滑块图标为 sb_press.jpg，滑块左边的滑动条为一种颜色，滑块右边的滑动条为另一种颜色，要实现这样的效果，可以按以下步骤操作。

图 8.1 拖动 SeekBar 效果

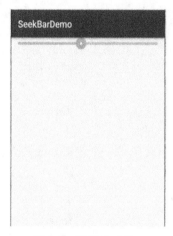

图 8.2 定制 SeekBar

（1）准备按钮滑块的图片和正常滑块的图片，本例中将 sb_normal.jpg 和 sb_press.jpg 图片保存到 mipmap 文件夹下。

（2）在 drawable 目录下创建文件 sb_thumb.xml 文件控制滑块图片，代码如下。

```
1    <?xml version ="1.0" encoding ="utf-8"? >
2    <selector xmlns:android ="http://schemas.android.com/apk/res/android" >
3        <item android:state_pressed =" true" android: drawable =" @mipmap/sb_press" />
4        <item android: state_pressed =" false" android: drawable =" @mipmap/sb_normal" />
5    </selector>
```

上述第 3 行代码表示按钮按下时，滑块加载 sb_press.jpg 图片；第 4 行代码表示按钮不按下时，滑块加载 sb_normal.jpg 图片。

（3）在 drawable 目录下创建文件 sb_bar.xml 文件控制滑动条效果，代码如下。

```
1    <?xml version ="1.0" encoding ="utf-8"? >
2    <layer-list xmlns:android ="http://schemas.android.com/apk/res/android" >
3        <item android:id ="@android:id/background" >
4            <shape > <solid android:color ="#FFFFD042" / > </shape >
5        </item >
6        <item android:id ="@android:id/secondaryProgress" >
7            <clip >
8                <shape > <solid android:color ="#FFFFFFFF" / > </shape >
9            </clip >
10       </item >
11       <item android:id ="@android:id/progress" >
12           <clip >
13               <shape > <solid android:color ="#FF96E85D" / > </shape >
14           </clip >
15       </item >
16   </layer-list >
```

（4）定义 SeekBar 的布局文件，代码如下。

```
1    <SeekBar
2        android:id ="@+id/sb_normal"
3        android: layout_width =" match_parent"
4        android: layout_height =" wrap_content"
5        android: maxHeight =" 5.0dp"
6        android: minHeight =" 5.0dp"
7        android: progressDrawable =" @drawable/sb_bar"
8        android: thumb =" @drawable/sb_thumb" >
9    </SeekBar>
```

上述第 7 行代码用于设置滑动条效果，第 8 行代码用于设置滑块效果。

8.2.2 音乐播放器的实现

1. 界面设计

【音乐播放器的实现】

当音乐播放器打开后，主界面以 ListView 方式显示每首歌曲的图标、歌曲名、歌曲存放位置及歌曲播放时间，运行效果如图 8.3 所示。为了实现这样的效果，本案例中定义了两个布局文件：一个布局文件作为主界面（main_layout.xml），另一个布局文件作为 ListView 列表中每一行显示效果布局（item_message.xml）。

主界面（main_layout.xml）的布局代码如下。

```
1    <?xml version ="1.0" encoding ="utf-8"? >
2    <LinearLayout xmlns:android ="http://schemas.android.com/apk/res/android"
3        android:layout_width =" match_parent"
4        android: layout_height =" match_parent"
5        android: orientation =" vertical" >
6        <ListView
7            android: id =" @+id/lvallname"
8            android: layout_width =" fill_parent"
9            android: layout_height =" fill_parent" / >
10   </LinearLayout >
```

ListView 列表中行显示效果（item_message.xml）的布局代码如下。

```
1   <RelativeLayout xmlns:android="http://schemas.android.com/apk/res/android"
2       xmlns:tools="http://schemas.android.com/tools"
3       android:layout_width=" match_parent"
4       android: layout_height=" wrap_content"
5       android: background=" #d7ced7"
6       android: padding=" 10dp" >
7       <ImageView
8           android: id=" @+id/im_iv_head"
9           android: layout_width=" 60dp"
10          android: layout_height=" 60dp"
11          android: scaleType=" fitXY"
12          android: src=" @mipmap/ic_launcher" />
13      <LinearLayout
14          android: layout_width=" wrap_content"
15          android: layout_height=" wrap_content"
16          android: layout_centerVertical=" true"
17          android: layout_marginLeft=" 10dp"
18          android: layout_toRightOf=" @+id/im_iv_head"
19          android: orientation=" vertical" >
20          <TextView
21              android: id=" @+id/im_tv_name"
22              android: layout_width=" wrap_content"
23              android: layout_height=" wrap_content"
24              android: text=" 歌名" />
25          <TextView
26              android: id=" @+id/im_tv_message"
27              android: layout_width=" wrap_content"
28              android: layout_height=" wrap_content"
29              android: layout_marginTop=" 10dp"
30              android: text=" 歌曲存放位置" />
31      </LinearLayout>
32      <TextView
33          android: id=" @+id/im_tv_time"
34          android: layout_width=" wrap_content"
35          android: layout_height=" wrap_content"
36          android: layout_alignParentRight=" true"
37          android: layout_marginTop=" 5dp"
38          android: text=" 歌曲播放时间" />
39  </RelativeLayout>
```

单击图 8.3 所示界面上的某一行歌曲信息时，打开图 8.4 所示播放控制界面，在这个界面上可以实现"上一首""下一首""播放""暂停""停止""音量+""音量-""顺序播放""随机播放""单曲循环""查看歌词"等操作功能，另外还可以显示正在播放的歌曲名称（用 TextView 实现）和显示播放的进度（用 SeekBar 实现）。界面比较简单，本处不做详述，读者可以参见 ALLAPP 代码包中 playmusic 文件夹里的内容（play_layout.xml）。

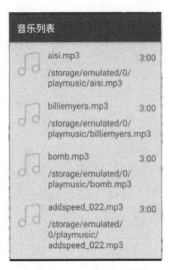

图8.3 音乐列表界面

图8.4 播放控制界面

2. 功能实现

(1) 音乐列表界面相关功能的实现。

本案例实现的音乐播放器需要将歌曲存放在 SD 卡的 playmusic 目录下，所以当应用程序运行时，首先需要判断 SD 卡的 playmusic 目标是否存在。如果 playmusic 目录不存在，就需要创建这个目录，如果 playmusic 目录存在，还要判断是否有音乐文件，并给用户相应提示。如果 playmusic 目录存在，并且在该目录中已经存放了音乐文件，那么就使用 ListView 和 SimpleAdapter 等将图片、歌名、歌曲存放位置和歌曲播放时间等显示在界面上。

- 定义变量。

```
1      protected Context mContext;//当前上下文
2      private String MEDIA_PATH = ""; //SD卡路径
3      private ListView listView;
4      private ArrayList<HashMap<String, Object> > listItems = new Array-
List<HashMap<String, Object> >(); //存放歌曲的相关信息
```

- 判断 SD 卡的 playmusic 目录及音乐文件功能。

```
1      if (Environment.getExternalStorageState().equals(Environment.MEDIA_MOUNTED)) {
2          MEDIA_PATH = Environment.getExternalStorageDirectory().toString();
3      } else {
4          Toast.makeText (mContext," 对不起，没有 SD 卡!", Toast.LENGTH_LONG)
.show();
5          finish();
6          return;
7      }
8      File folder = new File (MEDIA_PATH + " /playmusic/");
9      if (! folder.exists()) {
10         folder.mkdirs();
11         Toast.makeText (mContext, " 对不起，暂时没有音乐文件（SD卡的 playmusic
目录)!", Toast.LENGTH_LONG) .show();
```

```
12        finish();
13        return;
14    }
```

- 给 ListView 装配数据。

```
1   //定义音乐文件扩展名
2   private static final FileFilter filter = new FileFilter() {
3       public boolean accept(File f) {return ! f.isDirectory() && f.getName().matches("^.* ? \\ (mid |mp3 |wma) $");
4       }
5   };
6   //将满足扩展名的文件信息(图片、歌曲名、歌曲存放位置及歌曲播放时间先存入ArrayList)
7   File[] fileArr = folder.listFiles(filter);
8   HashMap <String, Object > map;
9   for (File file : fileArr) {
10      map = new HashMap <String, Object > ();
11      map.put("icon", R.mipmap.bendi);//图片
12      map.put("fileName", file.getName());//歌曲名
13      map.put("filePathName", file.getAbsolutePath());//歌曲存放位置
14      MediaPlayer mediaPlayer = new MediaPlayer();
15      try {
16          mediaPlayer.setDataSource(file.getAbsolutePath());
17          long mTime = mediaPlayer.getDuration();
18          String sTime = String.format("% d:% d",
19              TimeUnit.MILLISECONDS.toMinutes((long) mTime),
20              TimeUnit.MILLISECONDS.toSeconds((long) mTime)
21              - TimeUnit.MINUTES.toSeconds(TimeUnit.MILLISECONDS.
22              toMinutes((long) mTime)));
23          map.put("fileTime", sTime);//歌曲播放时间
24      } catch (IOException e) {
25          e.printStackTrace();
26      }
27      listItems.add(map);
28  }
29  //用 SimpleAdapter 将数据与 ListView 每行的布局组件匹配
30  SimpleAdapter simpleAdapter = new SimpleAdapter(this, listItems, R.layout.item_message, new String [] {" icon", " fileName", " filePathName", " fileTime"}, new int [] {R.id.im_iv_head, R.id.im_tv_name, R.id.im_tv_message, R.id.im_tv_time});
31  listView.setAdapter (simpleAdapter);
```

上述第 14~26 行代码使用 MediaPlayer 对象的 getDuration()方法取得歌曲总时长（以 ms 为单位），然后使用 TimeUnit 类中的方法得出分钟和秒数。

- 绑定 ListView 的单击事件。

单击 ListView 中的某一行内容时，用户界面跳转到图 8.4 所示的播放控制界面，并将当前歌曲的 index 和 playmusic 目录下的所有歌曲信息（保存在 listItems 中）用 Intent 传递给 PlayActivity。

```
1   listView.setOnItemClickListener(new AdapterView.OnItemClickListener() {
2       @Override
```

```
3      public void onItemClick(AdapterView <?> parent, View view,int position, long id) {
4          Intent intent = new Intent(mContext, PlayActivity.class);
5          intent.putExtra("index", position);
6          intent.putExtra("list", listItems);
7          startActivity(intent);
8      }
9  });
```

（2）播放控制界面相关功能的实现。

在单击如图8.3所示的音乐列表转到如图8.4所示的界面时，首先需要获取前一界面由 Intent 传递来的 index 和存放歌曲信息的 ArrayList，并播放 index 对应的歌曲。本案例中由于播放歌曲功能需要在多个地方调用，所以自定义了一个 playMusic（int index）方法。这些功能都需要在 PlayActivity 的 onCreate( )方法中实现。

- 定义播放音乐的 playMusic( )方法。

```
1   private void playMusic(int index) {
2       musicMap = musicList.get(index);//下标为 index 的那首歌
3       tvFileName.setText(musicMap.get("fileName").toString());//显示歌名
4       String path = musicMap.get("filePathName").toString();//歌曲存放位置
5       if (TextUtils.isEmpty(path)) {
6           return;
7       }
8       try {//正在播放,先停止
9           if (mediaPlayer.isPlaying()) {
10              mediaPlayer.stop();
11          }
12          mediaPlayer.release();//释放资源
13          mediaPlayer = null;
14          mediaPlayer = new MediaPlayer();
15          mediaPlayer.reset();//重置
16          mediaPlayer.setDataSource(path);//设置播放源
17          mediaPlayer.prepare();//就绪
18          mediaPlayer.setOnPreparedListener(new SetPreparedListener());//绑定音乐就绪的监听事件
19          mediaPlayer.setOnCompletionListener(new SetCompletionListener());//绑定播放完的监听事件
20          mediaPlayer.setOnErrorListener(new SetErrorListener());//绑定播放出错的监听事件
21      } catch (Exception e) {
22          e.printStackTrace();
23      }
24  }
```

- 自定义类 SetPreparedListener 实现音乐就绪的监听事件。

```
1   private class SetPreparedListener implements MediaPlayer.OnPreparedListener{
2       @Override
3       public void onPrepared(MediaPlayer mp) {
4           isStop = false;
```

```
5                    mediaPlayer.start();//开始播放
6                    new Thread(runnable).start();//启动进度更新线程
7                }
8            }
```

- 自定义类 SetCompletionListener 实现音乐播放完成的监听事件。

当一首歌曲播放完成时，根据用户选择的播放模式（定义一个 int 类型的 playMode，0—随机播放、1—顺序播放、2—单曲循环）实现对应的播放效果。

```
1       private class SetCompletionListener implements MediaPlayer.OnCompletionListener {
2           private Random random;
3           @Override
4           public void onCompletion(MediaPlayer mp) {
5               switch (playMode) {
6                   case 0://随机播放
7                       if (musicList != null) {
8                           if (random == null) {
9                               random = new Random();
10                          }
11                          index = random.nextInt(musicList.size());
12                          playMusic(index);
13                      }
14                      break;
15                  case 1://顺序播放
16                      if (musicList != null) {
17                          index ++;
18                          if (index == musicList.size()) {
19                              index = 0;
20                          }
21                          playMusic(index);
22                      }
23                      break;
24                  case 2://单曲循环
25                      mediaPlayer.seekTo(0);
26                      mediaPlayer.start();
27              }
28          }
29      }
```

上述第 6～14 行代码表示当传入的 playMode 为 0 时，该值由如图 8.4 所示的"随机播放"按钮控制，产生一个随机整数作为播放的下一首歌的 index，然后调用 playMusic( ) 方法；第 15～23 行代码表示当传入的 playMode 为 1 时，该值由如图 8.4 所示的"顺序播放"按钮控制，将刚刚播放完毕歌曲的 index 自增 1 作为 playMusic( ) 方法的参数；第 24～26 行代码表示当传入的 playMode 为 2 时，该值由如图 8.4 所示的"单曲循环"按钮控制，将刚刚播放完的歌曲进度设置为 0，然后再调用 start( ) 方法播放。

- 自定义类 SetErrorListener 实现音乐播放出错的监听事件。

音乐播放时可能会因为各种原因出错，为了在出错时能够及时释放相关的软硬件资源并改善用户体验，可以在这个自定义类中使 MediaPlayer 重新返回到 Idle 状态。

```
1    private class SetErrorListener implements MediaPlayer.OnErrorListener {
2        @Override
3        public boolean onError(MediaPlayer mp, int what, int extra) {
4            mp.reset();
5            try {
6                mp.setDataSource(musicMap.get("filePathName").toString());
7                mp.prepare();
8                mp.setOnPreparedListener(new SetPreparedListener());
9                mp.setOnCompletionListener(new SetCompletionListener());
10               mp.setOnErrorListener(new SetErrorListener());
11           } catch (IOException e) {
12               e.printStackTrace();
13           }
14           return false;
15       }
16   }
```

上述第6行代码表示重新给 MediaPlayer 对象设置数据源，以便实现出错后能够重新播放当前的歌曲。

- 实现"暂停""停止""播放"的监听事件。

```
1            btnStop.setOnClickListener(new View.OnClickListener() {
2                @Override
3                public void onClick(View v) {
4                    mediaPlayer.stop();//停止
5                    isStop = true;
6                }
7            });
8            btnPause.setOnClickListener(new View.OnClickListener() {
9                @Override
10               public void onClick(View v) {
11                   mediaPlayer.pause();//暂停
12                   isStop = true;
13               }
14           });
15           btnPlay.setOnClickListener(new View.OnClickListener() {
16               @Override
17               public void onClick(View v) {
18                   if(mediaPlayer.isPlaying()){
19                       return;
20                   }
21                   mediaPlayer.start();//播放
22                   isStop = false;
23               }
24           });
```

- 实现"上一首""下一首"的监听事件。

```
1    //下一首
2    btnNext.setOnClickListener(new View.OnClickListener() {
3        @Override
```

```
4       public void onClick(View v) {
5           index ++;
6           if (index >= musicList.size()) {
7               index = musicList.size()-1;
8               Toast.makeText(mContext,"已经是最后一首歌了",Toast.LENGTH_LONG).show();
9               return;
10          }
11          playMusic (index);
12      }
13  });
14  //上一首
15  btnPre.setOnClickListener (new View.OnClickListener() {
16      @Override
17      public void onClick (View v) {
18          index--;
19          if (index <0) {
20              index = 0;
21              Toast.makeText (mContext, "已经是第1首歌了", Toast.LENGTH_LONG).show();
22              return;
23          }
24          playMusic (index);
25      }
26  });
```

- 实现"音量+""音量-"的监听事件。

```
1   音量 +
2   btnVolAdd.setOnClickListener(new View.OnClickListener() {
3       @Override
4       public void onClick(View v) {
5           audioManager.adjustStreamVolume(AudioManager.STREAM_MUSIC,
6               AudioManager.ADJUST_RAISE, AudioManager.FLAG_SHOW_UI);;
7       }
8   });
9   音量 -
10  btnVolDec.setOnClickListener (new View.OnClickListener() {
11      @Override
12      public void onClick (View v) {
13          audioManager.adjustStreamVolume (AudioManager.STREAM_MUSIC,
14              AudioManager.ADJUST_LOWER, AudioManager.FLAG_SHOW_UI);
15      }
16  });
```

本案例的音量使用了系统提供的 AudioManager 类中的 adjustStreamVolume ( ) 方法实现，也可以使用 MediaPlayer 类中的 mediaPlayer. setVolume (float，float) 方法实现，读者可以自行编写实现。

- SeekBar 的进度更新实现。

SeekBar 的进度更新与 ProgressBar 类似，也需要使用 Handler 和 Thread 结合起来实现，即在线程中取出当前歌曲的播放进度和歌曲播放时间，分别对应 SeekBar 的当前进度值和进度最大值。为了保证"暂停""停止""播放"控制按钮对进度的控制，需要增加一个 boolean 类型的 isStop 变量，具体实现代码如下。

```
1    Handler handler = new Handler() {
2      @Override
3      public void handleMessage(Message msg) {
4        super.handleMessage(msg);
5        switch (msg.what) {
6          case 1111:
7            seekBar.setProgress(msg.arg2);//根据播放进度修改 seekBar 进度
8            seekBar.setMax(msg.arg1); //设置 seekBar 的最大进度值为歌曲播放时间
9            break;
10          }
11        }
12    };
13    Runnable runnable = new Runnable() {
14      @Override
15      public void run() {
16        while ((true)){
17          try {
18            Thread.sleep(200);
19            if (! isStop) {
20              if (null ! = mediaPlayer && mediaPlayer.isPlaying()) {
21                Message msg = Message.obtain();
22                msg.what = 1111;
23                msg.arg1 = mediaPlayer.getDuration();
24                msg.arg2 = mediaPlayer.getCurrentPosition();
25                handler.sendMessage(msg);
26              }
27            }
28          } catch (Exception e) {
29        }}
30      }
31    };
```

当用户拖动 SeekBar 时，播放位置能够根据拖动停止位置发生改变，此时可以给 SeekBar 设置事件，代码如下。

```
1    seekBar.setOnSeekBarChangeListener(new SeekBar.OnSeekBarChangeListener() {
2      @Override
3      public void onProgressChanged(SeekBar seekBar, int progress, boolean fromUser) {
4        //进度改变触发
5      }
6
7      @Override
8      public void onStartTrackingTouch(SeekBar seekBar) {//拖动开始触发
9
10      }
11      @Override
12      public void onStopTrackingTouch(SeekBar seekBar) {//拖动停止触发
```

```
13              //将播放进度设置为进度条当前位置值
14              mediaPlayer.seekTo(seekBar.getProgress());
15          }
16      });
```

- onCreate()方法实现功能。

前面已经阐述过,该界面一旦启动,就需要获取前面一个界面传递来的相关信息并进行相应处理,主要代码如下。

```
1   //音频管理器,用于调节音量
2   audioManager = (AudioManager) getSystemService(Service.AUDIO_SERVICE);
3   //获取从音乐列表页传递来的 ArrayList
4   musicList = (ArrayList<HashMap<String,Object>>)
getIntent().getSerializableExtra ("list");
5   //获取单击的歌曲对应的 Index
6   index = getIntent().getIntExtra ("index", 0);
7   playMusic (index); //根据歌曲的位置播放歌曲
8   //上面介绍的相应按钮的单击监听事件,此处略
```

另外,由于本案例中需要读取 SD 卡中的音乐文件,所以需要在 AndroidManifest.xml 配置文件中添加如下代码。

```
<uses-permission
android:name="android.permission.WRITE_EXTERNAL_STORAGE" />
```

至此,音乐播放器设计完成,感兴趣的读者可以在本案例的基础上增加快进、快退或歌词显示功能。本案例完整代码读者可以参见 ALLAPP 代码包中 playmusic 文件夹里的内容。

8.3 视频播放器的设计与实现

随着移动互联网的发展,用户在移动终端设备上观看视频已成为常态。本节将介绍基于 Android 平台的视频播放器的设计与实现方法,读者可根据本节介绍的内容设计一款适合自己的视频播放器。

8.3.1 预备知识

基于 Android 平台的视频播放器的实现有以下两种方式。

1. 使用 MediaPlayer

上一节已经介绍过,Android 平台提供了 android.media 包来管理各种音频和视频的媒体接口,该包中的 MediaPlayer 类(媒体播放器接口)用于控制音频文件、视频文件和流的回放。根据该类提供的方法来实现音频播放比较简单,但是由于 MediaPlayer 没有提供图像输出界面,所以需要借助其他组件(如 SurfaceView)来显示 MediaPlayer 播放的图像。

2. 使用 VideoView

为简化 MediaPlayer 播放视频的繁琐控制过程,android.widget 包中还提供了 VideoView(视频视图)组件专门用于播放视频文件。VideoView 组件是调用 MediaPlayer 实现视频播放的,其作用与 ImageView 类似。ImageView 用来显示图片,VideoView 用来播放视频。

Android 自带程序 Gallery 也是用 VideoView 实现的。VideoView 组件的常用方法及功能说明见表 8-3，常用事件及功能说明见表 8-4。

表 8-3　VideoView 组件的常用方法及功能说明

方　法　名	功　能　说　明
VideoView（Context context）	创建一个默认属性的 VideoView 实例
getCurrentPosition（）	得到当前播放位置
canPause（）	判断是否能够暂停播放视频
canSeekBackward（）	判断是否能够倒退
canSeekForward（）	判断是否能够快进
getDuration（）	得到文件的时间
isLooping（）	判断是否循环播放
isPlaying（）	判断是否正在播放
pause（）	暂停
resume（）	恢复挂起的播放器
release（）	释放 MediaPlayer 对象
reset（）	重置 MediaPlayer 对象
seekTo（int msec）	指定播放的位置（以毫秒为单位的时间）
setVideoPath（String path）	设置视频文件的路径名
setVideoURI（Uri uri）	设置视频文件的统一资源标识符
stopPlayback（）	停止回放视频文件
suspend（）	挂起视频文件的播放
setMediaController（MediaController controller）	设置媒体控制器
start（）	开始播放

表 8-4　VideoView 组件的常用事件及功能说明

事　件　名	功　能　说　明
setOnCompletionListener（MediaPlayer.OnCompletionListenerlistener）	播放结束监听事件
setOnErrorListener（MediaPlayer.OnErrorListener listener）	播放出错监听事件
setOnPreparedListener（MediaPlayer.OnPreparedListener l）	播放就绪监听事件

8.3.2　视频播放器的实现

1. MediaPlayer

MediaPlayer 类可以用于播放音频和视频，通过设置它的 setDataSource（）方法可以指定音频或视频的文件路径。与播放音频数据不同的是，视频播放需要设置显示视频内容的输出界面，此时可以使用 SurfaceView 控件，将它与 MediaPlayer 结合起来，就可以实现视频输出了。SurfaceView 类的常用方法及功能说明见表 8-5。

表 8-5　SurfaceView 类的常用方法及功能说明

方　法　名	功　能　说　明
public getHolder（）	得到 SurfaceHolder 对象用于管理 SurfaceView
public void setVisibility（int visibility）	设置是否可见，其值可以是 VISIBLE、INVISIBLE、GONE

为了管理 SurfaceView，Android 提供了一个 SurfaceHolder 接口。SurfaceView 用于显示，SurfaceHolder 用于管理显示的 SurfaceView 对象。SurfaceView 是视图（View）的一个继承类，每一个 SurfaceView 都内嵌封装一个 Surface（Surface 是 Android 的一个重要元素，用于 Android 界面的图形绘制）。通过调用 SurfaceHolder 可以调用 SurfaceView，控制图形的尺寸和大小，而 SurfaceHolder 对象是由 getHold()方法获得的，创建 SurfaceHolder 对象后，用 SurfaceHolder.Callback()方法回调 SurfaceHolder 对 SurfaceView 进行控制。实现步骤如下。

- 创建 MediaPlayer 对象，并设置加载的视频文件。
- 在界面布局文件中定义 SurfaceView 控件。
- 通过 MediaPlayer.setDisplay(SurfaceHolder sh) 来指定视频画面输出到 SurfaceView 之上。
- 通过 MediaPlayer 的其他一些方法控制播放视频。

（1）界面设计。

本案例的界面设计比较简单，上面放置了一个用于显示视频图像的 SurfaceView，下面放置了三个 Button 分别用于控制视频的"播放""暂停""停止"，其布局文件代码如下。

```
1   ?xml version="1.0" encoding="utf-8"?>
2   <LinearLayout xmlns:android="http://schemas.android.com/apk/res/android"
3       android:layout_width="fill_parent"
4       android:layout_height="fill_parent"
5       android:orientation="vertical">
6       <!--视频输出区域-->
7       <SurfaceView
8           android:id="@+id/surfaceView"
9           android:layout_width="fill_parent"
10          android:layout_height="360px"></SurfaceView>
11      <LinearLayout
12          android:layout_width="wrap_content"
13          android:layout_height="wrap_content"
14          android:orientation="horizontal">
15          <Button
16              android:id="@+id/startBtn"
17              android:layout_width="wrap_content"
18              android:layout_height="wrap_content"
19              android:text="播放"></Button>
20          <Button
21              android:id="@+id/pauseBtn"
22              android:layout_width="wrap_content"
23              android:layout_height="wrap_content"
24              android:text="暂停"></Button>
25          <Button
26              android:id="@+id/stopBtn"
27              android:layout_width="wrap_content"
28              android:layout_height="wrap_content"
29              android:text="停止"></Button>
30      </LinearLayout>
31  </LinearLayout>
```

(2) 功能实现。

功能实现中关于"暂停""停止"按钮的代码与 MediaPlayer 播放音频文件的方法完全一样，限于篇幅，此处只列出"播放"按钮的关键代码。

```
1     private MediaPlayer mPlayer = new MediaPlayer();
2     private SurfaceView sfv_show = (SurfaceView) this.findViewById (R.id.surfaceView);
3     private SurfaceHolder surfaceHolder;
4     try {
5         mediaPlayer.reset();
6         mediaPlayer.setAudioStreamType (AudioManager.STREAM_MUSIC);
7         //设置需要播放的视频文件
8         mediaPlayer.setDataSource (" /mnt/sdcard/lesson2.flv");
9         //初始化 SurfaceHolder 类，SurfaceView 的控制器
10        surfaceHolder = sfv_show.getHolder();
11        //显示的分辨率，不设置为视频默认
12        surfaceHolder.setFixedSize (320, 220);
13        //把视频画面输出到 SurfaceView
14        mediaPlayer.setDisplay (surfaceHolder);
15        mediaPlayer.prepare();
16        mediaPlayer.start(); // 播放
17    } catch (Exception e) {
18
19    }
```

上述第 8 行代码用于设置 MediaPlayer 的视频文件，此处也可以根据上一节中音乐播放器的设计思路来实现。感兴趣的读者可以在音乐播放器的基础上自己设计一个播放器，既可以播放视频，也可以播放音频。MediaPlayer 实现视频播放的效果如图 8.5 所示。

图 8.5 MediaPlayer 实现视频播放的效果

2. VideoView

用 VideoView 组件实现视频播放时，必须在布局文件中定义以下两个组件。

● VideoView 组件：用于视频输出。

● MediaController 组件：用于控制视频播放，即用于控制该视频文件的播放行为（如暂停、前进、后退和进度拖曳等）。

MediaController 类是 android.widget 包下的，包含控制 MediaPlayer 多媒体播放的组件，如"播放""暂停""快退""快进""进度条"等，常用方法及功能说明见表 8-6。

表 8-6 MediaController 类的常用方法及功能说明

方 法 名	功 能 说 明
MediaController(Context context, AttributeSetattrs)	通过 Context 对象和 AttributeSet 对象来创建 MediaController 对象
MediaController(Context context, boolean useFastForward)	通过 Context 对象指定是否允许用户控制进度，即是否有"快进""快退"按钮，若设置为 false，则不会显示
hide()	设置隐藏 MediaController
show()	设置显示 MediaController，3s 后自动消失
show(int timeout)	设置 MediaController 显示的时间（毫秒）
setMediaPlayer(VideoView player)	设置播放的 VideoView

使用 VideoView 组件实现视频播放的基本步骤如下。

● 在布局文件上定义 VideoView 组件。

● 调用 VideoView 的方法来加载指定的视频。setVideoPath(String path) 方法用于加载 path 路径下的视频文件；setVideoUri(Uri uri) 方法用于加载 Uri 指定的视频文件。

● 调用 VideoView 的 start()、stop()、pause() 方法来控制视频的播放。

（1）界面设计。

本案例的界面设计比较简单，只要放置一个 VideoView 组件，其布局文件代码如下。

```
1    <?xml version="1.0" encoding="utf-8"?>
2    <LinearLayout xmlns:android="http://schemas.android.com/apk/res/android"
3        android:layout_width="fill_parent"
4        android:layout_height="fill_parent"
5        android:orientation="vertical">
6        <VideoView
7            android:id="@+id/videoView"
8            android:layout_width="match_parent"
9            android:layout_height="wrap_content"></VideoView>
10   </LinearLayout>
```

（2）功能实现。

```
1        private VideoView videoView;
2        private MediaController mediaController;
3        videoView = (VideoView)this.findViewById(R.id.videoView);
4        mediaController = new MediaController(this);
5        videoView.setVideoPath("/mnt/sdcard/lesson3.mp4");
6        //设置 VideoView 与 MediaController 建立关联
7        videoView.setMediaController(mediaController);
8        //设置 MediaController 与 VideView 建立关联
```

```
9            mediaController.setMediaPlayer(videoView);
10          //让 VideoView 获取焦点
11           videoView.requestFocus();
12       //开始播放
13           videoView.start();
14       }
15  }
```

布局文件没有使用"播放""暂停"等按钮来实现对视频播放的控制,从上述第 4 行代码可以看出,本案例实例化了一个 MediaController 对象,该对象会在用户界面的最底部自动产生一个控制工具条,通过第 9 行代码的关联设置,就可以用这个控制工具控制播放的视频了。第 5 行代码指出了要加载播放的视频文件。本案例运行后的效果如图 8.6 所示。如果要让 MediaController 控制工具条上不会显示"快进"和"快退"两个按钮,则可以将上述第 4 行代码修改为如下代码。

```
mediaController = new MediaController(this,false);
```

图 8.6 VideoView 视频播放器

8.4 录音机的设计与实现

Android 多媒体框架支持对常见音频和视频的录制和编码,如果硬件支持,可以使用 MediaRecorder 类非常方便地实现录制音频和视频的功能。也就是说,只需要 Android 设备带麦克风就可以录制音频,带摄像头就可以录制视频了,本节介绍录制音频的方法。

8.4.1 预备知识

为了能在 Android 应用程序中录制音频,Android 提供了 MediaRecorder 类,该类的常用方法及功能说明见表 8-7。

表 8-7 MediaRecorder 类的常用方法及功能说明

方 法 名	功 能 说 明
MediaRecorder()	构造方法
prepare()	准备录音机
release()	释放 MediaRecorder 对象
reset()	重置 MediaRecorder 对象，使其为空闲状态
stop()	停止录制音频
setAudioEncoder()	设置音频编码格式
setAudioEncodingBitRate(int bitRate)	设置编码位数
setAudioSamplingRate(int samplingRate)	设置采样频率
setAudioSource()	设置录制的音频来源
setVideoSource()	设置录制的视频来源
setOutputFormat()	设置所录制的音频格式并保存
setOutputFile(String path)	设置录制的音频文件的保存位置

下面以录制后的音频文件存放到 SD 卡的 sounds 文件下并以当前时间命名的文件为例介绍录音机的实现过程。

8.4.2 录音机的实现

音频的录制比音频或者视频的播放复杂，一般按照如下步骤实现。

(1) 使用 android. media. MediaRecorder 类创建 MediaRecorder 实例。
(2) 使用 MediaRecorder. setAudioSource() 方法设置音频来源。
(3) 使用 MediaRecorder. setOutputFormat() 设置输出音频格式。
(4) 使用 MediaRecorder. setAudioEncoder() 设置输出编码格式。
(5) 使用 MediaRecorder. setOutputFile() 方法设置输出文件名。
(6) 在开始录制之前调用 MediaRecorder. prepare() 方法。
(7) 调用 MediaRecorder. start() 方法开始录制音频。
(8) 调用 MediaRecorder. stop() 方法停止录制音频。
(9) 录制完音频调用 MediaRecorder. release() 方法释放占用的相关资源。

在使用麦克风进行录音时，必须给应用程序打开录音的权限；保存文件到 SD 卡需要给应用程序打开读写 SD 卡的权限，即在 AndroidManifest. xml 配置文件中增加如下代码。

```
<uses-permission android:name ="android. permission. RECORD_AUDIO" />
<uses-permission android: name =" android. permission.WRITE_EXTERNAL_STOR-
AGE" />
```

本案例在界面上通过 ListView 显示音频文件列表，实现过程在前面用户界面设计的章节中已详细介绍，此处不再详述；单击"录制"按钮，开始录制音频，实现代码如下。

```
1    startBTN.setOnClickListener(new OnClickListener(){
2        @Override
3        public void onClick(View v){
```

```
4       File dir = new File(Environment.getExternalStorageDirectory(),"sounds");
5       if(!dir.exists()){
6            dir.mkdirs();
7       }
8       File soundFile = new File(dir,System.currentTimeMillis()+".amr");
9       if(!soundFile.exists()){
10          try{
11              soundFile.createNewFile();//创建文件
12          }catch(IOException e){
13              e.printStackTrace();
14          }
15      }
16      try{
17          //实例化 MediaRecorder 对象
18          mediaRecorder = new MediaRecorder();
19          //设置音频输入源——麦克风
20          mediaRecorder.setAudioSource(MediaRecorder.AudioSource.MIC);
21          //设置输出文件的格式
22          mediaRecorder.setOutputFormat(MediaRecorder.OutputFormat.DEFAULT);
23          //设置音频文件的编码
24          mediaRecorder.setAudioEncoder(MediaRecorder.AudioEncoder.DEFAULT);
25          //设置输出文件的路径
26          mediaRecorder.setOutputFile(soundFile.getAbsolutePath());
27          mediaRecorder.prepare();
28          mediaRecorder.start();
29      }catch(IOException e){
30          e.printStackTrace();
31      }
32  }
33 });
```

上述第 4～7 行代码表示首先在 SD 卡文件夹下判断有没有 sounds 文件夹，如果没有就创建；第 8～15 行代码表示在 SD 卡的 sounds 文件夹下创建以当前时间命名的 ".amr" 声音文件；最后按照步骤设置 MediaRecorder 对象的相关参数，并开始录音。

单击"停止"按钮，录制结束，实现代码如下。

```
1   stopBTN.setOnClickListener(new OnClickListener(){
2       @Override
3       public void onClick(View v){
4           if(mediaRecorder!=null){
5               mediaRecorder.stop();
6               mediaRecorder.release();
7               mediaRecorder.null;
8           }
9       }
10  });
```

录音机运行效果如图 8.7 所示。

第8章 多媒体应用开发

图 8.7 录音机运行效果

8.5 照相机的设计与实现

拍照功能已经成为移动终端设备的基本功能之一，拍照功能是通过摄像头实现的。摄像头的应用对于 Android 未来的发展有着至关重要的作用，用户认证、条形码、二维码、内容分享等方面都会涉及摄像头的应用。本节介绍开发者在自己开发的应用程序中增加照相机功能的方法。

8.5.1 预备知识

Android 为拍照提供了基本的编程接口，使得开发者可以非常容易地访问摄像头而不需要编写底层的代码。开发者在 Android 平台实现照相机功能有两种方式：第 1 种是调用系统自带的相机应用程序；第 2 种是使用 Camera API 自定义照相机。

【Intent启动Activity 获取返回值】

1. 调用系统自带相机应用程序实现拍照

用户 App 的 Activity 在调用系统相机拍照功能时，首先使用 Intent 启动安装在设备上的摄像头应用程序，代码如下。

```
Intent intent = new Intent (android.provider.MediaStore.ACTION_IMAGE_CAP-
TURE);
    startActivityForResult (intent, requesCode);
```

相机拍照后会返回一个 Intent，该 Intent 的 Extra 部分包含一个编码过的 Bitmap。此时，用户 App 中该 Activity 的 onActivityResult( )方法会收到这个 Intent，用户程序可以在这个方法中对收到的 Intent 中的 data 进行处理。其工作原理如图 8.8 所示。

与摄像头相关的 Intent 属性及功能说明见表 8-8。

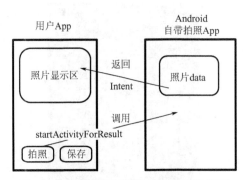

图 8.8 调用系统自带相机拍照功能原理

表 8-8 与摄像头相关的 Intent 属性及功能说明

属　　性	功 能 说 明
MediaStore.ACTION_IMAGE_CAPTURE	启动摄像头进行照片捕获
MediaStore.ACTION_IMAGE_CAPTURE_SECURE	返回照相机拍摄到的图像，设备被固定
MediaStore.ACTION_VIDEO_CAPTURE	调用已有的视频应用程序捕获视频
MediaStore.EXTRA_SCREEN_ORIENTATION	设置屏幕的方向：垂直或横向
MediaStore.EXTRA_FULL_SCREEN	控制 ViewImage 的用户界面（默认全屏）
MediaStore.INTENT_ACTION_VIDEO_CAMERA	启动摄像头进行视频捕获
MediaStore.EXTRA_SIZE_LIMIT	控制视频或图像捕获尺寸大小限制

2. Intent 启动 Activity 获取返回值

关于 Intent 的使用，本书在 4.2 节已经做了较为详细的介绍。使用 Intent 可以实现显示启动 Activity 和隐式启动 Activity，这两种方式都可以附加数据在 Activity 间进行传递。本节介绍一种实现 Activity 间传递数据的方法，即 Intent 启动其他 Activity 并获取返回值。

例如，要实现如图 8.9 所示的需求，即用户在 A 界面注册时需要到 B 界面选择籍贯，在 B 界面选择完后，能将选择结果立即返回到 A 界面指定位置。

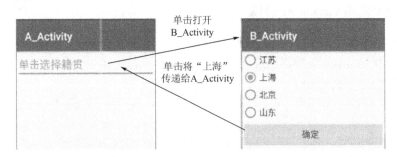

图 8.9 Intent 启动 Activity 获取返回值

图 8.9 中 A_Activity 与 B_Activity 的界面布局比较简单，限于篇幅，此处不详细列出布局文件的源代码，只介绍功能实现源代码。本例完整代码读者可以参见 FirstAPP 代码包中 a_b_activity 文件夹里的内容。

(1) A_Activity 中主要实现两个功能模块。

● 单击 EditText（图 8.9 中"单击选择籍贯"），使用 Intent 启动 B_Activity，并且要求返回值。其实现代码如下。

```
1    private final int codeRequest =1000;
2    private final int codeResult =2000;
3    edtJG.setOnClickListener(new View.OnClickListener() {
4        @Override
5        public void onClick(View v) {
6            Intent intent = new Intent(A_Activity.this, B_Activity.class);
7            A_Activity.this.startActivityForResult(intent,codeRequest);
8        }
9    });
```

上述第 1 行代码定义了一个 int 类型的 codeRequest 常量，用于标记本次 A_Activity 向其他 Activity（本例为 B_Activity）发出要求返回值的请求码。

● 重写 onActivityResult( ) 方法，用于根据"请求码"或另一个 Activity 返回的"结果码"对其返回来的 data（包含在 Intent 中）进行处理。其实现代码如下。

```
1    @Override
2    protected void onActivityResult(int requestCode, int resultCode, Intent data) {
3        super.onActivityResult(requestCode, resultCode, data);
4        switch (resultCode) {//根据结果码判断并处理
5            casecodeResult:
6                String jg = data.getStringExtra("jg");
7                edtJG.setText(jg);
8                break;
9        }
10   }
```

上述第 4～9 行代码表示如果该方法传回来的结果码（resultCode）与本 Activity 定义的结果码（codeResult）一样，那么就进行相应的处理（本例中将 B_Activity 传回的 data 数据处理后显示在"籍贯"位置处）。当然，也可以根据发出的请求码来辨别并加入功能处理。

(2) B_Activity 中主要实现一个功能模块，即单击"确定"按钮后将选中的籍贯通过 Intent 传递回调用的 A_Activity。

```
1    private final  int codeRseult =2000;
2    btnEnter.setOnClickListener(new View.OnClickListener() {
3        @Override
4        public void onClick(View v) {
5            String jg ="";
6            if(radBJ.isChecked()){
7                jg ="北京";
8            }
9            if(radSD.isChecked()){
10               jg ="山东";
11           }
```

```
12                    if(radSH.isChecked()){
13                         jg="上海";
14                    }
15                    if(radJS.isChecked()){
16                         jg="江苏";
17                    }
18                    Intent intent = getIntent();
19                    intent.putExtra("jg",jg);
20                    BActivity.this.setResult(codeResult,intent);
21                    finish();
22               }
23          });
```

上述第 20 行代码表示设置返回结果（含结果码 codeResult 和要传回的数据 Intent）。

从以上示例的实现过程可以看出，要实现在两个或两个以上的 Activity 间传递数据，必须使用以下步骤：

● 在 A_Activity 中使用 startActivityForResult(Intent intent，int requestCode) 方法启动 B_Activity。该方法的第 1 个参数为 Intent，即可以使用它绑定传递过去的值；第 2 个参数为请求码，即标识 A_Activity 发出的请求码，用于辨别 A_Actvity 发出的是哪一个请求。

● 在 A_Activity 中重写 onActivityResult(int requestCode，int resultCode，Intent data) 方法以获得 B_Activity 传回的值。该方法的第 1 个参数为请求码，即用于辨别 A_Activity 发出的是哪一个请求；第 2 个参数为结果码，即用于辨别是哪一个传回的结果；第 3 个参数为传回的用 Intent 绑定的值。

● 在 B_Activity 中调用 setResult(int resultCode，Intent data) 方法以返回结果给调用者（本案例中为 A_Activity）。该方法第 1 个参数为本次返回的结果码；第 2 个参数为要返回的用 Intent 绑定的数据。

3. 使用 Camera API 实现拍照

在实现自定义照相机时，需要使用 android.hardware.Camera 类，该类封装了所有摄像头相关的操作，如连接、断开摄像头服务，设置摄像头的各种参数，开始、结束摄像预览，抓取照片及连续抓取多个图像帧等。该类没有默认的构造函数，只能使用静态方法 open() 获得类对象，其常用方法及功能说明见表 8-9。

表 8-9 Camera 类的常用方法及功能说明

方 法 名	功 能 说 明
setPreviewDisplay()	用于设置预览窗口
startPreview()	Camera 对象把摄像头拍摄到的画面在对应的 View 对象上显示
stopPreview()	结束预览画面显示
takePicture()	用于真实拍摄，其参数全部是回调函数，完成拍摄后进行处理
release()	释放摄像头资源
autoFocus()	自动对焦
setParameters()	设置摄像头的参数

实现自定义相机除了需要使用 Camera 类外，还需要一个 SurfaceView 类来展示实时摄像头预览给用户，关于这个类的使用在 8.3 节已做过介绍。

8.5.2 照相机的实现

1. 调用系统相机

（1）界面设计。

【调用系统相机、自定义相机】

本案例的界面设计比较简单，一个 Button 按钮用于控制"拍照"，一个 ImageView 用于显示拍完后的照片。调用系统相机拍照的显示效果如图 8.10 所示。

图 8.10　调用系统相机拍照的显示效果

（2）功能实现。

- 给 Button 绑定监听事件

```
1   btnStart.setOnClickListener(new View.OnClickListener() {
2       @Override
3       public void onClick(View v) {
4           Intent intent = new Intent(MediaStore.ACTION_IMAGE_CAPTURE);
5           MainActivity.this.startActivityForResult (intent, 1000);
6       }
7   });
```

上述代码表示单击"拍照"按钮后，使用 Intent 打开摄像头，并调用 startActivityForResult()方法打开系统自带照相机界面（请求码为 1000）。

- 重写 Activity（本案例为 MainActivity.this.）的 onActivityResult()方法

```
1   @Override
2   protected void onActivityResult(int requestCode, int resultCode, Intent data) {
3       super.onActivityResult(requestCode, resultCode, data);
```

```
4       if (requestCode == 1000) {
5           Bitmap bp = (Bitmap) data.getExtras().get("data");
6           imgShow.setImageBitmap(bp);
7       }
8   }
```

上述第 4～7 行代码表示如果是 "1000" 的请求，就将系统照相机返回的 data 进行 Bitmap 处理，然后将这个 Bitmap 显示在界面上的 ImageView 上。

通过以上步骤实现的拍照功能存在两个方面的缺陷：一是没有保存到本地（当然也可以在界面上设计一个"保存"按钮，然后通过写文件的方式将 ImageView 显示的内容保存到指定文件中）；二是经过 Bitmap 处理的图片清晰度受影响。所以如果对照片质量要求较高，可以采取以下步骤实现。

● 给 Button 绑定监听事件

```
1   private File curImageFile = null; //定义一个保存图片的 File 变量
2   btnSave.setOnClickListener(new View.OnClickListener() {
3       @Override
4       public void onClick(View v) {
5           File dir = new File(Environment.getExternalStorageDirectory(),"pictures");
6           if(dir.exists()){
7               dir.mkdirs();
8           }
9           currentImageFile = new File(dir,System.currentTimeMillis() + ".jpg");
10          if(! curImageFile.exists()){
11              try {
12                  curImageFile.createNewFile();
13              } catch (IOException e) {
14                  e.printStackTrace();
15              }
16          }
17          Intent intent = new Intent(MediaStore.ACTION_IMAGE_CAPTURE);
18          intent.putExtra (MediaStore.EXTRA_OUTPUT, Uri.fromFile (curImageFile));
19          startActivityForResult (intent, 1000);
20      }
21  });
```

上述第 5～16 行代码表示如果 SD 卡中没有 pictures 文件夹，则创建；然后在该文件夹下创建一个以当前时间命名的 ".jpg" 文件用于存放照片文件。第 17～19 行代码表示用 Intent 启动摄像头并以指定文件名保存文件，然后调用 startActivityForResult()方法启动系统照相机。

● 重写 Activity（本案例为 MainActivity.this.）的 onActivityResult()方法

```
1   @Override
2   protected void onActivityResult(int requestCode, int resultCode, Intent data) {
3       super.onActivityResult(requestCode, resultCode, data);
4       if (requestCode == 1000) {
5           imgShow.setImageURI(Uri.fromFile(currentImageFile));;
6       }
7   }
```

需要特别说明的是：上述调用系统照相机并保存文件的方法在 Android 7.0 中调用可能会出现类似"android.os.FileUriExposedException：file:///storage/emulated/0/photo.jpeg exposed beyond app through ClipData.Item.getUri()"的错误信息，如果出现这种错误，读者可以在 Activity 的 onCreate()方法中加入如下代码。

```
1    StrictMode.VmPolicy.Builder builder = new StrictMode.VmPolicy.Builder();
2    StrictMode.setVmPolicy(builder.build());
3    builder.detectFileUriExposure();
```

调用系统相机是一种实现拍照功能的简单方法，它使用了系统内置的 Activity 完成拍照工作，所以开发者没有办法对它做任何调整，只能全部接受系统相机的各项设置。调用系统相机的方法虽然简单，但灵活性不够，所以也没有什么特色，适合把拍照作为辅助功能的应用程序采用。

2. 自定义相机

使用 Camera API 拍照或摄像，都需要用到预览。预览就要用到 SurfaceView，为此 Activity 的布局中必须有 SurfaceView 组件。通过 Camera 类进行拍照主要包含如下步骤。

- 调用 Camera 的 open()打开相机。
- 调用 Camera 的 getParameters()获取拍照的参数，该方法返回一个 Camera.parameters 对象。
- 调用 Camera.Parameters 对象对照相的参数进行设置。
- 调用 Camera.setParameters()，并将 Camera.Parameters对象作为参数传入，这样就可以对拍照的参数进行控制。
- 调用 Camera 的 startPreview()，开始预览取景，在预览取景之前需要调用 Camera 的 startPreviewDisplay(SufaceHolder holder) 设置使用哪个 SurfaceView 来显示取得的图片。
- 调用 Camera 的 takePicture()进行拍照。
- 结束程序时，调用 Camera 的 stopPreview()结束取景预览，并调用 release()释放资源。

(1) 界面设计。

图 8.11 所示操作界面的上部使用一个 SurfaceView 组件用于显示正在捕获的对象，中部使用三个 Button 分别实现"打开摄像头""拍照""关闭摄像头"功能，下部使用一个 ImageView 实现显示拍摄的照片。其代码如下。

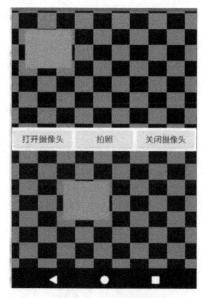

图 8.11 自定义照相机

```
1    <?xml version ="1.0" encoding ="utf-8"?>
2    <LinearLayout xmlns:android="http://schemas.android.com/apk/res/android"
3        android:layout_width =" fill_parent"
4        android: layout_height =" fill_parent"
5        android: orientation =" vertical" >
```

```xml
6      <SurfaceView
7          android:id="@+id/sufaceView"
8          android:layout_width="match_parent"
9          android:layout_height="0dp"
10         android:layout_weight="1"/>
11     <LinearLayout
12         android:layout_width="match_parent"
13         android:layout_height="0dp"
14         android:layout_weight="0.2" >
15         <Button
16             android:id="@+id/btnOpen"
17             android:layout_width="0dp"
18             android:layout_height="wrap_content"
19             android:layout_weight="1"
20             android:text="打开摄像头"/>
21         <Button
22             android:id="@+id/btnCapture"
23             android:layout_width="0dp"
24             android:layout_height="wrap_content"
25             android:layout_weight="1"
26             android:text="拍照"/>
27         <Button
28             android:id="@+id/btnClose"
29             android:layout_width="0dp"
30             android:layout_height="wrap_content"
31             android:layout_weight="1"
32             android:text="关闭摄像头"/>
33     </LinearLayout>
34     <ImageView
35         android:id="@+id/imgsShow"
36         android:layout_width="match_parent"
37         android:layout_height="0dp"
38         android:layout_weight="1"
39         android:scaleType="fitXY"
40         android:src="@mipmap/ic_launcher"/>
41 </LinearLayout>
```

（2）功能实现。
- 定义变量和实例化相关对象

```java
1   private SurfaceView sufaceView;//用于显示捕获对象
2   private SurfaceHolder surfaceHolder;
3   private ImageView imgShow;//用于显示拍摄的照片
4   private Button btnTake, btnOpen, btnClose;//拍照,打开、关闭摄像头
5   private Camera camera = null;//照相机对象
6   sufaceView = (SurfaceView) this.findViewById(R.id.sufaceView);
7   btnTake = (Button) this.findViewById(R.id.btnCapture);
8   btnOpen = (Button) this.findViewById(R.id.btnOpen);
9   btnClose = (Button) this.findViewById(R.id.btnClose);
10  imgShow = (ImageView) this.findViewById(R.id.imgsShow);
11  surfaceHolder = this.sufaceView.getHolder();
```

- 为 SurfaceHolder 对象添加一个回调监听器 SurfaceHolder.Callback，它用于接收发生在 SurfaceView 中变化的信息，通常需要重写三个方法，详细代码如下。

```
1    surfaceHolder.addCallback(new SurfaceHolder.Callback() {
2        @Override
3        public void surfaceCreated(SurfaceHolder holder) {
4            //当 surface 被创建时调用
5        }
6        @Override
7        public void surfaceChanged(SurfaceHolder holder, int format, int width, int height) {
8            //当 surface 的大小或尺寸变化的时候调用
9        }
10       @Override
11       public void surfaceDestroyed(SurfaceHolder holder) {
12           //当 surface 被毁坏时调用
13           if (camera != null) {
14               camera.release();//释放资源
15               camera = null;
16           }
17       }
18   });
```

- 绑定"打开摄像头"监听事件

```
1        btnOpen.setOnClickListener(new View.OnClickListener() {
2            @Override
3            public void onClick(View v) {
4                camera = Camera.open();
5                try {
6                    //此处可以增加 Camera 的参数设置
7                    camera.setPreviewDisplay(surfaceHolder);
8                } catch (IOException e) {
9                    e.printStackTrace();
10               }
11           }
12       });
```

上述第 6 行代码可以给摄像头设置参数，设置参数时先调用 getParameters()得到已打开的摄像头的配置参数 Parameters 对象，若要修改对象的参数，可以调用 setParameters()方法进行设置，代码如下。

```
                //获得相机参数对象
                Camera.Parameters parameters = mycamera.getParameters();
                //设置格式
                parameters.setPictureFormat(PixelFormat.JPEG);
                //设置预览大小,设置为 360px×480px
                parameters.setPreviewSize(360, 480);
                //设置自动对焦
                parameters.setFocusMode("auto");
                //设置图片保存时的分辨率大小
                parameters.setPictureSize(2592, 1456);
```

```
                //给相机对象设置刚才设定的参数
                camera.setParameters(parameters);
                //开始预览
                camera.startPreview();
```

- 绑定"拍照"按钮监听事件

```
1    btnTake.setOnClickListener(new View.OnClickListener() {
2        @Override
3        public void onClick(View v) {
4           if(camera==null){
5             Toast.makeText(CameraActivity.this,"打开摄像头再拍照!",Toast.LENGTH_SHORT).show();
6             return;
7           }
8           camera.takePicture (null, null, new Camera.PictureCallback() {
9             @Override
10            public void onPictureTaken (byte [] data, Camera camera) {
11              String path = "";
12              if ( (path = saveFile (data)) != null) {
13                 imgShow.setImageURI (Uri.fromFile (new File (path)));
14              } else {
15                 Toast.makeText (CameraActivity.this, "保存照片失败",Toast.LENGTH_SHORT).show();
16              }
17            }
18         });
19         camera.startPreview();
20        }
21    });
```

上述第8～18行代码启动拍照功能，并调用自定义的saveFile( )方法将照片以当时的时间为文件名保存到SD卡的pictures文件夹下，saveFile( )方法的代码如下。

```
1     //保存临时文件的方法
2     private String saveFile(byte[] bytes) {
3         File currentImageFile = null;
4         File dir = new File(Environment.getExternalStorageDirectory(), "pictures");
5         if (dir.exists()) {
6            dir.mkdirs();
7         }
8         currentImageFile = new File(dir, System.currentTimeMillis() + ".jpg");
9         if (! currentImageFile.exists()) {
10           try {
11              currentImageFile.createNewFile();
12           } catch (IOException e) {
13              e.printStackTrace();
14           }
15        }
16        try {
```

```
17              FileOutputStream fos = new FileOutputStream(currentImageFile);
18              fos.write(bytes);
19              fos.flush();
20              fos.close();
21              return currentImageFile.getAbsolutePath();
22          } catch (IOException e) {
23              e.printStackTrace();
24          }
25          return "";
26
27
```

- 绑定"关闭摄像头"按钮事件

```
1   btnClose.setOnClickListener(new View.OnClickListener() {
2       @Override
3       public void onClick(View v) {
4           camera.stopPreview();
5           camera.release();
6           camera = null;
7       }
8   });
```

为了在应用程序中使用摄像头，必须在 AndroidManifest.xml 文件中设置拍照的权限许可，即在配置文件中增加如下代码。

```
<uses-permission android:name="android.permission.CAMERA" />
```

如果系统中没有安装摄像头，则这个程序就无法安装。为了通知系统当前应用程序使用摄像头，可以在配置文件中加入如下代码。

```
<uses-feature android:name="android.hardware.camera" />
<uses-feature android:name="android.hardware.camera.autofocus" />
```

一般来说，拍照和摄像需要写到 SD 卡上，所以也需要在配置文件中加入如下代码。

```
<uses-permission android:name="android.permission.WRITE_EXTERNAL_STORAGE" />
```

开发具有拍摄视频功能的应用程序，需要用到音频录制和视频录制功能，所以又需要下面两个权限声明。

```
<uses-permission android:name="android.permission.RECORD_VIDEO" />
<uses-permission android: name=" android.permission.RECORD_AUDIO" />
```

读者可以在此应用的基础上进行应用程序的功能扩展。本案例的详细代码可以参见 FirstAPP 代码包中 MyCamera 文件夹里的内容。

本 章 小 结

本章结合实际示例项目的开发过程，介绍了 Android 中 MediaPlayer、VideoView、MediaRecorder、Camera 等多媒体组件的使用方法，让读者既能明白 Android 多媒体应用开发的流程，又能掌握相关技术。

习 题

一、选择题

1. 使用 MediaPlayer 播放保存在 SD 卡上的 mp3 文件时（   ）。
   A. 需要使用 MediaPlayer.create 方法创建 MediaPlayer
   B. 直接 new MediaPlayer 即可
   C. 需要调用 setDataSource 方法设置文件源
   D. 直接调用 start 方法，无须设置文件源

2. 在 Android 应用程序中，有些应用功能需要添加用户权限才能使用，下面选项中不需要添加用户权限就可以实现的功能是（   ）。
   A. 访问 SD 卡   B. 实现访问网络
   C. 使用手机拨号程序   D. 使用手机打电话的程序

3. 使用 Android 进行拍照用到的类是（   ）。
   A. SurfaceView   B. SurfaceHolder   C. Callback   D. Camera

4. 下列关于 MediaPlayer 类的叙述错误的是（   ）。
   A. 在开始播放之前，MediaPlayer 对象必须要进入 Prepared 状态
   B. 要开始播放，必须调用 start() 方法。当此方法成功返回时，MediaPlayer 的对象处于 Started 状态
   C. isPlaying() 方法可以被调用来测试某个 MediaPlayer 对象是否在 Started 状态
   D. 当 MediaPlayer 对象处于播放状态时不能调用 seekTo() 方法调整播放的位置

5. 下列哪个操作一定需要在配置文件中声明权限？（   ）
   A. 播放 mp3 文件   B. 访问 SD 卡
   C. 创建数据库   D. 访问网络

6. 以下属于调用摄像头硬件的权限的是（   ）。
   A. <uses-permission android:name = "android.permission.CAMERA"/>
   B. <uses-permission Android:name = "android.permission.MOUNT_UNMOUNT_FILESYSTEMS"/>
   C. <uses-permission Android:name = "android.permission.WRITE_EXTERNAL_STORAGE"/>
   D. <uses-permission android:name = "android.permission.INTERNET"/>

7. 下列哪个接口是 Camera 中处理快门关闭的接口？（   ）
   A. android.hardware.Camera.ShutterCallback
   B. android.hardware.Camera.PreviewCallback
   C. android.hardware.Camera.ErrorCallback
   D. android.hardware.Camera.AutoFocusCallback

二、填空题

1. Android 中专门用于录音的组件是_____。
2. VideoView 组件是调用_____实现视频播放的，它的作用与 ImageView 类似。
3. 如果启动一个 Activity，并且需要从该 Activity 中获取返回值，则需要使用_____方法。

4. _____类位于 Android.Media 包下，提供了音量控制与铃声模式相关操作。

5. 使用 MediaRecorder 录完音后，调用_____方法停止录制声音。

三、判断题

1. Android 界面上播放视频和音频都是调用 MediaPlayer 实现的。　　　　（　　）
2. Android 中可以应用 MediaRecorder 实现视频录制及音频录制的功能。（　　）
3. MediaRecorder 对象的 setAudioSource( )方法设置声音来源，一般传入 MediaRecorder.AudioSource.MIC 参数，指定录制来自麦克风的声音。（　　）
4. 使用 VideoView 来播放视频时，需要结合 SurfaceView 来实现图像输出。（　　）
5. 录制完声音调用 MediaRecorder.stop( )方法释放占用的相关资源。（　　）

【第8章参考答案】

第 9 章 网络应用开发

随着移动互联网技术的发展，越来越多的移动终端设备拥有了更为专业的网络性能。用户经常使用移动设备上网聊天、浏览页面及传送文件等，即用户可以在 Android 平台设备上实现数据上传、数据下载及数据浏览等功能，那么实现这些操作的应用程序是如何开发的呢？本章将用具体的案例详细介绍 Android 平台设备与网络进行数据交换的技术和实现方法。

教学目标

理解 Http、Web Service 和 WebView 这三种技术与网络交互数据的原理。

掌握利用 Http、Web Service 和 WebView 这三种技术进行 Android 平台的网络应用开发的方法。

教学要求

知识要点	能力要求	相关知识
概述	（1）理解 Http 的 GET 和 POST 两种请求方式 （2）理解 Web Service 的工作原理 （3）熟悉 WebView 的作用	网络请求
在线中英文互译工具的设计与实现	（1）掌握 HttpURLConnection 类的常用方法和使用步骤 （2）掌握使用 GET 和 POST 两种请求方式获取 HTML、图片和网页的方法 （3）了解 XML 格式数据的三种解析方式，掌握 PULL 解析器解析 XML 文件的方法	文件输入/输出流
股票即时查询工具的设计与实现	（1）了解第三方网络请求框架 OkHttp 的原理，并熟悉 OkHttp3 的相关核心类和方法 （2）掌握使用 OkHttp3 进行 GET/POST 请求的方法 （3）熟悉新浪网提供的股票信息 API，并掌握调用和解析方法	第三方服务

(续)

知识要点	能力要求	相关知识
快递单查询工具的设计与实现	（1）掌握 Android Studio 开发环境下加载第三方类库的方法 （2）熟悉 KSOAP2 包中的常用类及功能 （3）掌握调用 Web Service 服务的步骤及 SoapObject 对象的解析方法	第三方类库
WebView 的应用	（1）掌握 WebView 的常用方法和功能 （2）熟悉 WebSettings、WebChromeClient 和 WebViewClient 的常用方法和功能 （3）掌握使用 WebView 定制浏览器的开发方法 （4）掌握 Android 与 JavaScript 交互的开发方法	HTML 语言 JavaScript 语言

9.1 概　　述

Android 平台完全支持 JDK 本身的 TCP、UDP 网络通信 API，所以基于 Android 平台的网络应用开发与基于 Java 的网络应用开发几乎一样。本章主要讲述在 Android 平台上使用 Http 访问网络、使用 Web Service 与网络实现数据交换、使用 WebView 布局 UI 及浏览网页的开发技术。

1. Http

Http 是 TCP/IP 协议体系中的一个应用层协议，用于定义 Web 浏览器与 Web 服务器之间交换数据的过程。即 Web 浏览器（客户端）连上 Web 服务器后，若想获得 Web 服务器中的某个 Web 资源，就需要遵守一定的通信格式，Http 协议就是这样的通信格式。目前，该协议有 Http 1.0 与 Http 1.1 两个版本，Http 1.0 在客户端与 Web 服务器建立连接后，一个连接只能获取一个 Web 资源，而 Http 1.1 在客户端与 Web 服务器建立连接后，允许在一个连接上获取多个 Web 资源。

Http 通信技术是网络应用中最为常用的技术之一，客户端向服务器发出 Http 请求，服务器接收到客户端的请求后，处理客户端的请求，处理完成后再通过 Http 将应答传回给客户端。在 Java 网络编程中，客户端一般是浏览器，而 Android 平台的客户端是指安装了 Android 的智能终端，服务器一般是 Web 服务器。Http 请求通常有多种方式，实际开发中用得较多的是 GET 和 POST 两种方式。GET 方式在请求的 URL 地址后以 "?" 的形式带上交给服务器的数据，多个数据之间以 "&" 进行连接，但数据容量通常不能超过 2KB。例如，使用 "http://ie.nnutc.edu.cn? username = …&pawd = …" 这种样式访问 Web 服务器的就是 GET 方式。POST 方式可以在请求的实体内容中向 Web 服务器发送数据，传输没有数量限制。从它们的工作机制可以看出，CET 方式安全性非常低，POST 方式安全性比较高，但是 GET 方式的执行效率却比 POST 方式高。实际使用时，向 Web 服务器提出查询时一般使用 GET 方式，而进行数据的增加、删除和修改时通常使用 POST 方式。

在 Android 平台使用 Http 访问网络有 HttpURLConnection 和 HttpClient 两种传统方式。

HttpURLConnection 是一种多用途、轻量级的 Http 客户端，使用它进行 Http 操作适用于大多数的应用程序。HttpClient 是 Apache 的开源框架，封装了访问 Http 的请求头、参数、内容体和响应等，使用起来比较方便，也比较稳定。HttpURLConnection 是 Android SDK 的标准实现，自 Android 6.0 开始 Google 已经弃用 HttpClient，并推荐使用 HttpUrlConnection 代替 HttpClient。在实际 Android 网络开发中，越来越多的开发者开始使用另一个由 Square 公司提供的轻量级的开源框架——OkHttp。基于此本章只介绍用 HttpURLConnection 和 OkHttp 两种方式访问网络资源。

2. Web Service

Web Service（Web 服务）是一种跨编程语言和跨操作系统平台的远程调用技术。该技术能够让运行在不同机器上（可能是不同的操作系统）的不同应用程序（可能是用不同的编程语言）无须借助附加的、专门的第三方软件或硬件，就可以相互交换数据。也就是说，依据 Web Service 规范实施的应用程序之间，无论它们使用的语言、平台或内部协议是什么都可以方便地交换数据。

从表面上看，Web Service 就是一个应用程序向外界暴露的能够通过 Web 进行调用的 API，也就是说能够用编程的方法通过 Web 来调用这个应用程序。调用这个 Web Service 的应用程序称为客户端，提供这个 Web Service 的应用程序称为服务端。Web Service 的体系结构如图 9.1 所示。从深层次看，Web Service 是建立可互操作的分布式应用程序的标准平台，该平台需要一套协议实现分布式应用程序的创建，并提供一种标准来描述 Web Service 供客户调用，这种调用方式必须遵循远程过程调用协议（RPC）。

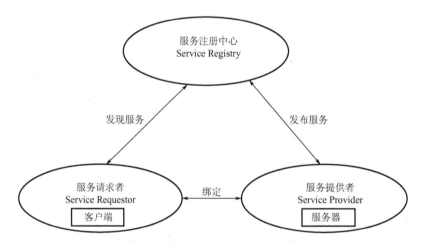

图 9.1　Web Service 的体系结构

构成 Web Service 平台的三大技术如下。

（1）XML 和 XSD。

XML 是 Web Service 平台中表示数据的基本格式，它解决了数据表示的问题，但没有定义一套标准的数据类型。W3C 制定的 XML Schema（XSD）就是专门解决这个问题的一套标准，它不仅定义了一套标准的数据类型，还给出了一种语言来扩展这套数据类型。Web Service 平台就是用 XSD 来作为其数据类型系统的。

（2）SOAP。

Web Service 通过 Http 协议发送请求和接收响应结果时，发送的请求内容和结果内容都采用 XML 格式封装，并增加了一些特定的 Http 消息头，用以说明 Http 消息的内容格式，这些特定的 Http 消息头和 XML 内容格式就是 SOAP（简单对象访问协议）。SOAP 提供了标准的 RPC 方法来调用 Web Service。SOAP 协议定义了 SOAP 消息的格式，它既基于 Http 协议，也基于 XML 和 XSD 的数据编码方式，即 SOAP 协议 = Http 协议 + XML 数据格式。

（3）WSDL。

Web Service 客户端要调用 Web Service 服务，就必须知道这个服务的地址和服务里提供了哪些方法可以调用。要知道这些信息，就需要 Web Service 服务器端通过一个 WSPL（Web Service 描述语言）文件来告诉开发者服务器上有哪些服务（包括服务的网络地址、服务中提供的方法、接收的参数及返回值）可以对外调用。WSDL 文件保存在 Web 服务器上，通过一个 URL 地址可以访问它。客户端在调用一个 Web Service 服务之前，就应该首先知道该服务的 WSDL 文件的地址。Web Service 服务提供商可以通过两种方式对外公开其 WSDL 文件地址，即注册到 UDDI（统一描述、发现和集成协议）服务器或直接告诉客户端调用者。UDDI 用于在网上自动查找 Web Service，一旦 Web Service 注册到 UDDI，客户端调用者就可以很方便地查找和定位其所需要的 Web Service。

由于 Android 平台目前没有提供 Web Service 客户端开发类库，所以只能借助第三方的 Web Service 客户端开发类库。KSOAP2 是目前在 Android 平台应用最为广泛的客户端开发类库，它是一个高效的、轻量级的 SOAP 开发包，读者可以到 http://ksoap2.sourceforge.net/ 下载。

3. WebView

WebView（网页视图）是 Android 中一个非常实用的组件，它和 Safari、Chrome 一样都是基于 Webkit 的网页渲染引擎，可以通过加载 HTML 数据的方式便捷地展现应用程序的界面。Android 4.3 及其以下版本，内部直接采用 Webkit 渲染引擎，从 Android 4.4 版本开始，内部采用 Chromium 渲染引擎来渲染 View 的内容。WebView 主要包含以下三个方面的功能。

（1）显示和渲染 Web 页面。

（2）直接使用 HTML 文件（文件可以存放在网络或本地 assets 目录下）作布局。

（3）与 JavaScript 交互调用。

9.2　在线中英文互译工具的设计与实现

英汉字典作为一种工具，能够查找英语单词的中文意思，方便用户学习英语。随着网络应用的发展，很多人已不习惯使用传统的纸质字典学习英语，英汉电子字典的使用正成为一种趋势。在线英汉双译字典主要通过 Http 通信在线实现英汉互译、单词发音的功能，极大方便了用户。

9.2.1　预备知识

HttpURLConnection 属于 Java API 的标准接口，包含在 java.net

【HttpURLConnection】

包中。HttpURLConnection是Java的标准类，继承自URLConnection类，在URLConnection的基础上进一步改进，增加了一些操作Http资源的快捷方法。由于URLConnection与HttpURLConnection都是抽象类，所以无法直接实例化对象，需要通过URL.openConnection()方法来获得，并需要进行强制类型转换。HttpURLConnetction类的常用方法及功能说明见表9-1。

表9-1 HttpURLConnetction类的常用方法及功能说明

方 法 名	功 能 说 明
setConnectTimeout(int timeout)	设置连接超时时长，如果超出，则放弃连接(单位为ms)
setDoInput(boolean newValue)	设置是否允许输入
setDoOutput(boolean newValue)	设置是否允许输出
getRequestMethod()	返回发送请求的方法
getResponseCode()	返回服务器的响应码
getResponseMessage()	返回服务器的响应消息
setRequestMethod(String method)	设置发送请求的方法
setUseCaches(boolean newValue)	设置是否允许使用网络缓存
setRequestProperty(String field,String newValue)	设置请求报文头，并且只对当前HttpURLConnection有效

使用HttpURLConnection的一般步骤如下。

- 创建一个URL对象。

```
URL url = new URL(http://www.baidu.com);
```

- 调用URL对象的openConnection()来获取HttpURLConnection对象实例。

```
HttpURLConnection conn = (HttpURLConnection) url.openConnection();
```

- 设置Http请求使用的方法（GET或者POST）。

```
conn.setRequestMethod("GET");//设置请求的方式为GET或者POST,默认方式为GET
```

- 设置其他一些参数（如连接主机超时时间、从主机读取数据超时时间，以及服务器希望得到的一些消息头等），这些参数根据具体开发需求进行设置。

```
conn.setConnectTimeout(6* 1000);
conn.setReadTimeout(6 * 1000);
```

- 调用getInputStream()方法获得服务器返回的输入流。

```
InputStream in = conn.getInputStream();
```

- 调用disconnect()方法关闭Http连接。

```
conn.disconnect();
```

由于需要访问网络资源，所以必须要用到Android中联网的权限，即在AndroidManifest.xml文件中添加如下内容。

```
<uses-permissionandroid:name ="android.permission.INTERNET"/>
```

1. GET方式

下面以实现图9.2所示效果为例介绍HttpURLConnection的用法。具体效果是当单击

"加载 HTML"按钮时,在界面上使用 ScrollView 控制 TextView 显示网页的 HTML 代码(图 9.2);单击"加载图片"按钮控制 ImageView 显示网页上的图片(图 9.3);单击"加载网页"按钮控制 WebView 显示网页内容(图 9.4)。

图 9.2　GET 方式获取 HTML 效果　　图 9.3　GET 方式获取图片效果　　图 9.4　GET 方式获取网页效果

(1) 界面实现。

要实现图 9.2～图 9.4 所示的界面效果,布局文件详细代码如下。

```
 1    <?xml version="1.0" encoding="utf-8"?>
 2    <LinearLayout xmlns:android="http://schemas.android.com/apk/res/android"
 3        android:layout_width="match_parent"
 4        android:layout_height="match_parent"
 5        android:orientation="vertical" >
 6        <LinearLayout
 7            android:layout_width="match_parent"
 8            android:layout_height="wrap_content" >
 9            <Button
10                android:id="@+id/btnHtml"
11                android:layout_width="0dp"
12                android:layout_height="wrap_content"
13                android:layout_weight="1"
14                android:text="加载 HTML" />
15            <Button
16                android:id="@+id/btnImg"
17                android:layout_width="0dp"
18                android:layout_height="wrap_content"
19                android:layout_weight="1"
20                android:text="加载图片" />
21            <Button
22                android:id="@+id/btnWebView"
23                android:layout_width="0dp"
```

```
24            android:layout_height="wrap_content"
25            android:layout_weight="1"
26            android:text="加载网页"/>
27    </LinearLayout>
28    <WebView
29        android:id="@+id/wvpage"
30        android:layout_width="match_parent"
31        android:layout_height="match_parent"
32        android:visibility="gone"/>
33    <ScrollView
34        android:id="@+id/scroll"
35        android:layout_width="match_parent"
36        android:layout_height="match_parent"
37        android:visibility="gone" >
38        <TextView
39            android:id="@+id/txthtml"
40            android:layout_width="wrap_content"
41            android:layout_height="wrap_content"/>
42    </ScrollView>
43    <ImageView
44        android:id="@+id/imgjpg"
45        android:layout_width="match_parent"
46        android:layout_height="match_parent"/>
47 </LinearLayout>
```

(2) 功能实现。

由于访问网络属于耗时操作，所以需要将访问网络的操作放在子线程中实现，本例中使用 Handler 和 Thread 相结合实现了从网络获取数据，通过 Handler 显示在用户界面上。

- 定义变量。

```
private TextView tvHtml;//显示 HTML 代码
private ScrollView scrollView;
private Button btnHtml, btnImg, btnWebView;
private ImageView imgJpg;//显示网络图片
private WebView wvPage;//显示网页
private String content = "";//存储网络读取的内容
private Bitmap bitmap;
```

- 定义 Handler 用于更新界面内容。

```
1  private Handler handler = new Handler() {
2      @Override
3      public void handleMessage(Message msg) {
4          super.handleMessage(msg);
5          switch (msg.what) {
6              case 1000://html 代码
7                  wvPage.setVisibility(View.GONE);
8                  imgJpg.setVisibility(View.GONE);
9                  scrollView.setVisibility(View.VISIBLE);
10                 tvHtml.setText(content);
11                 break;
```

```
12            case 2000://图片
13                wvPage.setVisibility(View.GONE);
14                imgJpg.setVisibility(View.VISIBLE);
15                scrollView.setVisibility(View.GONE);
16                imgJpg.setImageBitmap(bitmap);
17                break;
18            case 3000://网页
19                wvPage.setVisibility(View.VISIBLE);
20                imgJpg.setVisibility(View.GONE);
21                scrollView.setVisibility(View.GONE);
22                wvPage.loadDataWithBaseURL("", content, "text/html", "UTF-8", "");
23                Toast.makeText(MainActivity.this,"网页加载完毕", Toast.LENGTH_SHORT).show();
24                break;
25          }
26      }
27  };
```

上述第 7～10 行代码表示不显示 WebView、ImageView，ScrollView 用于显示"加载HTML"按钮传递过来的 content 内容；第 13～16 行代码表示不显示 WebView、ScrollView，ImageView 用于显示"加载图片"按钮传递过来的 bitmap 内容；第 19～22 行代码不显示 ScrollView、ImageView，WebView 用于显示"加载网页"按钮传递过来的 content 内容，其中使用了 WebView 组件，关于它的用法后面章节将做详细介绍，此处第 22 行代码调用 loadDataWithBaseURL（baseUrl, string, "text/html", "utf-8", null）方法加载页面，该方法的第 1 个参数表示要加载的 URL 页面，第 2 个参数表示页面的内容，第 3 个参数表示页面数据的 MIME 类型，第 4 个参数表示页面内容的编码格式。

- 定义从流中读取数据的 read( ) 方法。

```
1   public byte[] read(InputStream inStream) throws Exception {
2       ByteArrayOutputStream outStream = new ByteArrayOutputStream();
3       byte[] buffer = new byte[1024];
4       int len = 0;
5       while ((len = inStream.read(buffer)) != -1) {
6           outStream.write(buffer, 0, len);
7       }
8       inStream.close();
9       return outStream.toByteArray();
10  }
```

- 定义一个用 GET 方式获取网络图片数据的 getImage( ) 方法。

```
1   public byte[] getImage(String path) throws Exception {
2       URL url = new URL(path);
3       HttpURLConnection conn = (HttpURLConnection) url.openConnection();
4       //设置连接超时为 5s
5       conn.setConnectTimeout(5000);
6       //设置请求类型为 GET 类型
7       conn.setRequestMethod("GET");
8       //判断请求 URI 是否成功
9       if (conn.getResponseCode() != 200) {
```

```
10              throw new RuntimeException("请求url失败");
11          }
12          InputStream inStream = conn.getInputStream();
13          byte[] bt = read(inStream);
14          inStream.close();
15          conn.disconnect();
16          return bt;
17      }
```

上述第2～8行代码表示开启HttpURLConnection连接，并设置连接超时时间、请求获取方式（本例使用GET）等参数，然后第12～15行代码调用从流中读取数据的read()方法并返回给主调方法。

- 定义一个用GET方式获取网页的HTML源代码的getHtml()方法。

```
1   public String getHtml(String path) throws Exception {
2       URL url = new URL(path);
3       HttpURLConnection conn = (HttpURLConnection) url.openConnection();
4       conn.setConnectTimeout(5000);
5       conn.setRequestMethod("GET");
6       if (conn.getResponseCode() != 200) {
7           throw new RuntimeException("请求url失败");
8       }
9       InputStream inStream = conn.getInputStream();
10      byte[] data = read(inStream);
11      String html = new String(data, "UTF-8");
12      inStream.close()
13      conn.disconnect();
14      return html;
15  }
```

上述第2～10行代码的功能与上面的getImage()方法一样，第11行代码将从流中读取数据转换为"UTF-8"格式的字符串，并最后返回给主调方法。

- "加载HTML"按钮监听事件。

```
1   btnHtml.setOnClickListener(new View.OnClickListener() {
2       @Override
3       public void onClick(View v) {
4           new Thread() {
5               @Override
6               public void run() {
7                   try {
8                       String path = "http://news.nnutc.edu.cn/s2348.html";
9                       content = getHtml(path);
10                  } catch (Exception e) {
11                      e.printStackTrace();
12                  }
13                  handler.sendEmptyMessage(1000);
14              }
15          }.start();
16      }
17  });
```

第9章　网络应用开发

上述第 4～15 行代码表示开启子线程，在子线程调用获取网页源代码的 getHtml( ) 方法访问指定网页，然后通过 Handler 传送消息码 1000 给 Handler 进行处理。

- "加载图片"按钮监听事件。

```
1    btnImg.setOnClickListener(new View.OnClickListener() {
2      @Override
3      public void onClick(View v) {
4        new Thread(){
5          @Override
6          public void run() {
7            String path ="http://news.nnutc.edu.cn/upfile/201712/2017122763656621.jpg";
8            try {
9                byte[] data =  getImage(path);
10               bitmap = BitmapFactory.decodeByteArray(data, 0, data.length);
11           } catch (Exception e) {
12               e.printStackTrace();
13           }
14           handler.sendEmptyMessage(2000);
15         }
16       }.start();
17     }
18   });
```

上述第 4～16 行代码表示开启子线程，在子线程调用获取网页图片的 getImage ( ) 方法访问指定网页的图片，然后通过 Handler 传送消息码 2000 给 Handler 进行处理。

- "加载网页"按钮监听事件。

```
1    btnWebView.setOnClickListener(new View.OnClickListener() {
2      @Override
3      public void onClick(View v) {
4        new Thread(){
5          @Override
6          public void run() {
7            try {
8                String path = "http://news.nnutc.edu.cn/s2348.html";
9                content = getHtml(path);
10           } catch (Exception e) {
11               e.printStackTrace();
12           }
13           handler.sendEmptyMessage(3000);
14         }
15       }.start();
16     }
17   });
```

上述第 4～15 行代码表示开启子线程，在子线程调用获取网页源代码的 getHtml ( ) 方法访问指定网页，然后通过 Handler 传送消息码 3000 给 Handler 进行处理。

当然，在具体应用开发时有可能需要使用者输入类似用户名或密码等参数，这个时候就需要把对应的参数拼接到 URL 的尾部。例如，上述代码要访问的网页为"http://news.nnutc.edu.cn/show.asp?id=2351"这个形式，其中 id 为访问网页的参数，"2351"为传递给 id 的参数值。上面代码实现时就需要将 path 直接拼接为这种样式。这种方式能直接看出用户输入的参数值,所以 GET 方式并不安全。为此可以使用另外一种请求方式——POST 方式。

2. POST 方式

图 9.5 POST 方式获取网页效果

下面以用 POST 方式访问 http://fy.webxml.com.cn/webservices/EnglishChinese.asmx 网站上的资源为例介绍其实现过程。POST 方式获取网页效果如图 9.5 所示。如果要以 POST 方式访问某个网络资源,必须要保证这个网络资源支持 POST 方式。以这个网站为例,打开 EnglishChinese.asmx 页面后,单击 TranslatorString 链接可以看到的页面显示效果如图 9.6 所示。

图 9.6 中①②标注处表示 GET 方式访问该页面资源的 URL 格式,即 http://fy.webxml.com.cn/webservices/EnglishChinese.asmx/TranslatorString? wordKey = wide。"?"后面的 wordKey 是参数,wide 是要传递的参数值;③④标注处表示 POST 方式访问该页面资源的 URL 格式,即 http://fy.webxml.com.cn/webservices/EnglishChinese.asmx/TranslatorString,而⑤标注处表示要传递的参数名为 wordKey,"="后面表示要传递的参数值。

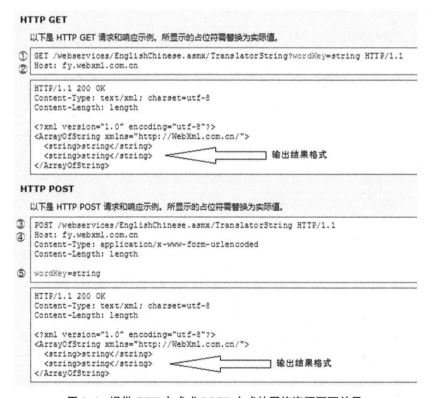

图 9.6 提供 GET 方式或 POST 方式的网络资源页面效果

(1) 界面设计。

图9.5所示界面包含一个 EditText 用于输入要翻译的词语,一个 Button 和一个 WebView用于显示POST方式的请求结果。代码比较简单,限于篇幅不再赘述。

(2) 功能实现。

- 定义一个使用POST方式获取网页HTML代码的loadByPost( )方法。

```
1    public String loadByPost(String path, String wordKey) throws Exception {
2        String msg = "";
3        URL url = new URL(path);
4        HttpURLConnection conn = (HttpURLConnection) url.openConnection();
5        conn.setRequestMethod("POST");
6        conn.setReadTimeout(5000);
7        conn.setConnectTimeout(5000);
8        conn.setDoOutput(true);
9        conn.setDoInput(true);
10       conn.setUseCaches(false); //POST方式不能缓存,需手动设置为false
11       data = "wordKey=" + URLEncoder.encode(wordKey, "UTF-8");
12       OutputStream out = conn.getOutputStream();//获取输出流
13       out.write(data.getBytes());
14       out.flush();
15       out.close();
16       if (conn.getResponseCode() == 200) {
17           //获取响应的输入流对象
18           InputStream is = conn.getInputStream();
19           //创建字节输出流对象
20           ByteArrayOutputStream message = new ByteArrayOutputStream();
21           int len = 0;
22           byte buffer[] = new byte[1024]; // 定义缓冲区
23           //按照缓冲区的大小,循环读取
24           while ((len = is.read(buffer)) ! = -1) {
25               //根据读取的长度写入os对象中
26               message.write(buffer, 0, len);
27           }
28           is.close(); //释放资源
29           message.close();
30           msg = new String(message.toByteArray());// 返回字符串
31           conn.disconnect();//关闭连接
32           return msg;
33       }
34       conn.disconnect();
35       return msg;
36   }
```

上述第11行代码表示要传递的参数(必须与图9.5的⑤标注处完全一样),第12~14行代码表示将参数通过POST方式传递出去。如果有多个参数,则需要使用"&"进行拼接。

- "POST 访问网络资源"按钮的监听事件。

```
1        btnWebView.setOnClickListener(new View.OnClickListener() {
2            @Override
3            public void onClick(View v) {
```

```
    4                new Thread() {
    5                    @Override
    6                    public void run() {
    7                        String path = "http://fy.webxml.com.cn/webservices/EnglishChinese.asmx/Translator";
    8                        String wordKey = edtContent.getText().toString();
    9                        try {
   10                            content = loadByPost(path,wordKey);
   11                            handler.sendEmptyMessage(3000);
   12                        } catch (Exception e) {
   13                            handler.sendEmptyMessage(4000);
   14                        }
   15                    }
   16                }.start();
   17            }
   18        });
```

上述代码块表示在子线程中调用自定义的 loadByPost( )方法连接网络。第7～8行代码表示要访问的URL和要传递的参数值。该参数值从 EditText（本案例对应 edtContent）输入，然后调用自定义的 loadByPost( )方法连接网络并获取网络数据资源，最后通过自定义的 Handler 处理结果。自定义的 Handler 与前面的 GET 方式类似，限于篇幅，不再赘述，读者可以参见 FirstAPP 代码包中 Httpuconn 文件夹里的内容（PostActivity.java）。

9.2.2 在线中英文互译工具的实现

【互译工具的实现】

1. 主界面的设计

主界面上放置了一个 EditText 用于输入要翻译的词语，两个 Button（一个用于单击后连接网络进行翻译，一个用于播放读音），三个 TextView 分别用于显示"读音""解释"和"例句"的文字，五个 TextView 分别用于显示翻译的读音内容、解释内容和三个例句。布局文件代码比较简单，代码不再列出，请读者参见 ALLAPP 代码包中 cetranslator 文件夹里的内容，运行后的效果如图 9.7 所示。

图 9.7　中英文互译主界面

2. 功能实现

(1) 定义变量。

```
1    private EditText edtContent;//输入翻译的词语
2    private Button btnOk, btnRead;//确定按钮、发音按钮
3    privateTextView tvDy, tvJs, tvLj1, tvLj2, tvLj3;//读音、解释、例句1、例句2、例句3
4    private String[] tempWord = new String[5];//存放翻译的词语内容
5    private String[] tempSentence = new String[3];//存放例句内容
```

(2) 定义 Handler 用于更新界面内容。

```
1    Handler handler = new Handler() {
2        @Override
3        public void handleMessage(Message msg) {
4            super.handleMessage(msg);
5            switch (msg.what) {
6                case 3000:
7                    tvDy.setText(tempWord[0]);//读音
8                    tvJs.setText(tempWord[3]);//解释
9                    tvLj1.setText("①" + tempSentence[0]);//例句1
10                   tvLj2.setText("②" + tempSentence[1]);//例句2
11                   tvLj3.setText("③" + tempSentence[2]);//例句3
12                   break;
13               case 4000:
14                   break;
15           }
16       }
17   };
```

(3) XML 格式数据的解析。

XML 类似于 HTML，通常用 XML 格式文件来存储数据，第 7 章介绍的 SharedPreferences 就是使用 XML 格式文件来保存信息的。SQLite 数据库底层其实也是一个 XML 文件，所以也可以将 XML 文件看成一个微型的数据库，在网络应用方面通常将数据包装成 XML 格式文件来进行传递。对 XML 格式文件的解析有 SAX（Simple API for XML）、DOM（Document Object Model）和 PULL（Android 推荐）三种方式。

本书只介绍 PULL 解析器解析 XML 文件的方法，对其他解析器感兴趣的读者可以参阅其他相关资料。XML 格式文件通常由文档开始、开始元素、属性、文本结点、结束元素和文档结束组成。对 XML 格式文件的解析通常按照下面的步骤实现。

- 获得 XmlPullParser 解析器。

```
XmlPullParserFactory factory = XmlPullParserFactory.newInstance();
XmlPullParser parser = factory.newPullParser();
```

- 为解析器对象提供 XML 流与编码格式。

```
parser.setInput(inputStream, "UTF-8");
```

- 获得事件的类型。

```
int eventType = parser.getEventType();
```

- 用 switch 对不同事件类型进行处理。

XmlPullParser 解析器的事件类型及含义见表 9-2

表 9-2 XmlPullParser 解析器的事件类型及含义

事件类型	含 义
START_DOCUMENT	开始文档解析，通常在这里完成初始化操作
START_TAG	解析元素，此处可以通过解析器的 getName( ) 方法获得元素名和 getAttributeValue（int i） 方法取出属性值；对于文本结点直接使用 parser. nextText( ) 获得结点内容
TEXT	解析文本
END_TAG	结束元素解析
END_DOCUMENT	结束文本解析

　　本案例开发中使用了中英文双向翻译永久免费的 Web 服务网站——http://fy. webxml. com. cn/webservices/EnglishChinese. asmx，该网站提供词典翻译、音标（拼音）、解释、相关词条、读音等功能。在线中英文互译工具实现了英汉互译、读音及例句三个功能。其中英汉互译及读音功能使用该服务网站提供的 TranslatorString( )方法，此方法输入参数 wordKey 后，返回 XML 格式的字符串内容，如图 9.8 所示，此格式字符串中，第 1 个 <string> 是输入的内容，第 2 个 <string> 是音标（拼音），第 4 个 <string> 是翻译释义，第 5 个 <string> 是对应词语读音的 mp3 文件名。例句功能使用该服务网站提供的 TranslatorSentenceString( )方法，此方法输入参数 wordKey 后，返回 XML 格式的字符串内容，如图 9.8 所示，此格式字符串中，第 1 个 <string> 是例句 1 内容，第 2 个 <string> 是例句 2 内容，第 3 个 <string> 是例句 3 内容。

```
<?xml version="1.0" encoding="UTF-8"?>
- <ArrayOfString xmlns="http://WebXml.com.cn/"
  xmlns:xsd="http://www.w3.org/2001/XMLSchema"
  xmlns:xsi="http://www.w3.org/2001/XMLSchema-instance">
    <string>good</string>
    <string>gud</string>
    <string/>
    <string>n. 善行,好处； adj. 好的,优良的,上等的； [pl.]商品</string>
    <string>1033.mp3</string>
  </ArrayOfString>
```

⇧

TranslatorString( )方法返回结果

```
<?xml version="1.0" encoding="UTF-8"?>
- <ArrayOfString xmlns="http://WebXml.com.cn/"
  xmlns:xsd="http://www.w3.org/2001/XMLSchema"
  xmlns:xsi="http://www.w3.org/2001/XMLSchema-instance">
    <string>My one good suit is at the cleaner's.|我那套讲究的衣服还在洗衣店里呢。
    </string>
    <string>He was very good to me when I was ill.|我生病时他帮了我的大忙。
    </string>
    <string>This is a good place for a picnic.|这是一个野餐的好地方。</string>
  </ArrayOfString>
```

⇧

TranslatorSentenceString( )方法返回结果

图 9.8　返回 XML 格式的字符串

为了将 XML 格式的字符串内容解析出来，此处自定义了方法 parsePullXML（String protocolXML，String fieldname，String［］temp）来实现，该方法实现时将 <string > 后的内容分别放入一个字符串数组 temp［i］中。实现代码如下。

```java
1    public void parsePullXML(String protocolXML, String fieldName, String[] temp) {
2      int index = 0;
3      try {
4        XmlPullParserFactory factory = XmlPullParserFactory.newInstance();
5        XmlPullParser parser = factory.newPullParser();
6        //把要解析的 XML 格式字符串转化为输入流,以便 PULL 解析器对它进行解析
7        InputStream inputStream = new ByteArrayInputStream(protocolXML.getBytes());
8        parser.setInput(inputStream, "UTF-8");
9        int eventType = parser.getEventType();
10       while (eventType != XmlPullParser.END_DOCUMENT) {
11         switch (eventType) {
12           case XmlPullParser.START_DOCUMENT:
13             break;
14           case XmlPullParser.START_TAG:
15             String tagName = parser.getName();
16             if (tagName.equals(fieldName)) {
17               temp[index] = parser.nextText();
18               index++;
19             }
20         }
21         eventType = parser.next();
22       }
23     } catch (XmlPullParserException e) {
24       e.printStackTrace();
25     } catch (IOException e) {
26       e.printStackTrace();
27     }
28   }
```

该方法使用 DOM 解析器解析 XML 格式文档，它有两个参数：protocolXML 表示要解析的 XML 格式字符串，fieldName 表示子结点名（本案例中从图 9.8 可以看出子结点名就是"string"）。详细步骤读者可参阅源代码中的注释，限于篇幅不再详述。

（4）定义一个使用 POST 方式获取网页 HTML 代码的 loadByPost（）方法，该方法内容与本节预备知识里讲述的内容完全一样。

（5）绑定"确定"按钮的监听事件。

```java
1    btnOk.setOnClickListener(new View.OnClickListener() {
2      @Override
3      public void onClick(View v) {
4        new Thread() {
5          @Override
6          public void run() {
7            String pathWord = "http://fy.webxml.com.cn/webservices/EnglishChinese.asmx/TranslatorString";
8            String pathSentence = "http://fy.webxml.com.cn/webservices/EnglishChinese.asmx/TranslatorSentenceString";
```

```
9                     String wordKey = edtContent.getText().toString();
10                     if (wordKey.trim().length() == 0) {
11                         Toast.makeText(MainActivity.this, "请输入翻译内容!",
Toast.LENGTH_LONG).show();
12                         return;
13                     }
14                     try {
15                         parsePullXML(loadByPost(pathWord, wordKey), "string",
tempWord);
16                         parsePullXML(loadByPost(pathSentence, wordKey), "string",
tempSentence);
17                         handler.sendEmptyMessage(3000);
18                     } catch (Exception e) {
19                         handler.sendEmptyMessage(4000);
20                     }
21                 }
22             }.start();
23         }
24     });
```

上述第 7、8 行代码分别定义访问 TranslatorString( )方法和 TranslatorSentenceString( )方法的 URL；第 9 行代码获取从 EditText 上输入的内容；第 16～17 行代码分别调用自定义访问网络资源的 loadByPost( )方法，将返回结果用 parseXml( )方法解析后分别存储到 tempWord、tempSentence 数组中，以便用 Handler 更新主界面。

（6）绑定"读音"按钮的监听事件。

```
1   btnRead.setOnClickListener(new View.OnClickListener() {
2       @Override
3       public void onClick(View v) {
4           if (tempWord[4].trim().length() == 0) {//读音文件名
5               Toast.makeText(MainActivity.this, "对不起,暂时没有该词的读音!",
Toast.LENGTH_LONG).show();
6               return;
7           }
8           String mp3URL = " http://fy.webxml.com.cn/sound/" + tempWord[4];
9           MediaPlayer mediaPlayer = new MediaPlayer();
10          try {
11              mediaPlayer.setDataSource(mp3URL);
12              mediaPlayer.prepare();
13              mediaPlayer.start();
14          } catch (IOException e) {
15              e.printStackTrace();
16          }
17      }
18  });
```

上述第 4 行代码中 tempWord[4] 存放的是单词的读音对应的 mp3 文件名（由 parseXml( )方法解析时存放在数组中的），然后使用 MediaPlayer 类中提供的方法播放音乐文件。

至此，在线中英文互译工具就已开发完成。读者可以在此案例项目基础上完善其功能,如将查找到的内容保存在本地数据库,下次使用时先查找本地数据库中的词语,以节省网络流量。

9.3 股票即时查询工具的设计与实现

在 Android 平台上使用 Http 访问网络除了可以使用前面介绍的 HttpURLConnection 方式外,目前开发中大多采用较为流行的第三方网络请求框架——OkHttp。本节以股票即时查询工具为例介绍 OkHttp 的用法。

9.3.1 预备知识

OkHttp 是由移动支付公司 Square 贡献的一款优秀的 Http 开源框架,它支持 GET 请求和 POST 请求,支持基于 Http 的文件上传和下载,支持加载图片,支持下载文件透明的 GZIP 压缩,支持响应缓存避免重复的网络请求,支持使用连接池来降低响应延迟问题。目前 OkHttp 有 2.0 版本和 3.0 以后的版本。这两个版本有的方法使用上有些不同,本书以 3.0 以后版本为例介绍它的用法。在使用 OkHttp3 之前,首先需要了解表 9-3 中的相关核心类(方法)及其功能。

【OkHttp】

表 9-3 OkHttp3 相关核心类(方法)及其功能

类 名	功 能
OkHttpClient	客户端对象
Request	访问请求,Post 请求中需要包含 RequestBody
RequestBody	请求数据,在 Post 请求中用到
Response	网络请求的响应结果
MediaType	数据类型,可以是 json、image、pdf 等一系列格式
newCall(request).execute()	同步的请求方法
newCall(request).enqueue(Callback callBack)	异步的请求方法,但 Callback 是执行在子线程中的,因此不能在此进行用户界面更新操作

1. 配置 OkHttp3

要使用 OkHttp3 进行项目开发,就必须首先在 Android 开发项目中添加第三方依赖库——OkHttp3。Android Studio 开发环境中添加第三方依赖库的方式有多种,下面先介绍在应用程序(Module)中通过 Gradle 抓取方式添加第三依赖库的步骤。

- 在 build.gradle 文件中添加以下代码。

```
dependencies {
    //其他依赖配置
    compile 'com.squareup.okhttp3:okhttp:3.2.0'
}
```

- 单击 Sync Now 进行网络同步,如图 9.9 所示。

需要特别说明的是:这种方式添加第三方依赖库必须在连接网络的状态。

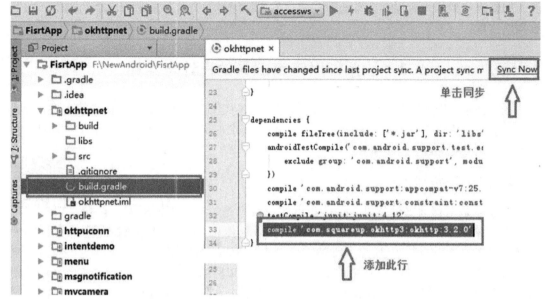

图 9.9　Gradle 抓取方式添加第三类库

2. Http GET 请求

(1) 同步 GET 请求。

同步 GET 请求的意思是一直等待 Http 请求，直到返回了响应。但是在等待期间可能会阻塞进程，所以这种方式只能在子线程中执行，否则会报错下面以实现如图 9.10 所示效果为例介绍其实现步骤。其界面设计比较简单，不再详细介绍。

- 自定义 getTBRequest( ) 方法使用 GET 请求获取网页资源数据。

```
1   public String getTBRequest(String url) {
2       //创建 OKHttpClient 对象
3       OkHttpClient client = new OkHttpClient();
4       //创建 Request 对象,至少要设置请求访问的 url
5       Request request = new Request.Builder().url(url).build();
6       try {
7           //创建一个包含新的 request 请求任务的 Call 对象
8           Call call = client.newCall(request);
9           //执行请求并获得 Http 请求的响应,可能会阻塞当前线程
10          Response response = call.execute();
11          if (response.isSuccessful()) {
12              return response.body().string();
13          }
14      } catch (IOException e) {
15          e.printStackTrace();
16      }
17      return "";
18  }
```

上述第 10 行代码表示启动同步 GET 请求；第 12 行代码中的 response.body( ) 是 ResponseBody 类，代表响应体（即请求响应返回的结果），可以通过 responseBody.string( )

获得字符串的表示形式。由于者通过 responseBody. bytes( )获得字节数组的表示形式。由于这两种形式都会把获得的文档内容（本例中就是返回 http://www. baidu. com 的主页内容）加载到内存中，所以 string( )方法只适用于获得小文档，而响应返回的文档太大（超过 1MB），实际开发时应避免使用这种方法。对于超过 1MB 的响应返回的文档，可以使用流的方式来处理，即可以通过 responseBody. charStream( )和 responseBody. byteStream( )返回流来进行处理。

图 9.10　**OkHttpClient 获取网页资源(GET)**

- 绑定"同步 GET"按钮的监听事件。

```
1    btnTBget.setOnClickListener(new View.OnClickListener() {
2        @Override
3        public void onClick(View v) {
4            new Thread() {
5                @Override
6                public void run() {
7                    String url = "http://www.baidu.com";
8                    content = getTBRequest(url);
9                    handler.sendEmptyMessage(1000);
10               }
11           }.start();
12       }
13   });
```

上述第 4～11 行代码开启一个子线程，调用 getTBRequest( )方法向网络发送 Http 的

GET 方式请求，并返回响应结果。
- 定义 Handler 处理界面内容更新。

```
1    Handler handler = new Handler() {
2        @Override
3        public void handleMessage(Message msg) {
4            super.handleMessage(msg);
5            switch (msg.what) {
6                case 1000:
7                    tvResult.setText(content);
8                    break;
9                //其他处理
10           }
11       }
12   };
```

（2）异步 GET 请求。

异步 GET 是指在另外的工作线程中执行 Http 请求，请求时不会阻塞当前的线程，所以可以在 Android 主线程中使用。

- 自定义 getYBRequest( ) 方法使用 GET 请求获取网页资源数据。

```
1    public String getYBRequest(String url) {
2        OkHttpClient client = new OkHttpClient();
3        Request request = new Request.Builder().url(url).build();
4        Call call = client.newCall(request);
5        //将请求加入队列,参数 Callback 监听
6        call.enqueue(new Callback() {
7            @Override
8            public void onFailure(Call call, IOException e) {
9                Toast.makeText(MainActivity.this, "异步 GET 失败", Toast.LENGTH_LONG).show();
10           }
11           @Override
12           public void onResponse (Call call, Response response) throws IOException {
13               final String responseStr = response.body().string();
14               runOnUiThread (new Runnable() {
15                   @Override
16                   public void run() {
17                       tvResult.setText (responseStr);
18                   }
19               });
20           }
21       });
22       return "";
23   }
```

上述第 6 行代码表示启动异步 GET 请求；第 8～10 行代码表示当请求失败的处理内容；第 12～20 行代码表示请求响应时的处理内容，此处代码直接开启一个子线程将响应结果显示在 txtResult 上。onResponse( )方法的回调参数 response 与同步请求方式一样，如果要获得返回的字符串，可以通过 responseBody.string( )获取；如果要获得返回的字节数组，则

可以调用 responseBody. bytes( ); 如果想获得 inputStream 流, 则可以调用 responseBody. byteStream( )。也就是说, 如果开发时需要实现大文件的下载, 可以使用 inputStream 通过 IO 流的方式写文件。但是 onResponse( )执行的线程并不是 UI 线程, 如果要在这儿更新界面上的内容, 一种方法是使用 Handler, 还有一种方法就参照上述第 14 ~ 19 行代码内容实现。

如果请求的网络资源需要带参数, 如要访问 http://news. nnutc. edu. cn/show. asp? id = 2350 这种格式的 URL, 可以直接将这个字符串作为自定义的 getYBRequest( )方法的参数, 也可以将上面的第 3 行代码替换为以下代码。

```
1    HttpUrl httpUrl = HttpUrl.parse(url).newBuilder().addQueryParameter("id","2350").build();
2    Request request = new Request.Builder().url(httpUrl).build();
```

在调用时, 直接使用 http://news. nnutc. edu. cn/show. asp 作为 getYBRequest( )方法的参数即可。

- 绑定"异步 GET"按钮的监听事件。

```
1    btnYBget.setOnClickListener(new View.OnClickListener() {
2        @Override
3        public void onClick(View v) {
4            String url = "http://www.baidu.com";
5            String content = getYBRequest(url);
6        }
7    });
```

此处调用自定义的异步请求访问网络并没有开启子线程, 而是直接在 Android 主线程中执行, 这就是它与同步 GET 请求的区别。

3. Http POST 请求

在 OkHttp3 中用 POST 方式向服务器发送一个请求体时, 请求体需要是一个 RequestBody。OkHttp3 中 POST 请求方式的类型及功能见表 9 - 4。

表 9 - 4   OkHttp3 中 POST 请求方式的类型及功能

类型	功能
key-value	键值对类型
String	字符串类型
Form	类似于 HTML 的表单数据提交
Stream	流类型
File	文件类型

(1) 同步 POST 请求。

- 自定义 postTBRequest( )方法使用 POST 请求获取网页资源数据。

```
1    public String postTBRequest(String url){
2        OkHttpClient client = new OkHttpClient();
3        //创建 Form 表单对象,可以 add 多个键值对(即参数)
4        FormBody formBody = new FormBody.Builder()
```

```
5                  .add("wordKey","chian")
6                  .build();
7          //创建Request对象,至少要设置请求访问的url和post对象
8          Request request = new Request.Builder()
9                  .url(url)
10                 .post(formBody)
11                 .build();
12         try{
13             Call call = client.newCall(request);
14             Response response = call.execute();
15             if(response.isSuccessful()){
16                 return response.body().string();
17             }
18         }catch(IOException e){
19             e.printStackTrace();
20         }
21         return "";
22     }
```

上述第5～6行代码表示该请求体的类型为FormBody表单类型,并传递wordKey的参数值为chian;第10行代码指明了请求方式为POST;其他代码与同步GET方式一样。

- 绑定"同步POST"按钮的监听事件。

```
1      btnTBpost.setOnClickListener(new View.OnClickListener() {
2          @Override
3          public void onClick(View v) {
4              new Thread(){
5                  @Override
6                  public void run() {
7                      String url =
"http://fy.webxml.com.cn/webservices/EnglishChinese.asmx/TranslatorString";
8                      content = postTBRequest(url);
9                      handler.sendEmptyMessage(2000);
10                 }
11             }.start();
12         }
13     });
```

- 定义Handler处理界面内容更新。

```
1      Handler handler = new Handler() {
2          @Override
3          public void handleMessage(Message msg) {
4              super.handleMessage(msg);
5              switch (msg.what) {
6                  case 2000:
7                      tvResult.setText(content);
8                      //其他处理
9                  }
10         }
11     };
```

第9章 网络应用开发

(2) 异步 POST 请求。

- 自定义 postYBRequest( )方法使用 POST 请求获取网页资源数据。

```
1       public String postYBRequest(String url) {
2           OkHttpClient client = new OkHttpClient();
3           FormBody formBody = new FormBody.Builder()
4                   .add("wordKey","chian")
5                   .build();
6           Request request = new Request.Builder()
7                   .url(url)
8                   .post(formBody)
9                   .build();
10          Call call = client.newCall(request);
11          call.enqueue(new Callback() {
12              @Override
13              public void onFailure(Call call, IOException e) {
14                  Toast.makeText(MainActivity.this, "异步GET失败", Toast.LENGTH_LONG).show();
15              }
16              @Override
17              public void onResponse (Call call, Response response) throws IOException {
18                  final String responseStr = response.body().string();
19                  runOnUiThread (new Runnable() {
20                      @Override
21                      public void run() {
22                          tvResult.setText (responseStr);
23                      }
24                  });
25              }
26          });
27          return "";
28      }
```

- 绑定"异步 POST"按钮的监听事件。

```
1       btnYBpost.setOnClickListener(new View.OnClickListener() {
2           @Override
3           public void onClick(View v) {
4               String url = "http://fy.webxml.com.cn/webservices/EnglishChinese.asmx/TranslatorString";
5               String content = postYBRequest(url);
6           }
7       });
```

在上面的同步 POST 和异步 POST 方式提交请求的代码中都用了 post(formBody) 进行 Request 的请求属性设置,也就表示了该请求为 POST 方式。上面两个 POST 请求方式代码的运行效果如图 9.11 所示。

图 9.11　OkHttpClient 获取网页资源（POST）

9.3.2　股票即时查询工具的实现

1. 主界面的设计

股票即时查询工具的主界面分三个区域，上面区域用来显示"上证指数""深圳成指""创业板指"；中间区域用来提供给用户输入、添加股票代码；下面区域用来显示用户添加股票的"股票名称""股票代码""最新报价""涨幅""涨跌"等数据。从图 9.12 可以看出，该主界面的整个布局从上到下是一个线性布局方式。

【查询工具的实现】

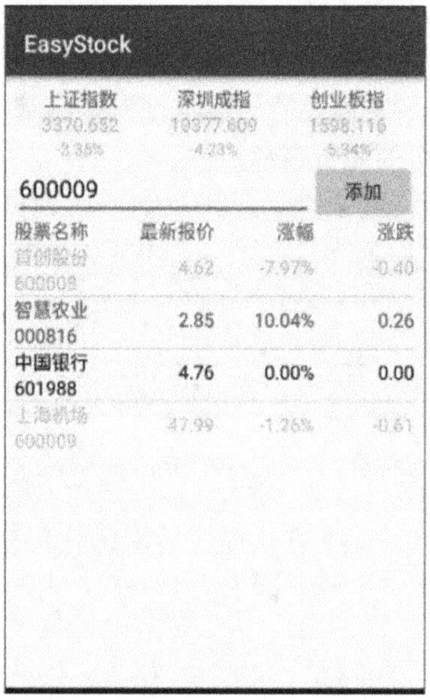

图 9.12　股票即时查询工具的主界面

（1）上面区域的布局。

该区域从左到右分为三个部分，每个部分占用的水平方向空间相等，所以将三个部分的 layout_weight 的属性值设为 "1"，代码如下。

```
1    <LinearLayout
2        android:layout_width=" match_parent"
3        android: layout_height =" wrap_content"
4        android: orientation =" horizontal" >
5        <LinearLayout
6            android: layout_width =" 0dp"
7            android: layout_height =" wrap_content"
8            android: layout_weight =" 1"
9            android: gravity =" center"
10           android: orientation =" vertical" >
11           <TextView
12               android: layout_width =" wrap_content"
13               android: layout_height =" wrap_content"
14               android: text =" 上证指数" / >
15           <TextView
16               android: id =" @ + id/txtshzs"
17               android: layout_width =" wrap_content"
18               android: layout_height =" wrap_content" / >
19           <TextView
20               android: id =" @ + id/txtshzf"
21               android: layout_width =" wrap_content"
22               android: layout_height =" wrap_content"
23               android: textSize =" 12sp" / >
24       </LinearLayout >
25       <LinearLayout
26           android: layout_width =" 0dp"
27           android: layout_height =" wrap_content"
28           android: layout_weight =" 1"
29           android: gravity =" center"
30           android: orientation =" vertical" >
31           <TextView
32               android: layout_width =" wrap_content"
33               android: layout_height =" wrap_content"
34               android: text =" 深圳成指" / >
35           <TextView
36               android: id =" @ + id/txtszzs"
37               android: layout_width =" wrap_content"
38               android: layout_height =" wrap_content" / >
39           <TextView
40               android: id =" @ + id/txtszzf"
41               android: layout_width =" wrap_content"
42               android: layout_height =" wrap_content"
43               android: textSize =" 12sp" / >
44       </LinearLayout >
45       <LinearLayout
46           android: layout_width =" 0dp"
```

```
47              android:layout_height=" wrap_content"
48              android: layout_weight=" 1"
49              android: gravity=" center"
50              android: orientation=" vertical" >
51              <TextView
52                  android: layout_width=" wrap_content"
53                  android: layout_height=" wrap_content"
54                  android: text=" 创业板指" />
55              <TextView
56                  android: id=" @+id/txtcyzs"
57                  android: layout_width=" wrap_content"
58                  android: layout_height=" wrap_content" />
59              <TextView
60                  android: id=" @+id/txtcyzf"
61                  android: layout_width=" wrap_content"
62                  android: layout_height=" wrap_content"
63                  android: textSize=" 12sp" />
64          </LinearLayout>
65      </LinearLayout>
```

(2) 中间区域的布局。

该区域的布局比较简单,水平放置一个 EditText 用于输入股票代码,一个 Button 用于向下面区域添加股票,代码如下。

```
1   <LinearLayout
2       android:layout_width=" match_parent"
3       android: layout_height=" wrap_content"
4       android: orientation=" horizontal" >
5       <EditText
6           android: text=" 600008"
7           android: hint=" 请输入股票代码"
8           android: id=" @+id/edtstockId"
9           android: layout_width=" wrap_content"
10          android: layout_height=" wrap_content"
11          android: layout_weight=" 1"
12          android: inputType=" number"
13          android: maxLength=" 6" />
14      <Button
15          android: id=" @+id/btnadd"
16          android: layout_width=" wrap_content"
17          android: layout_height=" wrap_content"
18          android: text=" 添加" />
19  </LinearLayout>
```

(3) 下面区域的布局。

下面区域分两个部分,第 1 行用水平方式的线性布局放置"股票名称""最新报价""涨幅""涨跌"等四个 TextView,第 2 行开始用 ListView 显示添加的股票信息,代码如下。

```
1   <LinearLayout
2       android:layout_width=" match_parent"
```

```
3          android:layout_height =" wrap_content"
4          android: orientation =" horizontal" >
5          <TextView
6              android: layout_width =" 0dp"
7              android: layout_height =" wrap_content"
8              android: layout_weight =" 1"
9              android: text =" 股票名称" />
10         <TextView
11             android: layout_width =" 0dp"
12             android: layout_height =" wrap_content"
13             android: layout_weight =" 1"
14             android: gravity =" right"
15             android: text =" 最新报价" />
16         <TextView
17             android: layout_width =" 0dp"
18             android: layout_height =" wrap_content"
19             android: layout_weight =" 1"
20             android: gravity =" right"
21             android: text =" 涨幅" />
22         <TextView
23             android: layout_width =" 0dp"
24             android: layout_height =" wrap_content"
25             android: layout_weight =" 1"
26             android: gravity =" right"
27             android: text =" 涨跌" />
28     </LinearLayout >
29     <ListView
30         android: id =" @ +id/listView"
31         android: layout_width =" wrap_content"
32         android: layout_height =" wrap_content" />
```

（4）整体布局。

整体布局就是一个垂直方向的线性布局，代码如下。

```
1   <?xml version ="1.0" encoding ="utf-8"? >
2   <LinearLayout xmlns:android ="http://schemas. android. com/apk/res/android"
3       android:layout_width =" match_parent"
4       android: layout_height =" match_parent"
5       android: orientation =" vertical"
6       android: padding =" 8dp" >
7       <!-- 上面区域代码-->
8       <!-- 中间区域代码-->
9       <!-- 下面区域代码-->
10  </LinearLayout >
```

上述代码中的 ListView 用于显示添加的每个股票的名称、代码、最新价格、涨幅和涨跌。要想获得如图 9.12 所示的样式，就需要再定义一个显示 ListView 每一行内容的布局文件（本案例的布局文件为 show_layout.xml），该布局文件中需要将"股票名称"和"股票代码"显示在同列，然后依次向右显示"最新报价""涨幅""涨跌"。详细代码如下。

```xml
1   <?xml version="1.0" encoding="utf-8"?>
2   <LinearLayout xmlns:android="http://schemas.android.com/apk/res/android"
3       android:layout_width="match_parent"
4       android:layout_height="match_parent"
5       android:orientation="horizontal">
6       <RelativeLayout
7           android:layout_width="0dp"
8           android:layout_height="wrap_content"
9           android:layout_weight="1"
10          android:orientation="vertical">
11          <TextView
12              android:id="@+id/txtname"
13              android:layout_width="match_parent"
14              android:layout_height="wrap_content"/>
15          <TextView
16              android:id="@+id/txtid"
17              android:layout_below="@+id/txtname"
18              android:layout_width="match_parent"
19              android:layout_height="wrap_content"/>
20      </RelativeLayout>
21      <TextView
22          android:id="@+id/txtnewprice"
23          android:layout_marginTop="8dp"
24          android:layout_width="0dp"
25          android:layout_height="wrap_content"
26          android:gravity="right"
27          android:layout_weight="1"/>
28      <TextView
29          android:gravity="right"
30          android:id="@+id/txtzf"
31          android:layout_marginTop="8dp"
32          android:layout_width="0dp"
33          android:layout_height="wrap_content"
34          android:layout_weight="1"/>
35      <TextView
36          android:gravity="right"
37          android:id="@+id/txtzd"
38          android:layout_marginTop="8dp"
39          android:layout_width="0dp"
40          android:layout_height="wrap_content"
41          android:layout_weight="1"/>
42  </LinearLayout>
```

2．功能实现

（1）定义变量。

```
1   //上证指数、上证涨幅、深圳指数、深圳涨幅、创业板指、创业涨幅
2   private TextView txtshzs, txtshzf, txtszzs, txtszzf, txtcyzs, txtcyzf;
3   //股票名称、股票代码、股票最新价格、涨幅、涨跌
4   private TextView tvname, tvid, tvprice, tvzf, tvzd;
```

```
5        private EditText edtstockid;//股票代码
6        private Button btnadd;//添加
7        private ListView lv;//显示添加股票的相关信息
8        private String content;//返回网络响应结果字符串
9        //用键值对格式保存添加的股票信息
10       List<Map<String,Object>> listitem = new ArrayList<Map<String,Object>>();
11       //ListView的自定义适配器
12       private MySimpleAdapter adapter = null;
```

上述第12行代码定义了一个继承自SimpleAdapter类的自定义适配器。因为本案例中ListView每一行显示的股票信息会根据该行股票涨跌情况分别显示不同的颜色,当涨跌值为0时股票信息用黑色显示;当涨跌值大于0时股票信息用红色显示;当涨跌值小于0时股票信息用绿色显示。该类的详细代码如下。

```
1    public class MySimpleAdapter extends SimpleAdapter {
2        List<? extends Map<String, ?>> mdata;
3        Context context;
4        public MySimpleAdapter(Context context, List<? extends Map<String, ?>> data, int resource, String[] from, int[] to) {
5            super(context, data, resource, from, to);
6            this.mdata = data;
7            this.context = context;
8        }
9        @Override
10       public View getView(int position, View convertView, ViewGroup parent) {
11           if (convertView == null) {
12               convertView = LinearLayout.inflate(context, R.layout.show_layout, null);
13           }
14           //此处的TextView是R.layout.show_layout里面的,也就是需要重新设置的字体颜色
15           TextView textView1 = (TextView) convertView.findViewById(R.id.txtname);
16           TextView textView2 = (TextView) convertView.findViewById(R.id.txtid);
17           TextView textView3 = (TextView) convertView.findViewById(R.id.txtnewprice);
18           TextView textView4 = (TextView) convertView.findViewById(R.id.txtzf);
19           TextView textView5 = (TextView) convertView.findViewById(R.id.txtzd);
20           String ss = (String) mdata.get(position).get("zd");//取出涨跌值
21           if (Float.parseFloat(ss) <0) {//涨跌值小于0,则绿色
22               textView1.setTextColor(Color.rgb(00, 255, 00));
23               // textView2、textView3、textView4、textView5设置一样,此处略
24           } else if (Float.parseFloat(ss) ==0) {//黑色
25               textView1.setTextColor(Color.rgb(0, 0, 0));
26               //textView2、textView3、textView4、textView5设置一样,此处略
27           } else  {//红色
28               textView1.setTextColor(Color.rgb(255, 0, 0));
29               //extView2、textView3、textView4、textView5设置一样,此处略
30           }
31           return super.getView(position, convertView, parent);
32       }
33   }
```

上述第 12 行代码使用 inflate( ) 方法将 ListView 的每一行布局文件转换为 View；第 15 ～ 19 行代码表示实例化的 TextView，以便对它们重新进行属性设置，从而控制它们的显示效果；第 20 行代码表示取出每一行匹配的 Map <String,？> > 数据中的涨跌值（键——zd，此处存放的值为 String 类型）；第 21 ～ 30 行代码表示将 String 类型的涨跌值转换为 Float 类型，并对其判断。如果涨跌值小于 0，将该行中对应的全部 TextView 设置为绿色；如果涨跌值等于 0，将该行中对应的全部 TextView 设置为黑色；如果涨跌值大于 0，将该行中对应的全部 TextView 设置为红色。

（2）自定义获取三大指数数据的 showInfo( ) 方法。

为了获取股票实时信息，本案例使用了新浪网提供的 URL（http://hq.sinajs.cn/list = 代码），其中代码由传递的参数值决定。例如，http://hq.sinajs.cn/list = sz399001 表示要访问的是深圳成指的信息；http://hq.sinajs.cn/list = sh600008 表示要访问的是上海的股票——600008（首创股份）的信息。用这个 URL 访问网络资源时返回的数据格式如图 9.13 所示。从图中可以看出，其返回的字符串由许多数据拼接在一起，不同含义的数据用逗号隔开。按照程序员的思路，这些拼接的数据含义见表 9 - 5。

图 9.13　股票返回数据格式

表 9 - 5　股票数据含义

下标	含义	下标	含义	下标	含义
0	股票名字	9	成交金额（万元单位）	20	卖一股数
1	今日开盘价	10	买一申请股数	21	卖一报价
2	昨日收盘价	11	买一报价	22	卖二股数
3	当前价格	12	买二申请股数	23	卖二报价
4	今日最高价	13	买二报价	24，25	卖三股数、报价
5	今日最低价	14	买三申请股数	26，27	卖四股数、报价
6	竞买价，即"买一"报价	15	买三报价	28，29	卖五股数、报价
7	竞卖价，即"卖一"报价	16，17	买四申请股数、报价	30	日期
8	成交的股票数（100 单位）	18，19	买五申请股数、报价	31	时间

所以，在实现这一功能时需要先将如图 9.13 所示的字符串进行解析，然后再将解析结果按照表 9 - 5 对应的含义进行相应处理后放置在界面的不同位置上，详细代码如下。

```
1    void showInfo(String stockid, final TextView tv1, final TextView tv2) {
2        String url = "http://hq.sinajs.cn/list =" + stockid;
3        OkHttpClient okHttpClient = new OkHttpClient();
4        Request request = new Request.Builder().url(url).build();
```

```
5              Call call = okHttpClient.newCall(request);
6              call.enqueue(new Callback() {
7                  @Override
8                  public void onFailure(Call call, IOException e) {
9                  }
10                 @Override
11                 public void onResponse(Call call, final Response response) throws IOException {
12                     runOnUiThread(new Runnable() {
13                         @Override
14                         public void run() {
15                             try {
16                                 String stockinfo = response.body().string();
17                                 String[] temp = stockinfo.split(",");
18                                 float zf = (Float.parseFloat(temp[3]) - Float.parseFloat(temp[2])) / Float.parseFloat(temp[2]) * 100;
19                                 if (zf <0) {//涨幅小于0,则绿色显示
20                                     tv1.setTextColor(0xff00ff00);
21                                     tv2.setTextColor(0xff00ff00);
22                                 } else if(zf ==0) {//涨幅等于0,则黑色显示
23                                     tv1.setTextColor(0xff000000);
24                                     tv2.setTextColor(0xff000000);
25                                 }else {////涨幅大于0,则红色显示
26                                     tv1.setTextColor(0xffff0000);
27                                     tv2.setTextColor(0xffff0000);
28                                 }
29                                 DecimalFormat decimalFormat = new DecimalFormat("0.00");
30                                 String p = decimalFormat.format(zf);
31                                 tv1.setText(Float.parseFloat(temp[3]) + "");
32                                 tv2.setText(decimalFormat.format(zf) + "%");
33                             } catch (IOException e) {
34                                 e.printStackTrace();
35                             }
36                         }
37                     });
38                 }
39             });
40         }
```

上述第1行代码的三个参数分别表示要访问的股票代码值、当前报价和涨幅显示的TextView；第6行代码表示以异步GET方式执行Request请求；第12～39行代码表示在onResponse()方法中对返回的response.body().string()数据（格式如图9.12所示）进行解析，即首先将字符串用","分隔后放到temp数组中，其中第18行代码表示按"（当前报价－昨日收盘价）/昨日收盘价*100"公式算出涨幅zf，然后第19～28行代码根据涨幅zf的值判断是用红色显示、绿色显示还是黑色显示；最后使用DecimalFormat将数据格式化后显示到指定对象上（当前报价显示在tv1上，涨幅显示在tv2上）。

应用程序一旦运行，首先就需要向网络请求信息，即获取三大指数的值，并在界面的指定位置显示，所以该方法的调用应该放置在主界面对应的Activity（本案例为MainActivity）

的 onCreate( )方法中，调用代码如下。

```
1   showInfo("sh000001", txtshzs, txtshzf);// (上证指数、指数/当前报价、涨幅)
2   showInfo("sz399001", txtszzs, txtszzf);//(深圳成指、指数/当前报价、涨幅)
3   showInfo("sz399006", txtcyzs, txtcyzf);//(创业板指数、指数/当前报价、涨幅)
```

（3）自定义获取股票相关数据的 getData( )方法。

在输入股票代码并单击"添加"按钮后，使用同步 GET 请求方式获取网络资源，其代码如下。

```
1   String getData(String url) {
2       OkHttpClient okHttpClient = new OkHttpClient();
3       Request request = new Request.Builder().url(url).build();
4       Call call = okHttpClient.newCall(request);
5       try {
6           Response response = call.execute();
7           return response.body().string();
8       } catch (IOException e) {
9           e.printStackTrace();
10      }
11      return "";
12  }
```

上述第 1 行代码的 url 参数格式为 "http://hq.sinajs.cn/list=股票代码参数"，由于股票代码的格式服务器需要使用类似 "sh600008" 这样的形式，所以在添加时需要根据输入的代码判断在代码前加上 "sh" 还是 "sz"，这个功能本案例是直接在"添加"按钮监听事件中实现的。

（4）绑定"添加"按钮监听事件。

```
1   btnadd.setOnClickListener(new View.OnClickListener() {
2       @Override
3       public void onClick(View v) {
4           final String stockid = edtstockid.getText().toString();
5           String findstock = "sh";
6           if (stockid.trim().length() == 0 || stockid.trim().length() <6) {
7               Toast.makeText(MainActivity.this, "请输入股票代码", Toast.LENGTH_LONG).show();
8               return;
9           }
10          String preid = stockid.trim().substring(0, 1);
11          switch (preid) {
12              case "6":
13                  findstock = "sh"; //上海股票
14                  break;
15              case "0":
16              case "3":
17                  findstock = "sz"; //深圳股票
18          }
19          final String url = "http://hq.sinajs.cn/list=" + findstock + stockid;
20          new Thread() {
```

第9章 网络应用开发

```
21                    @Override
22                    public void run() {
23                        content = getData(url);
24                        String[] temp = content.split(",");
25                        Map<String, Object> showitem = new HashMap<String, Object>();
26                        String[] name = temp[0].split("\"");
27                        showitem.put("name", name[1]);
28                        showitem.put("id", stockid);
29                        showitem.put("price", Float.parseFloat(temp[3]));
30                        float zf = (Float.parseFloat(temp[3]) - Float.parseFloat(temp[2])) / Float.parseFloat(temp[2]) * 100;
31                        DecimalFormat decimalFormat = new DecimalFormat("0.00");//构造方法的字符格式,这里如果小数不足2位,会以0补足.
32                        String p = decimalFormat.format(zf);
33                        showitem.put("zf", decimalFormat.format(zf) + "% ");
34                        float zd = Float.parseFloat(temp[3]) - Float.parseFloat(temp[2]);
35                        showitem.put("zd", decimalFormat.format(zd));
36                        if (! listitem.contains(showitem)){
37                        listitem.add(showitem);}//如果不重复
38                        handler.sendEmptyMessage(3000);
39                    }
40                }.start();
41            }
42        });
```

上述第4行代码表示从输入框中取出输入的股票代码（第6～9行代码对输入的代码合理性进行判断），因为使用http://hq.sinajs.cn/list=sz399001地址时，list参数值的最前面两个字母sz表示深圳、sh表示上海，这个字母值需要根据输入股票代码的最前一位字符进行判断，即6开头对应sh，0或3开头对应sz。第23行代码表示调用getData()方法从网络获得数据；第24行代码根据","对获得的字符串进行分隔，并保存在temp数组中。因为temp[0]中的元素值包含股票名称，所以可根据"\"将股票名称解析出来，然后将股票名称、股票代码、最新报价、涨幅和涨跌分别以name、id、price、zf和zd为键保存到Map，并添加到ArrayList中（本案例为listitem）；最后通知Handler更新界面。

（5）定义Handler更新界面内容。

```
1    Handler handler = new Handler() {
2        @Override
3        public void handleMessage(Message msg) {
4            super.handleMessage(msg);
5            switch (msg.what) {
6                case 3000://单击"添加"按钮后执行
7                    String[] from = new String[]{"name", "id", "price", "zf", "zd"};
8                    int[] to = new int[]{R.id.txtname, R.id.txtid, R.id.txtnewprice, R.id.txtzf, R.id.txtzd};
9                    MySimpleAdapter adapter = new MySimpleAdapter(MainActivity.this, listitem, R.layout.show_layout, from, to);
```

```
10                lv.setAdapter(adapter);;
11                break;
12           }
13       }
14  };
```

上述第 7～9 行代码表示使用自定义的 SimpleAdapter 为 ListView 填充数据。

到此,股票即时查询工具就已经开发完成。如果在正常股票交易时间需要自动更新 ListView 上添加的股票,可以创建一个子线程,每隔一段时间使用自定义的 showInfo() 和 getData() 方法到网络中获取实时数据,然后对获得的数据按照前面介绍的方法进行解析并处理后,就可以定时刷新股票数据了。感兴趣的读者可以自行完成此项功能。本案例的实现代码读者可以参见 ALLAPP 代码包中 easystock 文件夹里的内容。

9.4 快递单查询工具的设计与实现

随着移动互联网和移动终端设备的发展,基于 Android 移动设备端的查询需求越来越多,快递单查询就是其中最常用的应用之一。本案例通过开发一个基于 Android 平台的快递单查询系统介绍 Android 客户端使用 SOAP 技术实现与 Web Service 服务器进行交互的方法。

9.4.1 预备知识

1. 加载第三方类库

【第三方类库KSOAP2 (WebService访问)】

在 Android SDK 中并没有提供调用 Web Service 的类库,所以目前开发者都在使用效率较高的第三方类库(KSOAP2)来调用 Web Service。KSOAP2 是 Enhydra.org 的一个开源作品,在使用前需要先下载 ksoap2-android-assembly-3.6.0-jar-with-dependencies.jar(本书案例使用该版本)或更高版本的包,然后需要在 Android Studio 开发环境中导入后才能在 Android 项目开发中使用。这种下载到本地的第三方依赖库(包)的导入步骤如下。

● 将 ksoap2-android-assembly-3.6.0-jar-with-dependencies.jar 文件复制到 Project→App (Module) 文件夹下的 libs 文件夹下(如果没有该文件夹,用户可以自己创建)。

● 右键单击 ksoap2-android-assembly-3.6.0-jar-with-dependencies.jar 文件,在弹出的菜单上选择 Add As Library…,弹出图 9.14 所示的 Create Library 对话框,单击该对话框上的 OK 按钮即可。

如果 KSOAP2 库导入成功,则会在 Project→App(Module) 文件夹下的 build.gradle 文件中的 dependencies 依赖设置区域显示如下代码。

```
1  dependencies {
2      //其他设置
3      compile files('libs/ksoap2-android-assembly-3.6.0-jar-with-dependencies.jar')
4  }
```

导入 KSOAP2 类库后,org.ksoap2 包中如表 9-6 所示的类在访问 Web Service 时是必须使用的。

第9章 网络应用开发

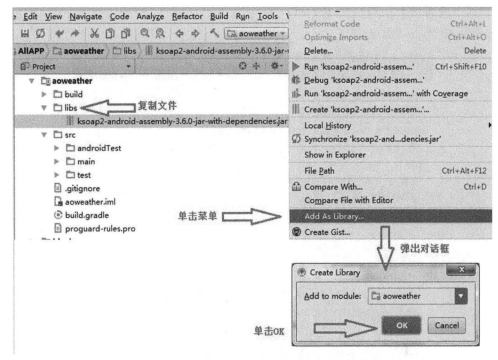

图 9.14 导入第三方类库

表 9-6 KSOAP2 的常用类及功能说明

类　　名	功　能　说　明
org.ksoap2.SoapObject	用于创建 SOAP 对象，实现 SOAP 调用
org.ksoap2.SoapEnvelope	实现了 SOAP 标准中的 SOAP Envelope，封装了 head 对象和 body 对象
org.ksoap2.SoapSerializationEnvelope	是 KSOAP2 中对 SOAP Envelope 的扩展，支持 SOAP 序列化（Serialization）格式规范，可以对简单对象自动进行序列化
org.ksoap2.Transport.HttpTransport	用于进行 Internet 访问，以获取 Web Service 服务器 SOAP

2. 调用 Web Service 的步骤

下面以访问 http://www.webxml.com.cn 网站手机号码归属地查询服务为例（运行效果如图 9.15 所示）介绍 KSOAP2 访问 Web Service 的实现步骤。

（1）查阅 WSDL 文档获取访问 Web Service 服务的重要参数。

在访问 Web Service 服务时需要 URL、NameSpace、Method 和 SoapAction 四个重要参

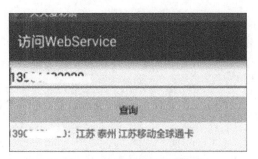

图 9.15 手机号码归属地查询

数，这四个重要参数中后三个参数可以直接参照图 9.16 获得，SoapAction 的值通常由 NameSpace 和 Method 组成。

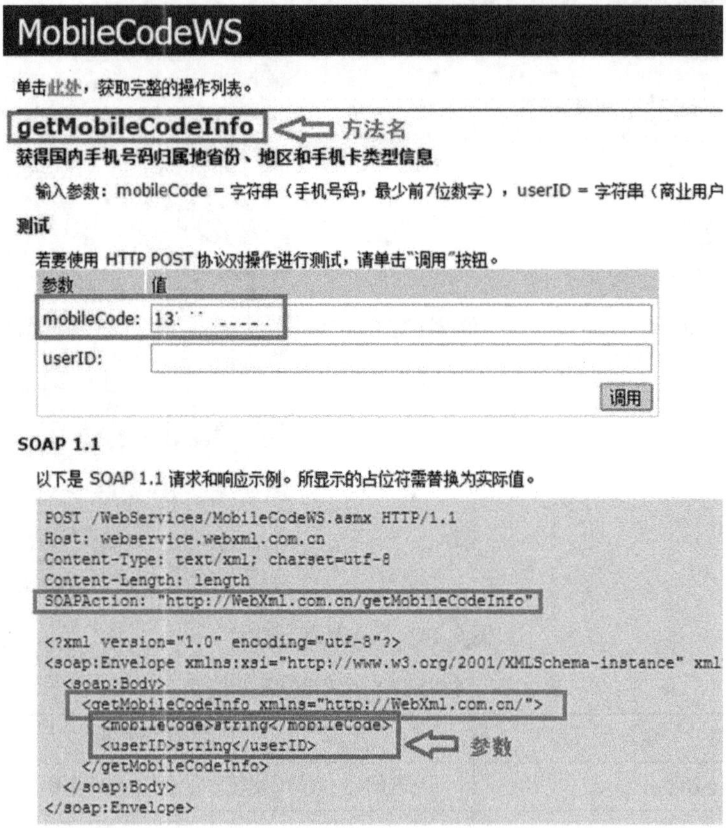

图 9.16 Web Service 页面

（2）实例化 SoapObject 对象，指定 Web Service 的命名空间、URL 和服务调用方法名称，代码如下。

```
1    //归属地查询相关参数
2    private  String  nameSpace = "http://WebXml.com.cn/";//命名空间
3    private  String  url = "http://ws.webxml.com.cn/WebServices/MobileCodeWS.asmx";
4    private  String  method = "getMobileCodeInfo";//调用方法
5    private  String  soapAction = nameSpace + method;
6    SoapObject request = new SoapObject(nameSpace, method); //实例化 SoapObject 对象
```

（3）设置调用方法的参数和参数值（若没有参数，此步骤可以省略），代码如下。

```
request.addProperty("参数名称1","参数值1");
request.addProperty("参数名称2","参数值2");
//……
```

要注意的是，addProperty()方法的参数名与 Web Service 服务端提供的名称必须完全一样。

（4）设置 SOAP 请求信息。参数部分为 SOAP 协议版本号，与要调用的 Web Service 中

的版本号一致，如图 9.16 所示的 SOAP1.1，代码如下。

```
//获得序列化的 Envelope
SoapSerializationEnvelope envelope = new SoapSerializationEnvelope(SoapEn-
velope.VER11);
envelope.bodyOut = request;
envelope.dotNet = true;//.Net 环境必须加
```

创建 SoapSerializationEnvelope 对象时需要通过 SoapSerializationEnvelope 类的构造方法设置 SOAP 协议的版本号，该版本号需要根据 Web Service 服务端的版本号设置，对应版本号见表 9-7。在创建 SoapSerializationEnvelope 对象后，一定要设置 SOAPSoapSerialization Envelope 类的 bodyOut 属性，该属性的值就是在第 1 步创建的 SoapObject 对象。如果提供 Web Service 的服务器是 .NET 环境，还必须在此处增加语句"envelope.dotNet = true;"。

表 9-7　SOAP 协议的版本号

SOAP 规范	版 本 号
SOAP 1.0	SoapEnvelope.VER10
SOAP 1.1	SoapEnvelope.VER11
SOAP 1.2	SoapEnvelope.VER12

（5）通过传递 SOAP 数据的目标地址实例化 HttpTransportsSE 传输对象，代码如下。

```
HttpTransportSE transport = new HttpTransportSE(url,2000);
transport.debug = true;
```

（6）使用 call( ) 方法调用 Web Service 方法（其中第 1 个参数为 SoapAction 值，通常由命名空间+方法名称组成，第 2 个参数为 Envelope 对象），代码如下。

```
transport.call(serviceNameSpace + methodName, envelope);
```

（7）获得 Web Service 服务返回的结果，可以使用以下方法获得返回结果，代码如下。

```
SoapObject result = (SoapObject)envelope.bodyIn;
```

（8）解析 SoapObject 对象，代码如下。

```
for (int i = 0; i < result.getPropertyCount(); i ++){
    System.out.println("result.getProperty(" + i + ")" + result.getProperty(i));
}
```

此处以获得 String 类型的获取目标为例，通常返回的 String 类型有单一 String、一个一维 String 数组或一个二维 String 数组三种情况。

对于单一 String 返回值，envelope.bodyIn 就不能强制转换成 SoapObject 类型，否则会出现转换类型错误提示信息。此时 envelope.bodyIn 可以直接作为一个 Object 类型的对象返回，也可以直接调用其 toString( ) 方法转换为字符串。

对于一个一维 String 数组返回值，envelope.bodyIn 就需要强制转换成 SoapObject 类型，然后调用其 getProperty(int index) 方法即可获得数据，参数对应一维数组的下标，上面代码就是这种方式。

对于一个二维 String 数组，getProperty(int index) 方法返回的仍然是 SoapObject 类型对象，此时需要两次调用 getProperty(int index) 方法才能获得最终数据。

(9) 设置访问网络的权限。在 AndroidManifest.xml 文件中加入 uses-permission 项，代码如下。

```
<uses-permission android:name ="android.permission. INTERNET" > </uses-permission >
```

图 9.15 的界面实现比较简单，限于篇幅，不再赘述。功能实现时按照如下步骤执行。

● 定义变量。

```
1       private TextView tvInfo;
2       private Button btnFind;
3       private EditText edtNo;
4       private SoapObject soapObject;
5       private  String  nameSpace = "http://WebXml.com.cn/";
6       private String content ="";
7       private  String  url =
"http://ws.webxml.com.cn/WebServices/MobileCodeWS.asmx";
8       private  String  method = "getMobileCodeInfo";
9       private  String  soapAction = nameSpace +  method;
```

● 定义获取 Web Service 数据的 getMoblieLand（String nameSpace，String method，String url，String tel）方法，详细代码如下。

```
1     public String getMoblieLand( String nameSpace,String method,String url,String tel) {
2         String result = "网络连接失败!";
3         SoapObject soapObject = new SoapObject(nameSpace, method);
4         soapObject.addProperty("mobileCode",tel);
5         SoapSerializationEnvelope envelope = new
SoapSerializationEnvelope(SoapEnvelope.VER11);
6         envelope.bodyOut = soapObject;
7         envelope.dotNet = true;
8         envelope.setOutputSoapObject(soapObject);
9         HttpTransportSE httpTransportSE = new HttpTransportSE(url);
10        try {
11            httpTransportSE.call( nameSpace +method, envelope);
12            SoapObject object = (SoapObject) envelope.bodyIn;
13            result = object.getProperty(0).toString();
14            return result;
15        } catch (Exception e) {
16            e.printStackTrace();
17        }
18        return result;
19    }
```

● 绑定"查询"按钮监听事件。

```
1     btnFind.setOnClickListener(new View.OnClickListener() {
2         @Override
3         public void onClick(View v) {
4             new Thread(){
5                 @Override
6                 public void run() {
7                     String tel = edtNo.getText().toString();
8                     content = getMoblieLand( nameSpace, method, url, tel);
```

```
 9                    handler.sendEmptyMessage(1000);
10                }
11            }.start();
12        }
13    });
```

- 定义 Handler 更新界面内容。

```
 1  private Handler handler = new Handler(){
 2      @Override
 3      public void handleMessage(Message msg) {
 4          super.handleMessage(msg);
 5          switch (msg.what){
 6              case 1000:
 7                  tvInfo.setText(content);
 8                  break;
 9          }
10      }
11  };
```

9.4.2 快递单查询工具的实现

1. 主界面的设计

主界面上放置了一个 Spinner 组件,用于显示快递公司名称;一个 EditText 组件用于输入快递单号;一个 ScrollView 组件和一个 TextView 组件可以上下滚动显示快递单跟踪信息。整个布局文件代码比较简单,不再列出,请读者参见 ALLAPP 代码包中 findexpress 文件夹里的内容,运行后效果如图 9.17 所示。

【联动效果的实现】

图 9.17 快递单查询效果

2. 功能实现

获取快递单信息的 Web Service 如图 9.18 所示。从该网站的信息可以看到使用 Web Service 服务的相关重点参数 URL、SoapAction、NameSpace、Method 的值及调用时需要传递的参数名及参数类型。

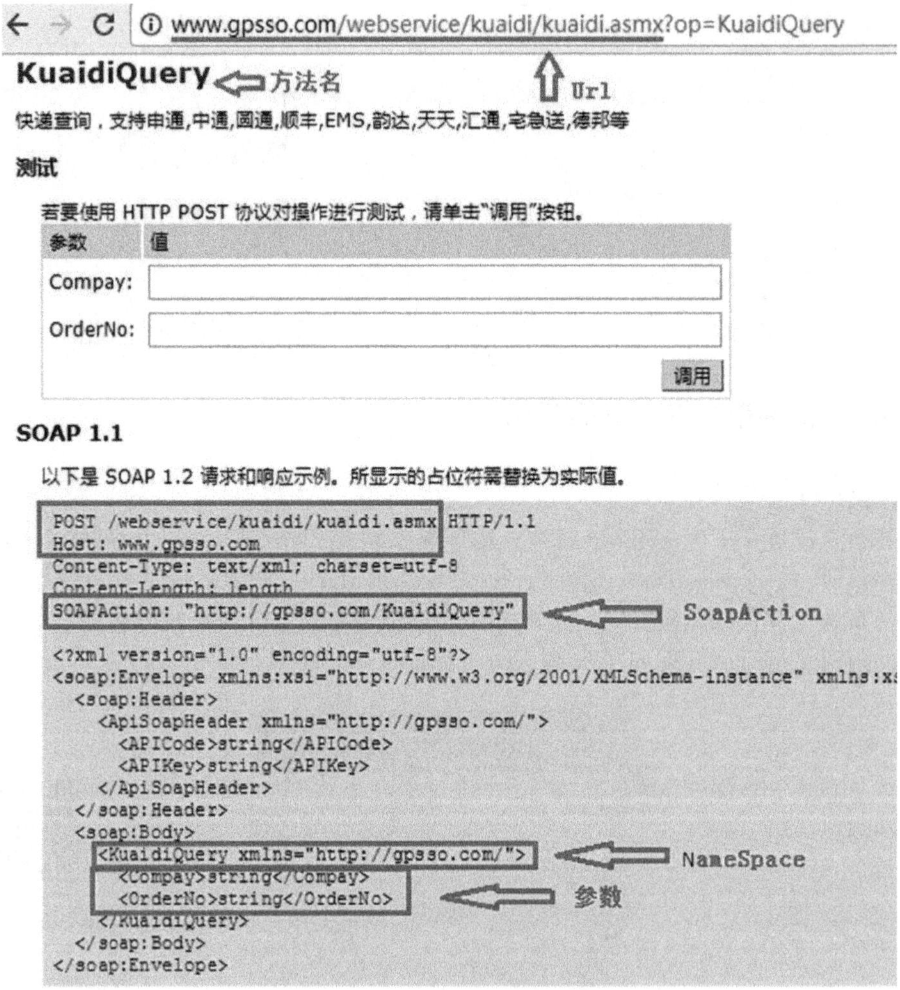

图 9.18 快递查询 Web Service

(1) 定义变量。

```
1    private String nameSpace ="http://gpsso.com/";//命名空间
2    private String methodName ="KuaidiQuery";//方法名
3    //Web Service 的 Url
4    private String url ="http://www.gpsso.com/webservice/kuaidi/kuaidi.asmx";
5    private TextView tvInfo;
6    private Button btnFind;
7    private EditText editNo;
8    private Spinner spinner;
```

```
9       private String[] company = new String[]{"申通","中通","圆通","顺丰",
"EMS","韵达","天天","汇通","宅急送","德邦"};
10      private String cpy = "";//存放 Spinner 选择的快递公司名
11      String content ="";//存放网络访问的返回结果
```

（2）给 Spinner 绑定数据并设置监听事件。

应用程序运行后，就直接将快递公司的数据绑定到 Spinner 组件，即直接在 onCreate( ) 方法中加入如下代码。

```
1    ArrayAdapter arrayAdapter = new ArrayAdapter(this,android.R.layout.simple_list_item_1, company);
2    spinner.setAdapter (arrayAdapter);
3    spinner.setOnItemSelectedListener (new AdapterView.OnItemSelectedListener() {
4        @Override
5        public void onItemSelected (AdapterView<?> parent, View view, int position, long id) {
6            cpy = company [position];
7        }
8        @Override
9        public void onNothingSelected (AdapterView<?> parent) {
10       }
11   });
```

上述第 6 行代码表示选中 Spinner 列表中的某项后，直接将该项的内容赋值给 cpy 变量，以便在查询快递信息时传递给访问 Web Service 的自定义方法 getOrderInfo( )。

（3）自定义获取快递单信息的 getOrderInfo( )方法。

```
1    public SoapObject getOrderInfo (String nameSpace, String method, String company,String orderNo) {
2        SoapObject soapObject = new SoapObject(nameSpace, method);
3        soapObject.addProperty("Compay",company);
4        soapObject.addProperty("OrderNo",orderNo);
5        SoapSerializationEnvelope envelope = new SoapSerializationEnvelope(SoapEnvelope.VER11);
6        envelope.bodyOut = soapObject;
7        envelope.dotNet = true;
8        envelope.setOutputSoapObject(soapObject);
9        HttpTransportSE httpTransportSE = new HttpTransportSE(url,2000);
10       try {
11           httpTransportSE.call(nameSpace + method, envelope);
12           SoapObject object = (SoapObject) envelope.bodyIn;
13           return object;
14       } catch (Exception e) {
15           e.printStackTrace();
16       }
17       return null;
18   }
```

该方法有四个参数，第 1 个参数为命名空间，第 2 个参数为方法名，第 3 个参数为传递的公司名称，第 4 个参数为传递的快递单号。

(4)"查询"按钮绑定监听事件。

```
1    btnFind.setOnClickListener(new View.OnClickListener() {
2        @Override
3        public void onClick(View v) {
4            final String no = editNo.getText().toString();
5            new Thread(){
6                @Override
7                public void run() {
8                    SoapObject soapObject = getOrderInfo(nameSpace, methodName,cpy,no);
9                    content = soapObject.toString();
10                   handler.sendEmptyMessage(1000);
11               }
12           }.start();
13       }
14   });
```

由于单击按钮后需要访问网络资源,所以调用 Web Service 需要在子线程中实现,并将返回结果通过 Handler 传递到主线程,最后更新 TextView 中的内容。

(5)Handler 更新主界面内容。

```
1    Handler handler = new Handler(){
2        @Override
3        public void handleMessage(Message msg) {
4            switch (msg.what){
5                case 1000:
6                    tvInfo.setText(content);
7                    break;
8            }
9        }
10   };
```

要实现在线查询快递单信息必须要用到 Android 中联网的权限,即在 AndroidManifest.xml 文件中添加如下内容。

```
<uses-permission android:name="android.permission.INTERNET"/>
```

至此,快递单查询工具的基本功能已经实现,读者可以在此项目示例的基础上进行扩展,如将获得的快递单信息解析出更符合平时阅读习惯的样式,这样既可以巩固前面介绍的知识,还能增加软件的友好性。

9.5 WebView 的应用

现在诸如淘宝、京东、苏宁易购等电商平台的 App 里都内置了 Web 网页,它们都是通过 WebView 来实现的。WebView 的开发使用比较灵活,如果想要高效快速地展现商家的信息,只需要修改网页代码即可,并不需要升级客户端程序,所以一些需要经常变化的页面可以用 WebView 这种方式加载到客户端。例如,在不同的节日商家推荐的商品页面是不一样的,此时程序员只需要修改 HTML 页面就可以了。

9.5.1 预备知识

使用 WebView 进行应用程序开发时需要使用到 WebSettings、WebChromeClient 和 WebViewClient 三个类。WebView 的常用方法及功能说明见表 9-8。

表 9-8 WebView 的常用方法及功能说明

方 法 名	功 能 说 明
getSettings()	返回一个 WebSettings 对象用来控制 WebView 的属性设置
onResume()	激活 WebView 为活跃状态,能正常执行网页的响应
onPause()	当页面失去焦点被切换到后台不可见状态时,需要执行 onPause
pauseTimers()	当应用程序(存在 WebView)被切换到后台时,暂停所有 WebView 的 layout、parsing、javascripttimer,降低 CPU 功耗
resumeTimers()	恢复 pauseTimers 状态
canGoBack()	是否可以后退
goBack()	后退网页
canGoForward()	是否可以前进
goForward()	前进网页
clearCache(boolean)	清除网页访问留下的缓存
clearHistory()	清除当前 WebView 访问的历史记录
clearFormData()	清除自动完成填充的表单数据并不会清除 WebView 存储到本地的数据
goBackOrForward(int s)	s 为负数则后退,为正数则前进
loadUrl(String url)	加载指定的 URL
loadData(String data, String mimeType, String encoding)	加载指定的 Data 到 WebView 中。使用"data:"作为标记头,该方法不能加载网络数据,其中 mimeType 为数据类型,如 text/html、image/jpeg;encoding 为字符的编码方式
loadDataWithBaseURL(String baseUrl, String data, String mimeType, String encoding, String historyUrl)	加载指定的 Data 到 WebView 中
setWebViewClient(WebViewClient client)	为 WebView 指定一个 WebViewClient 对象,WebViewClient 可以辅助 WebView 处理各种通知、请求等事件
setWebChromeClient(WebChromeClient client)	为 WebView 指定一个 WebChromeClient 对象,WebChromeClient 专门用来辅助 WebView 处理 JavaScript 的对话框、网站标题、网站图标、加载进度条等

WebSettings 用于管理 WebView 状态配置，当 WebView 第 1 次被创建时，WebView 包含着一个默认的配置。WebSettings 的常用方法及功能说明见表 9-9。

表 9-9 WebSettings 的常用方法及功能说明

方 法 名	功 能 说 明
setJavaScriptEnabled(boolean)	设置 WebView 是否允许执行 JavaScript 脚本，默认是 false，不允许
setJavaScriptCanOpenWindowsAutomatically(boolean)	设置脚本是否允许自动打开弹窗，默认是 false，不允许
setDefaultTextEncodingName(String)	设置 WebView 加载页面文本内容的编码，默认是"UTF-8"
setUseWideViewPort(boolean)	设置是否将图片调整到适合 WebView 窗口大小
setLoadWithOverviewMode(boolean)	设置是否缩放到屏幕大小
setGeolocationEnabled(boolean)	设置是否开启定位功能，默认是 true，开启定位
setGeolocationDatabasePath(String)	设置 WebView 保存地理位置信息数据的路径，指定的路径 Application 具备写入权限
setAppCachePath(String)	设置当前 Application 缓存文件路径
setAppCacheEnabled(boolean)	设置 Application 缓存 API 是否开启，默认是 false
setLoadsImagesAutomatically(boolean)	设置 WebView 是否自动加载图片资源，默认是 true，自动加载图片
setBlockNetworkImage(boolean)	设置 WebView 是否以 http、https 方式访问从网络加载图片资源，默认是 false
setSupportZoom(boolean)	设置 WebView 是否支持使用屏幕控件或手势进行缩放，默认是 true，支持缩放
setMediaPlaybackRequiresUserGesture(boolean)	设置 WebView 是否通过手势触发播放媒体，默认是 true，需要手势触发
setBuiltInZoomControls(boolean)	设置 WebView 是否使用其内置的缩放机制，该机制集合屏幕缩放控件使用，默认是 false，不使用内置缩放机制
setDisplayZoomControls(boolean)	设置 WebView 使用内置缩放机制时，是否展现在屏幕缩放控件上，默认是 true，展现在控件上
setAllowFileAccess(boolean)	设置在 WebView 内部是否允许访问文件，默认是允许访问

(续)

方 法 名	功 能 说 明
setAllowContentAccess(boolean)	设置 WebView 是否允许访问 URL，默认是 true
setSaveFormData(boolean)	设置 WebView 是否保存表单数据，默认是 true，保存数据
setSupportMultipleWindows(boolean)	设置 WebView 是否支持多屏窗口，默认是 false，不支持
setMinimumFontSize(int)	设置 WebView 字号最小值，默认值是 8，取值 1 到 72
setDefaultFontSize(int)	设置 WebView 默认值字号值，默认值是 16，取值 1 到 72
setDatabaseEnabled(boolean)	设置是否开启数据库存储 API 权限，默认是 false，不开启
setDatabasePath(String)	设置数据库存储路径

WebViewClient 用于辅助 WebView 处理各种通知、请求等事件。WebViewClient 的常用方法及功能说明见表 9 - 10。

表 9 - 10 WebViewClient 的常用方法及功能说明

方 法 名	功 能 说 明
onPageStared(WebView view, String url)	开始载入页面调用的，可以设定一个 loading 进度页面，告诉用户程序在等待网络响应
onPageFinished(WebView view, String url, Bitmap favicon)	在页面加载结束时调用，可以关闭 loading 进度条，切换程序动作
doUpdateVisitedHistory(WebView view, String url, boolean isReload)	更新历史记录
onLoadResource(WebView view, String url)	在加载页面资源时会调用，每一个资源（例如图片）的加载都会调用一次
onScaleChanged(WebView view, float oldScale, float newScale)	WebView 的缩放发生改变时调用
shouldOverrideKeyEvent(WebView view, KeyEvent event)	控制 WebView 是否处理按键时间，返回 true 则 WebView 不处理，返回 false 则处理

(续)

方 法 名	功 能 说 明
shouldOverrideUrlLoading（WebView view, String url）	当新的 URL 被加载时，也就是用户单击了 view 内容里面的一个链接时会调用该方法
onReceivedError（WebView view, int errorCode, String description, String failingUrl）	加载出现错误时调用，在其中可以进行错误处理，如再请求加载一次或提示 404 错误
onReceivedSslError（WebView view, SslErrorHandler handler, SslError error）	接收到 https 错误时调用，在其中可以进行错误处理

WebChromeClient 用于辅助 WebView 处理 JavaScript 的对话框、网站图标、网站标题等，它的常用方法及功能说明见表 9-11。

表 9-11 WebChromeClient 的常用方法及功能说明

方 法 名	功 能 说 明
onJsAlert（WebView view, String url, String message, JsResult result）	处理 JavaScript 中的 Alert（警告）对话框
onJsConfirm（WebView view, String url, String message, JsResult result）	处理 JavaScript 中的 Confirm（确认）对话框
onJsPrompt（WebView view, String url, String message, String defaultValue, JsPromptResult result）	处理 JavaScript 中的 Prompt（输入）对话框
onProgressChanged（WebView view, int newProgress）	当加载进度条发生改变时调用，可以获得网页的加载进度并显示
onReceivedIcon（WebView view, Bitmap icon）	获得网页的 icon
onReceivedTitle（WebView view, String title）	获得网页的标题

9.5.2 WebView 定制浏览器

1. 个性化浏览器

使用 WebView 组件可以开发一个具有个性的浏览器，当在浏览器的地址栏输入网址后，不会打开系统自带的浏览器，而用自己的浏览器浏览网页。运行效果如图 9.19 所示。

图 9.19 所示的界面设计比较简单，最上一行放置了一个 EditText 组件用于输入网址、一个 Button 组件用于开始打开网页；下面放了一个 WebView 组件用于显示网页内容。"开始浏览"按钮的监听事件代码如下。

第9章 网络应用开发

```
1    btnEnter.setOnClickListener(new View.OnClickListener() {
2        @Override
3        public void onClick(View v) {
4            String url = edtUrl.getText().toString();
5            webView.setWebViewClient(new WebViewClient());
6            webView.getSettings().setJavaScriptEnabled(true);
7            webView.loadUrl(url);
8        }
9    });
```

此时，单击 Android 设备的"回退"按钮后会退出应用程序，如果需要退回到前一个网页页面，就需要重写 Activity 的 onBackPressed() 方法，其代码如下。

```
1    public void onBackPressed() {
2        if (webView.canGoBack()) {
3            webView.goBack();
4        }
5    }
```

2. 带工具条的浏览器

图 9.20 所示的浏览器是在图 9.19 浏览器的基础上增加了一行工具条，该工具条可以实现"回退""收藏""显示当前网页的标题""回到页面顶端""刷新"功能，其他的功能，如输入网址、开始浏览和显示网页与前面的个性化浏览器一样。

图 9.19 个性化浏览器

图 9.20 带工具条的浏览器

- 工具条的布局代码。

```
1       <LinearLayout
2           android:layout_width=" match_parent"
3           android: layout_height=" wrap_content"
4           android: orientation=" horizontal" >
5           <LinearLayout
6               android: layout_width=" 0dp"
7               android: layout_height=" wrap_content"
8               android: layout_weight=" 1" >
9               <ImageButton
10                  android: id=" @ +id/ibret"
11                  android: layout_width=" 40dp"
12                  android: layout_height=" 40dp"
13                  android: src=" @mipmap/ret" />
14              <ImageButton
15                  android: id=" @ +id/ibsc"
16                  android: layout_width=" 40dp"
17                  android: layout_height=" 40dp"
18                  android: src =" @mipmap/shouc" />
19              <TextView
20                  android: id=" @ +id/tvtitle"
21                  android: layout_width=" wrap_content"
22                  android: layout_height=" wrap_content"
23                  android: layout_marginTop=" 10dp"
24                  android: lines=" 1"
25                  android: text=" " />
26          </LinearLayout>
27          <ImageButton
28              android: id=" @ +id/ibtop"
29              android: layout_width=" 40dp"
30              android: layout_height=" 40dp"
31              android: src=" @mipmap/top" />
32          <ImageButton
33              android: id=" @ +id/ibfresh"
34              android: layout_width=" 40dp"
35              android: layout_height=" 40dp"
36              android: src=" @mipmap/refresh" />
37      </LinearLayout>
```

上述第 5～26 行代码实现了将当前页面的"标题"显示在工具条的"返回""收藏""回顶""刷新"按钮中间区域。

- 绑定"开始浏览"按钮监听事件。

```
1       btnEnter. setOnClickListener(new View. OnClickListener() {
2           @Override
3           public void onClick(View v) {
4               String url = edtUrl. getText(). toString();
5               webView. loadUrl(url);
6               webView. setWebChromeClient(new WebChromeClient() {
7                   @Override
```

第9章 网络应用开发

```
8                    public void onReceivedTitle(WebView view, String title) {
9                        super.onReceivedTitle(view, title);
10                       txtTitle.setText(title);//将显示内页的标题显示在TextView上
11                       //将打开页面的链接地址显示在EditText上
12                       edtUrl.setText(webView.getOriginalUrl());
13                   }
14               });
15               //设置后不使用系统默认浏览器打开链接
16               webView.setWebViewClient(new WebViewClient());
17           }
18       });
```

上述第 6 ~ 13 行代码给 WebView 设置了 WebChromeClient 属性,并且重写了 WebChromeClient 对象的 onReceivedTitle( ) 方法,该方法当接收到浏览网页标题时触发。其中第 10 行代码表示将网页标题显示在工具条的 TextView 上;第 11 行代码表示将当前访问页面的链接地址显示在 EditText 上。

- 绑定"返回"按钮监听事件。

```
1       imgReturn.setOnClickListener(new View.OnClickListener() {
2           @Override
3           public void onClick(View v) {
4               webView.goBack();//返回到前一页
5           }
6       });
```

- 绑定"回到顶端"按钮监听事件。

```
1       imgTop.setOnClickListener(new View.OnClickListener() {
2           @Override
3           public void onClick(View v) {
4               webView.setScrollY(0);//设置滚动轴Y坐标为0
5           }
6       });
```

- 绑定"刷新"按钮监听事件。

```
1       imgRefresh.setOnClickListener(new View.OnClickListener() {
2           @Override
3           public void onClick(View v) {
4               webView.reload();
5           }
6       });
```

"收藏"功能的实现属于读写文件(数据库)的内容,即当用户单击"收藏"按钮时,将 EditText 中的内容和 TextView 中的内容作为键值对保存到文件中;如果要使用收藏的链接,可以从文件中将这个键值对信息读出来放置在菜单中。限于篇幅,此处不再列出代码,读者可以自行设计实现。

如果要让自定义的浏览器能够获得更好的页面显示效果,可以增加下面的代码以保证页面内容自适应屏幕大小。

```
1    WebSettings settings = webView.getSettings();
2    settings.setUseWideViewPort(true);//设定支持 viewport
3    settings.setLoadWithOverviewMode(true);//自适应屏幕
4    settings.setBuiltInZoomControls(true);
5    settings.setDisplayZoomControls(false);
6    settings.setSupportZoom(true);//设定支持缩放
```

要实现图9.21所示的带页面加载进度的效果,需要重写 WebChromeClient 对象的 onProgressChanged()方法和 WebViewClient 对象的 onPageFinished()方法。代码如下。

```
1    @Override
2    public void onProgressChanged(WebView view, int newProgress) {
3        prgress = newProgress;
4        if(pd==null){
5            pd = new ProgressDialog(CustomActivity.this); //创建进度条
6        }
7        pd.setMessage("正在载入,请稍后..."+prgress+"% "); //提示信息
8        pd.show(); //显示进度条
9        super.onProgressChanged(view, newProgress);
10   }
```

图9.21 带加载进度条效果

WebChromeClient 对象的 onProgressChanged()方法在加载页面的进度发生改变时触发,上述第7行代码表示在进度条对话框显示相应信息。

```
1    @Override
2    public void onPageFinished(WebView view, String url) {
3        pd.cancel();
```

第9章 网络应用开发

```
4        super.onPageFinished(view, url);
5    }
```

WebViewClient 对象的 onPageFinished() 方法在页面加载结束时触发，上述第 3 行代码表示在页面加载结束时取消进度条对话框的显示。

9.5.3 Android 与 JavaScript 交互

Android 与 JavaScript 交互是通过 WebView 实现的，实际上包含两个层面的调用，即 Android 调用 JavaScript 代码和 JavaScript 调用 Android 代码。

1. Android 调用 JavaScript 代码

Android 通过 WebView 调用 JavaScript 代码的方法有 loadUrl() 和 evaluateJavascript() 两种。

- 将需要调用的 JavaScript 代码以 .html 格式存放到 src/main/assets 文件夹里。如果当前项目下没有 assets 文件夹，则需要在项目名处右击鼠标，在弹出的菜单中分别选择 new→folder→Assets folder 菜单命令，在弹出的对话框中输入 assets 文件夹名称（必须是这个文件夹名称），也可以在 assets 文件夹创建 .html 格式的文件。本示例直接复制了一个 course.html 文件到 assets 文件夹下，详细代码如下。

```
1    <html>
2      <head>
3        <title>课程调查</title>
4        <script>
5          function callJS(){
6            alert("Android调用了JS的callJS()方法");
7          }
8        </script>
9      </head>
10     <body>
11       <font >请选择你喜欢的课程名称</font>
12       <select>
13         <option value="jsj">计算机应用基础</option>
14         <option value="sjjg">数据结构</option>
15         <option value="jsjzc">计算机组成原理</option>
16         <option value="sjk">数据库及应用</option>
17       </select>
18     </body>
19   </html>
```

- 在 Android 里通过 WebView 设置调用 JavaScript 代码。

```
1    //先载入JS代码,格式必须为"file:///android_asset/文件名.html"
2    webView.loadUrl (" file: ///android_asset/course.html");
3    WebSettings webSettings = webView.getSettings();
4    webSettings.setJavaScriptEnabled (true); // 设置与JS交互的权限
5    //设置允许JS弹窗
6    webSettings.setJavaScriptCanOpenWindowsAutomatically (true);
7    btnCallJs.setOnClickListener (new View.OnClickListener() {
```

405

```
8       @Override
9       public void onClick(View v) {
10          // Android 版本变量
11          final int version = Build.VERSION.SDK_INT;
12          //因为该方法在 Android 4.4 版本才可使用，所以使用时需进行版本判断
13          if (version <18) {
14              webView.loadUrl (" javascript: callJS()");
15          } else {
16              webView.evaluateJavascript (" javascript: callJS()",
new ValueCallback<String>() {
17                  @Override
18                  public void onReceiveValue (String value) {
19                      //此处为 JS 返回的结果
20                  }
21              });
22          }
23      }
24  });
```

上述第 4～16 行代码表示如果使用 Android 4.4 之前版本，就需要使用 loadUrl( )方法调用 JavaScript，否则使用 evaluateJavascript( )方法调用 JavaScript。Android 调用 JS 效果如图 9.22 所示。

图 9.22 Android 调用 JS 效果

2. JavaScript 调用 Android 代码

JavaScript 通过 WebView 调用 Android 代码的方法有 WebView 的 addJavascriptInterface( )对象映射、WebViewClient 的 shouldOverrideUrlLoading ( )方法回调拦截和 WebChromeClient 的 onJsAlert( )、onJsConfirm( )、onJsPrompt( )方法回调拦截 JavaScript 的 alert( )、confirm( )及

prompt( )方法的对话框。

(1) 通过 WebView 的 addJavascriptInterface( )方法进行对象映射。

● 定义一个与 JavaScript 对象具有映射关系的 Android 类——ShowInfo.java，其代码如下。

```
1    public class ShowInfo {
2        private Context context;
3        public ShowInfo(Context context){
4            this.context = context;
5        }
6        //定义JS需要调用的方法,被JS调用的方法必须加入@JavascriptInterface注解
7        @JavascriptInterface
8        public voidshowToast(String msg) {
9            Toast.makeText(context,msg,Toast.LENGTH_LONG).show();
10       }
11   }
```

上述第 8～10 行代码定义了一个可以被 JavaScript 调用的 showToast( )方法，用于在界面上用 Toast 显示信息。

● 将需要调用的 JavaScript 代码以 .html 格式放到 src/main/assets 文件夹里——submit.html，其代码如下。

```
1    <html>
2        <head>
3            <title>提交页面</title>
4        </head>
5        <body>
6            <input type="submit" value="提交" onclick="mysubmit.showToast('真的成功了')"/>
7        </body>
8    </html>
```

上述第 6 行代码表示在单击 HTML 页面上的"提交"按钮时，调用 ShowInfo 对象中的 showToast( )方法。此处的 mysubmit 即为在 Android 代码中映射的 ShowInfo 对象的名称。

● 在 Android 里通过 WebView 设置 Android 类与 JS 代码的映射。

```
1    btnSubmit.setOnClickListener(new View.OnClickListener() {
2        @Override
3        public void onClick(View v) {
4            WebSettings webSettings = webView.getSettings();
5            //设置与JS交互的权限
6            webSettings.setJavaScriptEnabled(true);
7            ShowInfo showInfo = new ShowInfo((SubmitActivity.this));
8            //通过addJavascriptInterface()将Java对象映射到JS对象
9            //ShowInfo类对象映射到JS的mysubmit对象
10           webView.addJavascriptInterface(showInfo, "mysubmit");
11           webView.loadUrl("file:///android_asset/submit.html");
12       }
13   });
```

上述第 11 行代码非常重要，addJavascriptInterface( )方法的第 1 个参数是要被 JavaScript 调用的 Android 类对象，第 2 个参数是映射到 JavaScript 中的对象名（此名称必须与上面

HTML文件代码中第6行的对象名一致)。

(2) 通过 WebViewClient 的 shouldOverrideUrlLoading()方法回调拦截指定的 Url。

● 在 JavaScript 代码中约定所需要的 Url 协议,即将 JavaScript 代码以.html 格式放到 src/main/assets 文件夹里——geturl.html(本例文件名)。

```
1    <html>
2      <head>
3        <meta charset="utf-8">
4        <title>Carson_Ho</title>
5        <script>
6          function callAndroid() {
7            document.location ="linkjs://www.myweb?username=1&userpwd=2";;
8          }
9        </script>
10     </head>
11     <!--单击按钮则调用callAndroid方法-->
12     <body>
13       <button type="button" id="button1" onclick="callAndroid()">调用Android代码</button>
14     </body>
15   </html>
```

● 在 Android 代码中重写 WebViewClient 的 shouldOverrideUrlLoading()方法。

```
1    WebSettings webSettings = webView.getSettings();
2    webSettings.setJavaScriptEnabled(true);
3    webSettings.setJavaScriptCanOpenWindowsAutomatically(true);
4    webView.loadUrl("file:///android_asset/geturl.html");
5    webView.setWebViewClient(new WebViewClient() {
6      @Override
7      public boolean shouldOverrideUrlLoading(WebView view, String url) {
8        Uri uri = Uri.parse(url);
9        if (uri.getScheme().equals("linkjs")) {//协议格式判断
10         if (uri.getAuthority().equals("www.myweb")) {//协议名判断
11           Toast.makeText(GetUrlActivity.this,"恭喜你调用成功!",Toast.LENGTH_LONG).show();
12           HashMap<String, String> params = new HashMap<>();
13           Set<String> collection = uri.getQueryParameterNames();
14         }
15         return true;
16       } else {
17         Toast.makeText(GetUrlActivity.this,"Url 中的协议有误,请核查!",Toast.LENGTH_LONG).show();
18       }
19       return super.shouldOverrideUrlLoading(view, url);
20     }
21   });
```

上述第 9~15 行代码表示根据协议判断是否为所需要的 Url,如果是所需要的 Url,那么就执行第 11~13 行的逻辑处理代码,否则执行第 17 行代码。在进行协议判断时一般根据 scheme（协议格式）和 authority（协议名），本例中传入进来的"url = linkjs：//www.myweb? username = 1&userpwd = 2"，该 url 的协议格式为"linkjs"，协议名为 www.myweb。第 12～13 行代码表示可以在协议上带有参数（本例中参数名有 username 和 userpwd 两个）并传递到 Android 代码。

(3) 通过 WebChromeClient 的 onJsAlert()、onJsConfirm()、onJsPrompt()方法回调拦截 JavaScript 的 alert()、confirm()和 prompt()方法的对话框。

JavaScript 中有三个常用对话框，见表 9-12。

表 9-12 JavaScript 中的常用对话框

对话框名	方法名	返回值	备注
警告框	alert()	无	在文本中可加入\n换行
确认框	confirm()	两个返回值	返回逻辑值，true 表示确认，false 表示取消
输入框	prompt()	任意设置返回值	确认返回输入框中的值，取消返回 null

● 将需要调用的 JavaScript 代码以 .html 格式放到 src/main/assets 文件夹里——getdialog.html。

```
1    <html>
2        <head>
3            <title>三种不同对话框</title>
4            <script language="JavaScript">
5                function alertFun()
6                { alert("这是一个警告对话框!");}
7                function confirmFun()
8                {
9                    if(confirm("这是一个确认对话框——访问网易吗?"))
10                   {location.href = "http://www.163.com";}
11                   else alert("你选择的取消访问!");
12               }
13               function promptFun()
14               {
15                   var s = prompt("这是一个输入对话框——请输入你的年龄","");
16                   if(s)
17                   { alert("你的年龄为:" + s)
18                   }else{alert("对不起,你没有输入年龄!");}
19               }
20           </script>
21       </head>
22       <body>
23           <p>
24               <input type="submit" name="Submit1" value="Alert 对话框" onclick="alertFun()"/>
25           </p>
26           <p>
27               <input type="submit" name="Submit2" value="Confirm 对话框" onclick="confirmFun()"/>
28           </p>
29           <p>
```

```
30          <input type ="submit" name ="Submit3" value ="Prompt 对话框"
onclick ="promptFun()"/>
31        </p>
32     </body>
33  </html>
```

- 创建 Prompt 对话框的布局文件——prompt_layout.xml。

```
1   <?xml version ="1.0" encoding ="utf-8"? >
2   <LinearLayout xmlns:android ="http://schemas.android.com/apk/res/android"
3       android:layout_width =" match_parent"
4       android: layout_height =" match_parent"
5       android: orientation =" vertical" >
6       <TextView
7           android: id =" @ +id/tvprompt"
8           android: layout_width =" match_parent"
9           android: layout_height =" wrap_content" />
10      <EditText
11          android: id =" @ +id/edtprompt"
12          android: layout_width =" match_parent"
13          android: layout_height =" wrap_content" />
14  </LinearLayout >
```

- 创建继承于 WebChromeClient 的类——MyWebChromeClient.java，并重写 onJsAlert( )、onJsConfirm( )和 onJsPrompt( )方法。

```
1   class MyWebChromeClient extends WebChromeClient {
2       @Override
3       public boolean onJsAlert(WebView view, String url, String
message,final JsResult result) {
4           new AlertDialog. Builder(DialogActivity.this).setTitle("Alert 对话
框"). setMessage(message). setPositiveButton("确定", new DialogInterface. OnClick-
Listener() {
5           @Override
6           public void onClick(DialogInterface dialog, int which) {
7               result. confirm();
8           }}). setCancelable(false). show();
9           return true;
10      }
11      @Override
12      public boolean onJsConfirm(WebView view, String url, String message,
final JsResult result) {
13          new AlertDialog. Builder(DialogActivity.this).setTitle("Confirm 对
话框"). setMessage(message). setPositiveButton("确定", new DialogInterface. On-
ClickListener() {
14          @Override
15          public void onClick(DialogInterface dialog, int which) {
16              result. confirm();
17          }}). setNegativeButton("取消", new
DialogInterface. OnClickListener() {
```

```
18            @Override
19            public void onClick(DialogInterface dialog, int which) {
20                result.cancel();
21            }}).setCancelable(false).show();
22            return true;
23        }
24        @Override
25        public boolean onJsPrompt(WebView view, String url, String message,
String defaultValue, final JsPromptResult result) {
26            //获得一个LayoutInflater对象factory,加载指定布局成相应对象
27            final LayoutInflater inflater =
LayoutInflater.from(DialogActivity.this);
28            final View myview = inflater.inflate(R.layout.prompt_layout, null);
29            //设置TextView对应网页中的提示信息，edit设置来自于网页的默认文字
30            TextView tprompt = ((TextView) myview.findViewById (R.id.tvprompt));
31            final EditText eprompt = ((EditText) myview.findViewById (R.id.edtprompt));
32            tprompt.setText (message);
33            eprompt.setText (defaultValue);
34            new AlertDialog.Builder (DialogActivity.this).setTitle (" Prompt
对话框").setView (myview)
35            .setPositiveButton (" 确定", new DialogInterface.OnClickListener() {
36                @Override
37                public void onClick (DialogInterface dialog, int which) {
38                    //单击确定后取得输入的值，传给网页处理
39                    String value = eprompt.getText().toString();
40                    result.confirm (value);
41                }})
42            .setNegativeButton (" 取消", new DialogInterface.OnClickListener() {
43                @Override
44                public void onClick (DialogInterface dialog, int which) {
45                    result.cancel();
46                }}).show();
47            return true;
48        }
49    }
```

上述第3～10行代码表示重写onJsAlert()方法，该方法用于在Android中创建一个对话框来显示网页中的警告对话框。第12～23行代码表示重写onJsConfirm()方法，该方法用于在Android中创建一个对话框来显示网页中的确认对话框。第24～48行代码表示重写onJsPrompt()方法，该方法用于在Android中创建一个输入对话框来显示网页中的消息对话框，但这个输入对话框需要定义对话框的格式，即在布局文件中定义对话框布局后，使用LayoutInflater实现自定义对话框；其中第32行代码表示将网页中输入对话框的message值传入Android自定义对话框的TextView对象，第33行代码表示将网页中输入对话框的defaultValue值传入Android自定义对话框的EditText对象；第39～40行代码表示将Android对话框中输入的内容传递给网页显示。上述代码运行效果如图9.23、图9.24、图9.25、图9.26所示。

图9.23 加载网页效果

图9.24 警告对话框效果

图9.25 确认对话框效果

图9.26 输入对话框效果

9.5.4 在HTML页面显示Android通讯录联系人信息

下面以图9.27所示的显示效果为例,从Android终端设备的通讯录中读出联系人编号、姓名和联系电话,并将其显示在HTML页面上。

图9.27 显示联系人信息

1. 定义 JavaScript 文件

将需要调用 Android 代码的 JavaScript 文件以 .html 格式存放到 src/main/assets 文件夹里——showdetail.html，其详细代码如下。

```
1   <html>
2       <head>
3           <meta http-equiv="Content-Type" content="text/html; charset=utf-8" />
4           <title>显示获取的联系人列表</title>
5           <script language="JavaScript">
6               function show(jsondata)
7               {
8                   //将传递过来的Json转换为对象
9                   var jsonobjs = eval(jsondata);
10                  //获取body中定义的表格
11                  var table = document.getElementById("PersonTable");
12 //遍历上面的Json对象,将每个对象添加为表格中的一行,而它的每个属性作为每列内容
13                  for(var i = 0;i <jsonobjs.length;i ++)
14                  {
15                      //插入一行,每行包含三个单元格
16                      var tr = table.insertRow(table.rows.length);
17                      var td1 = tr.insertCell(0);
18                      td1.align = "center";
19                      var td2 = tr.insertCell(1);
20                      td2.align = "center";
21                      var td3 = tr.insertCell(2);
22 //设置单元格的内容和属性,其中innerHTML为设置或者获取位于对象起始和结束标签内的HTML,jsonobjs[i]为对象数组中的第i个对象
23                      td1.innerHTML = jsonobjs[i].pid;
24                      td2.innerHTML = jsonobjs[i].pname;
25  //为显示内容添加超链接,超链接调用Android代码中的call方法并且把内容作为参数传递过去
26                      td3.innerHTML = "<a href = 'javascript:tanstoJS.call(\""+jsonobjs[i].ptel + "\")'>"+jsonobjs[i].ptel + "</a>";
27                  }
28              }
```

```
29        </script>
30      </head>
31      <!-- onload 指定该页面被加载时调用的方法,这里调用的是Android代码中的
personlist()方法-->
32      <body style ="margin:0px; background-color:#FFFFFF; color:#000000;"
onload = "tanstoJS.personlist()" >
33        <table border = "1" width = "100%" id = "PersonTable" cellspacing = "0">
34          <tr>
35            <td width = "30%" >联系人编号</td>
36            <td align = "center" >联系人姓名</td>
37            <td width = "30%" >电话号码</td>
38          </tr>
39        </table>
40      </body>
41    </html>
```

上述第 5~28 行代码表示定义一个让 Android 代码调用的 JavaScript 方法,该方法将 Android 代码中传递过来的 Json 格式数据显示在 HTML 定义的表格中。第 32~38 行代码定义了一个表格,用于显示 Android 代码传递过来的数据。

2. 定义联系人类

按照图 9.27 的显示效果,一个联系人包含编号、姓名和电话号码三个属性,所以定义了一个 Person 类——Person.java,代码如下。

```
1     public class Person {
2         private int pId;
3         private String pName;
4         private String pTel;
5         public Person(int pid, String pname, String ptel) {
6             this.pId = pid;
7             this.pName = pname;
8             this.pTel = ptel;
9         }
10
11        public int getpId() {
12            return pId;
13        }
14        public String getpName() {
15            return pName;
16        }
17        public String getpTel() {
18            return pTel;
19        }
20    }
```

3. 定义用于 Android 代码与 JavaScript 代码交互的类

为了能够实现将 Android 代码获取的联系人信息传递给 JavaScript,此处定义了一个类——TransToJs.java,该类中包含以下四个方法。

(1) getPersons()方法,用于从 Android 的通讯录 ContentProvider 中取出_ID(编号)、

DISPLAY_NAME（姓名）和 NUMBER（电话号码），并存放到 List 中。

（2）buildJson(List <Person>)方法，用于将 List 中的联系人信息封装成 Json 格式，以便 JavaScript 解析。

（3）personlist()方法，用于让 JavaScript 代码调用该方法，同时该方法中又调用了 JavaScript 代码中定义的方法，所以这个方法既实现了 JavaScript 代码调用 Android 代码（JavaScript 代码的第 31 行），又实现了 Android 代码调用 JavaScript 代码。但是，自 Android 4.2 版本（API 18）开始，在 Android 代码中调用 JavaScript 代码必须在子线程中，所以下面的代码使用 handler. post()开启了一个子线程。

（4）call(String phone)方法，用于在 HTML 页面上单击电话号码链接就可以直接拨号。

```
1     class TransToJs {
2         @RequiresApi(api = Build.VERSION_CODES.KITKAT)
3         @JavascriptInterface
4         public void personlist() {
5             handler.post (new Runnable() {
6               @Override
7                 public void run() {
8                   String json = null;
9                   try {
10                      json = buildJson (getPersons());
11                      final int version = Build.VERSION.SDK_INT;
12                      if (version <18) {
13                        webView.loadUrl (" javascript: show ('" + json + "')");
14                      } else {
15                        webView.evaluateJavascript (" javascript: show ('" + json + "')", null);
16                      }
17                  } catch (Exception e) {
18                      e.printStackTrace();
19                  }
20              }
21          });
22      }
23      @JavascriptInterface
24      public void call (String phone) {
25          Intent it = new Intent (Intent.ACTION_CALL, Uri.parse (" tel:" + phone));
26          startActivity (it);
27      }
28      @JavascriptInterface
29      public String buildJson (List <Person> persons) throws Exception {
30          JSONArray array = new JSONArray();
31          for (Person person : persons) {
32              JSONObject jsonObject = new JSONObject();
33              jsonObject.put (" pid", person.getpId());
34              jsonObject.put (" pname", person.getpName());
35              jsonObject.put (" ptel", person.getpTel());
36              array.put (jsonObject);
```

```
37        }
38        return array.toString();
39    }
40    @JavascriptInterface
41    public List<Person> getPersons() {
42        List<Person> persons = new ArrayList<Person>();
43        ContentResolver resolver = getContentResolver();
44        Uri uri = ContactsContract.CommonDataKinds.Phone.CONTENT_URI;
45        Cursor cursor = resolver.query (uri, null, null, null, null);
46        while (cursor.moveToNext()) {
47            int pid = cursor.getInt (cursor.getColumnIndex (ContactsContract.Contacts._ID));
48            String pname = cursor.getString (cursor.getColumnIndex (ContactsContract.CommonDataKinds.Phone.DISPLAY_NAME));
49            String ptel = cursor.getString (cursor.getColumnIndex (ContactsContract.CommonDataKinds.Phone.NUMBER));
50            Person person = new Person (pid, pname, ptel);
51            persons.add (person);
52        }
53        cursor.close();
54        return persons;
55    }
56 }
```

运行效果如图 9.27 所示，本案例的实现代码读者可以参见 ALLAPP 代码包中 webgetandroid 文件夹里的内容。

本 章 小 结

本章结合实际示例项目的开发过程介绍了 Android 中 Http、WebService 和 WebView 访问网络的使用方法，详细阐述了这三种技术的基本原理，让读者既明白了进行 Android 中网络应用开发的流程，也掌握了相关技术。

习　题

一、选择题

1. 下列叙述中错误的是（　　）。

A. Android 中提供了 HttpURLConntection 和 HttpClient 接口实现与服务器的 Http 通信

B. Android 与服务器进行 Http 通信有 POST 和 GET 两种方式

C. Android 与服务器的 POST 请求方式中，openConnection( ) 方法既可以创建实例，也进行真正的连接操作

D. HttpClient 接口有一个实现类 DefaultHttpClient

2. 下列关于 Web Service 的叙述错误的是（　　）。

A. Web Service 使用基于 XML 的消息处理，可以消除不同操作系统和编程语言之间的差异

B. Web Service 的主要技术包括 SOAP、WSDL 和 UDDI

C. Android SDK 中直接提供了调用 Web Service 的类库

D. 要实现调用 Web Service 来获得服务器数据必须要添加联网的权限

3. 下列属于 SAX 解析 XML 文件的优点的是（　　）。

A. 将整个文档读入在内存中，便于操作，支持删除、修改、重新排列等多种功能

B. 不用事先调入整个文档，占用资源少

C. 整个文档调入内存，浪费时间和空间

D. 不是长久驻留在内存，数据不是持久的，事件过后，若没有保存数据，数据就会消失

4. Hanlder 是线程与 Activity 通信的桥梁，如果线程处理不当，机器就会变得越来越慢，那么线程销毁的方法是（　　）。

A. onDestroy( )　　　B. onClear( )　　　C. onFinish( )　　　D. onStop( )

5. WebView 中可以用来处理 JavaScript 中警告、确认等对话框的是（　　）。

A. WebSettings　　　B. WebViewClient　　C. WebChromeClient　D. WebViewChrome

6. Android 解析 XML 的方法中，将整个文件加载到内存中进行解析的是（　　）。

A. SAX　　　　　B. PULL　　　　　C. DOM　　　　　D. JSON

7. 在 AsyncTask 中，下列哪个方法是负责执行那些很耗时的后台计算工作的？（　　）

A. run　　　　　B. execute　　　　C. doInBackground　D. onPostExecute

8. 使用 HttpUrlConnection 实现移动互联时，设置读取超时属性的方法是（　　）。

A. setTimeout( )　　　　　　　　B. setReadTimeout( )

C. setConnectTimeout( )　　　　　D. setRequestMethod( )

9. 使用 HttpURLConnection 的 GET 方式请求数据时，下列（　　）属性是必须设置的。

A. connection.setDoOutput(true)

B. connection.connect( )

C. connection.setRequestMethod("POST")

D. connection.setDoInput(true)

10. 使用 HttpClient 的 GET 方式请求数据时，可以用（　　）类来构建 Http 请求。

A. Get　　　　　B. URLConnection　　C. HttpGet　　　　D. HttpPost

11. 若希望在单击网页中超链接时，在当前 WebView 中显示该网页，则需要覆盖 WebViewClient 类的（　　）方法。

A. shouldOverrideUrlLoading( )　　　B. onPageStarted( )

C. loadUrl( )　　　　　　　　　　　D. show( )

12. Android 中不是 SAX 方式解析 XML 需要用的类是（　　）。

A. SAXParserFactory　B. XMLReader　　C. PullParser　　D. DefaultHandler

13. 关于 Socket 通信正确的是（　　）。

A. 服务器端需要 ServerSocket，需要绑定端口号

B. 服务器端需要 ServerSocket，需要绑定端口号和 IP 地址

C. 客户端需要 Socket，需要绑定端口号

D. 客户端需要 ServerSocket，需要绑定端口号

二、填空题

1. Android 中与服务器实现 Http 通信，有 POST 和 _____ 这两种请求方式。

2. Android 平台使用 Http 访问网络有 HttpURLConnection 和_____两种传统方式。

3. _____是一种跨编程语言和跨操作系统平台的远程调用技术，该技术能让不同平台的不同应用间相互交换数据。

4. Web service 的标准通信协议是_____。

5. httpClient 中发送请求的方法是_____。

6. 给 ListView 设置适配器的方法是_____。

三、判断题

1. 在调用远程服务时，需要使用 AIDL 语言定义跨进程服务的接口。　　　　（　　）

2. Android 提供了 HttpURLConnection 接口和 Apache 接口实现与服务器的 Http 通信，并有 POST 和 GET 两种请求方法。　　　　（　　）

3. 在 Android 4.0 版本以后，编写应用时处理上网的代码只能放在子线程中执行，不能放在主线程中执行。　　　　（　　）

4. Andriod 中 android.permission.INTERNET 权限是允许打开网络套接字。　　　（　　）

5. HttpClient 是一种多用途、轻量级的 Http 客户端，使用它进行 Http 操作适用于大多数的应用程序。　　　　（　　）

【第9章参考答案】

第 10 章
传感器与位置服务应用开发

现在几乎所有的移动设备都配置有不同类型的传感器。开发者可以利用不同类型的传感器进行一些耳目一新的功能开发,如微信中的摇一摇抽红包、摇一摇切歌等;当然也可以把传感器的应用与地图结合起来实现定位、导航等功能。本节将结合传感器和百度地图开发技术介绍 Android 平台下的地图与传感器应用开发。

教学目标

理解 Android 中传感器的分类和功能。
理解 Android 中位置服务的原理和方式。
掌握 Android 应用开发中传感器的常用方法和使用步骤。
掌握百度地图在 Android 中的应用开发过程。

教学要求

知识要点	能力要求	相关知识
概述	(1) 掌握传感器的概念、分类和功能 (2) 掌握位置服务的概念和方式	硬件访问接口
指南针的设计与实现	(1) 掌握 Android 应用开发中传感器的使用步骤 (2) 掌握加速度传感器、磁场传感器的工作原理 (3) 掌握 SensorManager 类的作用和常用方法 (4) 了解使用 RotateAnimation 类实现图片动态旋转的方法	摇一摇实现
百度地图在 Android 中的应用	(1) 掌握 Android Studio 环境下获取 SHA1 的方法 (2) 掌握将百度地图加入 Android 项目中的方法 (3) 掌握利用百度地图进行地图切换、定位和添加覆盖物等基本的地图应用的扩展方法	百度地图开放平台开发者注册步骤

10.1 概　　述

Android 平台具有强大的应用层编程接口和丰富的传感器功能，其开发平台有利于开发者开发出适合在移动设备上运行的各类应用软件。现在很多第三方平台利用 Android 的这一特点向开发者提供了诸如传感器开发、地图开发等不同类型的开发包及 API，以便开发者利用移动设备配置的硬件设备和 Android 平台提供的 API 开发具有计步、定位和导航等功能的实用 App。

1. 传感器

传感器能够探测、感受外界信号，并可以将探知的外界信息传递给其他设备或器官，如光、热、湿度等物理信号，烟雾、毒气等化学信号。Android 平台支持的传感器可以分为动作传感器、位置传感器和环境传感器三大类。Android 平台支持的传感器的详细类别及功能说明见表 10-1。

表 10-1　Android 平台支持的传感器的详细类别及功能说明

名　称	类型常量（整数值）	功　能　说　明
加速度传感器	TYPE_ACCELEROMETER(1)	以 m/s^2 为单位测量设备三个物理轴线方向（x、y 和 z）的加速度
磁场传感器	TYPE_MAGNETIC_FIELD(2)	测量周围的三个物理轴线方向（x、y 和 z）的磁场
方向传感器	TYPE_ORIENTATION(3)	测量设备三个物理轴线方向（x、y 和 z）的旋转角度
陀螺仪传感器	TYPE_GYROSCOPE(4)	以 rad/s 为单位测量设备三个物理轴线方向（x、y 和 z）的旋转速度
光线传感器	TYPE_LIGHT(5)	以 lx 为单位测量周围光线级别
气压传感器	TYPE_PRESSURE(6)	测量周围空气气压
温度传感器	TYPE_TEMPERATURE(7)	检测设备的温度
距离传感器	TYPE_PROXIMITY(8)	检测物体与设备的距离
重力传感器	TYPE_GRAVITY(9)	测量重力
线性加速度传感器	TYPE_LINEAR_ACCELERATION(10)	检测沿着一个轴向的加速度
旋转矢量传感器	TYPE_ROTATION_VECTOR(11)	检测运动和旋转
相对湿度传感器	TYPE_RELATIVE_HUMIDITY(12)	检测周围空气相对湿度
心率传感器	TYPE_HEART_RATE(21)	检测心率

（1）动作传感器。

动作传感器用于监视设备动作，包括加速度传感器（Accelerometer Sensor）、陀螺仪传感器（Gyroscope Sensor）、重力传感器（Gravity Sensor，Android 2.3 引入）、线性加速传感器（Linear Acceleration Sensor，Android 2.3 引入）和旋转矢量传感器（Rotation Vector Sensor，Android 2.3 引入）。其中前两种是基于硬件的传感器，后三种可以是基于硬件的传感器，也可以是基于软件的传感器。基于软件的传感器在不同的 Android 设备中回传的数据可能来自不同的硬件传感器，所以基于软件的同一种传感器在不同的设备中的精确度、使用范围可能会有所不同。

（2）位置传感器。

位置传感器用于确定设备的位置，包括磁场传感器（Magnetic Field Sensor）、距离传感器（Proximity Sensor）和方向传感器（Orientation Sensor）。其中前两种是基于硬件的传感器，后一种是基于软件的传感器。

（3）环境传感器。

环境传感器用于检测不同的外部环境，包括光线传感器（Light Sensor）、气压传感器（Pressure Sensor）、温度传感器（Temperature Sensor）和相对湿度传感器（Relative Humidity Sensor，Android 4.0 引入）。

另外，自 Android 5.0 之后新增了一种心率传感器（Heart Rate Sensor），该传感器用于返回佩戴设备的人每分钟的心跳次数。该传感器返回的数据准确性可以通过 SensorEvent 的 accuracy 进行判断，如果该属性值为 SENSOR_STATUS_UNRELIABLE 或 SENSOR_STATUS_NO_CONTACT，则表明传感器返回的数据是不太可靠的，应该丢弃。

2. 位置服务

位置服务（Location Based Services，LBS）又称定位服务，是指通过无线电通信网络或 GPS 卫星定位等方式，获取各种终端的地理坐标（经度和纬度），在电子地图平台的支持下为用户提供基于位置导航和查询的一种信息业务。

基于地理定位服务的核心就是确定用户所在的位置，通常有以下三种实现方式。

（1）GPS 定位。

GPS 定位是通过终端设备（手机、平板电脑等）内置的 GPS 硬件与卫星交互来获取当前设备所在位置的经纬度坐标。该定位方式速度快、精度高，可在无网络情况下使用；但是首次连接时间长，且由于室内无法接收卫星信号，所以只能在室外使用。

（2）网络定位。

网络定位分为基站定位和 WiFi 定位。基站定位通常是根据 Android 终端设备（手机、平板电脑等）附近的三个基站进行三角定位，由于每个基站的位置是固定的，利用电磁波在这三个基站间中转所需要时间来算出设备所在位置的坐标。这种方式受环境影响较小，只要有基站的地方就能用，缺点是需要消耗流量，精度也相对较低。WiFi 定位是根据一个固定的 WifiMAC 地址，通过收集到的该 WiFi 热点的位置，然后访问网络上的定位服务器以获得经纬度坐标。这种方式和基站定位一样，优势在于受环境影响较小，只要有 WiFi 的地方就可以使用，缺点是需要有 WiFi，精度也不高。

(3) AGPS 定位。

AGPS（Assisted GPS，A-GPS，网络辅助 GPS）定位技术结合了 GPS 定位和蜂窝基站定位的优势，借助蜂窝网络的数据传输功能，可以达到很高的定位精度和很快的定位速度，但是它的硬件要求很高，造价也高。

10.2 指南针的设计与实现

为了应对城市复杂的交通情况和达到野外定向穿越的目标，指南针是一个不可或缺的工具。本节将以简易指南针的实现为例详细介绍基于 Android 平台的传感器的应用开发方法。

【传感器和计步实现】

10.2.1 预备知识

1. 传感器的使用步骤

(1) 获得传感器管理器。

Android 的所有传感器都是由传感器管理器（SensorManager）管理，获取传感器管理器的代码如下。

```
String service_name = Context.SENSOR_SERVICE;
SensorManager sensorManager = (SensorManager) getSystemService (service_name);
```

(2) 获得需要的传感器。

从传感器管理器中获取其中某个或者某些传感器的方法有如下 3 种。

• 返回某个指定类型的传感器，此处以重力传感器为例（其他传感器读者可以参见表 10-1）。

```
Sensor gravitySensor = sensorManager.getDefaultSensor(Sensor.TYPE_GRAVITY);
```

• 返回某个指定传感器列表。

```
List <Sensor> gravitySensorList = sensorManager.getSensorList(Sensor.TYPE_GRAVITY);
```

• 返回所有传感器列表。

```
List <Sensor> allSensors = sensorManager.getSensorList(Sensor.TYPE_ALL);
```

传感器的常用方法和功能说明见表 10-2。

表 10-2 传感器的常用方法和功能说明

方 法 名	功 能 说 明
getMaximumRange()	获得最大取值范围
getName()	获得设备名称
getPower()	获得功率
getResolution()	获得精度
getType()	获得传感器类型
getVentor()	获得设备供应商
getVersion()	获得设备版本号

(3) 实现 SensorEventListener 接口,并重写相关方法。

```
class MySensorListener implements SensorEventListener{
    @Override
    public void onSensorChanged(SensorEvent event) {
        final float[] eData = event.values;
    }
    @Override
    public void onAccuracyChanged(Sensor sensor, int accuracy) {

    }
}
```

onSensorChanged()方法在传感器的值发生变化时调用,传感器的数据来源于 SensorEvent 类,该类中有一个 float[] 数组类型的 values 变量,变量中最多包含三个元素,对于不同的传感器,对应元素代表的含义也不相同。常用传感器对应元素代表的含义见表10-3。onAccuracyChanged()方法在传感器的精度发生变化时调用。

表10-3 常用传感器对应元素代表的含义

传感器名称	数组元素	含义
方向传感器	values[0]	绕 z 轴转过的角度
	values[1]	绕 x 轴转过的角度
	values[2]	绕 y 轴转过的角度
陀螺仪传感器	values[0]	绕 x 轴转过的角速度
	values[1]	绕 y 轴转过的角速度
	values[2]	绕 z 轴转过的角速度
磁场传感器	values[0]	x 轴方向磁场强度
	values[1]	y 轴方向磁场强度
	values[2]	z 轴方向磁场强度
重力传感器	values[0]	x 轴方向的重力
	values[1]	y 轴方向的重力
	values[2]	z 轴方向的重力
线性加速度传感器	values[0]	x 轴方向线性加速度
	values[1]	y 轴方向线性加速度
	values[2]	z 轴方向线性加速度
加速度传感器	values[0]	x 轴方向加速度
	values[1]	y 轴方向加速度
	values[2]	z 轴方向加速度
温度传感器	values[0]	当前温度
光线传感器	values[0]	当前光的强度
气压传感器	values[0]	当前气压

(4)为传感器注册监听器。

调用 SensorManager 的 registerListener()方法来注册 SensorEventListener 使其生效,代码如下。

```
MySensorListener myListener = new MySensorListener();
sensorManager.registerListener(myListener,gravitySensor,
SensorManager.SENSOR_DELAY_UI);
```

其中第 1 个参数为传感器监听事件对象,第 2 个参数为传感器对象,第 3 个参数为传感器的更新速率。其中第 3 个参数有 4 个可选值,见表 10-4。

表 10-4 传感器的更新速率

常 量 名	说 明
SENSOR_DELAY_FASTEST	最低时延,以最快的速度获得传感器数据(延时 0ms),消耗大量电量
SENSOR_DELAY_GAME	以较快的速度获得传感器数据(延时 20ms),适用于实时性较高的游戏
SENSOR_DELAY_NORMAL	默认值,以一般的速度获得传感器数据(延时 60ms),适用于一般游戏
SENSOR_DELAY_UI	以较低速度获得数据(延时 200ms),适用于传感器更新 UI

以上更新速率依次递减,当然低延时意味着更频繁的检测,也会消耗更多的电量,实际开发中如果不是要求精度非常高,建议不使用太高精度,通常用 SENSOR_DELAY_UI 较多。

(5)为传感器注销监听器。

用完传感器后,一般需要注销传感器监听器,实现时通常在 Activity 或者 Service 的 onDestroy()方法中调用 SensorManager 的 unregisterListener()方法,其代码如下。

```
@Override
protected void onDestroy() {
    super.onDestroy();
    if (sensorManager ! = null) {
        sensorManager.unregisterListener(myListener);
    }
}
```

2. 加速度传感器

加速度是一种用于描述物体运行速度改变快慢的物理量,以 m/s^2 为单位。而 Android 中的加速度传感器则是提供了一种机制,使得开发者能够在应用程序中获取到设备当前的加速信息,合理利用这些信息可以开发出有趣、实用的功能。

设备静止时,加速度传感器返回的值为地表静止物体的重力加速度(约为 $9.8m/s^2$)。加速度传感器输出的信息存放在 SensorEvent 的 values 数组中,该数组中的三个值分别代表设备在 x 轴、y 轴和 z 轴方向上的加速度信息。因为重力的作用方向永远是向下的,所以当设备竖直放置时,重力作用在 y 轴;当设备水平放置时,重力作用在 z 轴;当设备横立放置时,重力作用在 x 轴。加速度传感器坐标系如图 10.1 所示。

下面以摇一摇计数为例介绍加速度传感器的用法,运行效果如图 10.2 所示。界面设计比较简单,读者可以参见 FirstAPP 代码包中 YYSensor 文件夹里的内容(main_layout.xml)。

第10章 传感器与位置服务应用开发

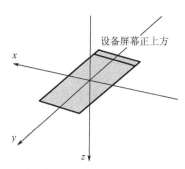

图10.1 加速度传感器坐标系

图10.2 摇一摇效果图

(1) 定义加速度传感器监听事件类。

```
1    class MySensorListener implements SensorEventListener {
2        @Override
3        public void onSensorChanged(SensorEvent event) {
4            //加速度可能会是负值,所以要取它们的绝对值
5            float xValue = Math.abs(event.values[0]);
6            float yValue = Math.abs(event.values[1]);
7            float zValue = Math.abs(event.values[2]);
8            if (xValue > 15 || yValue > 15 || zValue > 15) {
9                //认为用户摇动了手机,触发摇一摇逻辑
10               count ++;//计数器加1
11               tvCount.setText("你已经摇了"+count +"次!");//显示在 TextView 上
12               Toast.makeText(MainActivity.this,"摇一摇", Toast.LENGTH_SHORT).show();
13           }
14       }
15       @Override
16       public void onAccuracyChanged (Sensor sensor, int accuracy) {
17       }
18   }
```

(2) 注册传感器监听事件。

```
1    sensorManager = (SensorManager) this.getSystemService(Context.SENSOR_SERVICE);
2    Sensor sensor = sensorManager.getDefaultSensor (Sensor.TYPE_ACCELEROMETER);
3    MySensorListener listener = new MySensorListener();
4    sensorManager.registerListener (listener, sensor, SensorManager.SENSOR_DELAY_NORMAL);
```

(3) 注销传感器监听事件。

```
1    @Override
2    protected void onDestroy() {
3        super.onDestroy();
```

```
    4        if (sensorManager ! = null) {
    5            sensorManager.unregisterListener(listener);
    6        }
    7    }
```

3. 磁场传感器

磁场传感器主要用于感应周围的磁感应强度。即使周围没有任何直接的磁场，移动设备也始终会处于地球磁场中。随着移动设备摆放状态的改变，周围磁场在移动设备的 x 轴、y 轴和 z 轴方向上会发生改变。

磁场传感器与加速度传感器一样也会返回三个数据，三个数据分别代表周围磁场分解到 x 轴、y 轴和 z 轴三个方向上的磁场分量。

4. SensorManager

Android 的方向传感器［TYPE_ORIENTATION，最新版的 SDK 已经提示过期，并建议用 SensorManager.getOrientation( )替代］不是实际物理存在的，它实际上是通过磁场传感器和加速度传感器数据整合构成的。即方位数据是由这两个传感器的数据通过一定的算法得到的，这个算法现在已经封装成了 API，开发者直接使用即可。

SensorManager 是 Android 中的一个类，该类的 getRotationMatrix( )方法可以计算出旋转矩阵，然后通过 getOrientation( )方法求得设备的方向（方向角、倾斜角、旋转角）。

getRotationMatrix(float[ ] R,float[ ] I,float[ ] gravity,float[ ] geomagnetic)方法有以下四个参数。

• 第 1 个参数 R 是一个长度为 9 的 float 数组，该方法计算出的数据将会存放到这个数组中（保存旋转矩阵 R 的数据，可以为 null）。

• 第 2 个参数 I 是一个用于将磁场向量转换成重力坐标的旋转矩阵，通常为 null。

• 第 3 个参数 gravity 是加速度传感器输出的值。

• 第 4 个参数 geomagnetic 是磁场传感器输出的值。

getOrientation(float[ ] R,float[ ]values)方法有以下两个参数。

• 第 1 个参数是 R 数组，通过 getRotationMatrix( )方法获取。

• 第 2 个参数是长度为 3 的 float 数组，移动设备在各个方向上的旋转数据都会存放在该数组中，其中 values[0] 表示移动设备围绕 z 轴的旋转弧度，values[1] 表示移动设备围绕 x 轴的旋转弧度，values[2] 表示移动设备围绕 y 轴的旋转弧度。values[0] 代表方向角，它用磁场和加速度感应器得到的数据是 −180°~180°，即 0°表示正北，90°表示正东，180°/−180°表示正南，−90°表示正西；而直接通过方向感应器得到的数据是 0°~359°，即 360°/0°表示正北、90°表示正东、180°表示正南、270°表示正西。values[1] 代表倾斜角，即由静止状态开始，前后翻转，手机顶部往上抬起 0°~−90°，手机尾部往上抬起 0°~90°。values[2] 代表旋转角，即由静止状态开始，左右翻转，手机左侧抬起 0°~90°，手机右侧抬起 0°~−90°。

10.2.2 指南针的实现

1. 主界面的设计

根据指南针的功能，在主界面上用一个 TextView 显示方向信息，一

【指南针的实现】

个 ImageView 用于显示指南针罗盘（compass. png），一个 ImageView 用于显示指南针箭头（arrow. png），如图 10.3 所示。主界面布局文件代码如下。

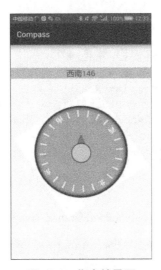

图 10.3　指南针界面

```
1   <?xml version ="1.0" encoding ="utf-8"? >
2   <RelativeLayout xmlns:android ="http://schemas.android.com/apk/res/android"
3       android:layout_width =" match_parent"
4       android: layout_height =" match_parent" >
5       <TextView
6           android: id =" @ +id/tvInfo"
7           android: layout_width =" match_parent"
8           android: layout_height =" wrap_content"
9           android: layout_marginTop =" 60dp"
10          android: background =" #e1d274"
11          android: gravity =" center_horizontal"
12          android: text =""
13          android: textColor =" #c64646"
14          android: textSize =" 20sp" />
15      <ImageView
16          android: id =" @ +id/compass_img"
17          android: layout_width =" 250dp"
18          android: layout_height =" 250dp"
19          android: layout_centerInParent =" true"
20          android: src =" @mipmap/compass" />
21      <ImageView
22          android: id =" @ +id/arrow_img"
23          android: layout_width =" 60dp"
24          android: layout_height =" 110dp"
25          android: layout_centerInParent =" true"
26          android: src =" @mipmap/arrow" />
27  </RelativeLayout >
```

2. 功能实现

(1) 定义变量。

```
1    private TextView tvInfo;
2    private SensorManager sm = null;
3    private Sensor aSensor = null;
4    private Sensor mSensor = null;
5    private ImageView imgCompass;
6    float[] aValues = new float[3];//记录加速度传感器输出数据
7    float[] mValues = new float[3];//记录磁场传感器输出数据
8    float lrDegree;//记录最后旋转角度
9    MySensorListener myListener;
```

上述第 9 行代码的 MySensorListener 是自定义的传感器监听类, 用于监听传感器的数据变化并执行相关操作。

(2) 定义传感器监听事件类 (MySensorListener.java)。

```
1    class MySensorListener implements SensorEventListener {
2        @Override
3        public void onSensorChanged(SensorEvent event) {
4            //判断当前是加速度传感器还是磁场传感器
5            if (event.sensor.getType() == Sensor.TYPE_ACCELEROMETER) {
6                //必须调用 clone 方法进行赋值, 否则 aValues 和 mValues 将会指向同一个引用
7                aValues = event.values.clone();
8            } else if (event.sensor.getType() == Sensor.TYPE_MAGNETIC_FIELD) {
9                mValues = event.values.clone();
10           }
11           float [] R = new float [9];
12           float [] values = new float [3];
13           SensorManager.getRotationMatrix (R, null, aValues, mValues);  //为 R 数组赋值
14           SensorManager.getOrientation (R, values);    //为 values 数组赋值
15           float rotateDegree = -(float) Math.toDegrees (values [0]);
16           int temp = (int) rotateDegree;
17           if (temp >=90&&temp <=180) {
18               if (temp ==90) {
19                   tvInfo.setText (" 正西");
20               } else if (rotateDegree ==180) {
21                   tvInfo.setText (" 正南");
22               } else {
23                   tvInfo.setText (" 西南" +temp);
24               }
25           }
26           if (temp >= -180&&temp <= -90) {
27               if (temp == -90) {
28                   tvInfo.setText (" 正东");
29               } else if (rotateDegree == -180) {
30                   tvInfo.setText (" 正南");
31               } else {
```

```
32                tvInfo.setText("东南"+temp);
33            }
34        }
35        if(temp>=-90&&temp<=0){
36            if(temp==-90){
37                tvInfo.setText("正东");
38            }else if(rotateDegree==0){
39                tvInfo.setText("正北");
40            }else {
41                tvInfo.setText("东北"+temp);
42            }
43        }
44        if(temp>=0&&temp<=90){
45            if(temp==90){
46                tvInfo.setText("正西");
47            }else if(rotateDegree==0){
48                tvInfo.setText("正北");
49            }else {
50                tvInfo.setText("西北"+temp);
51            }
52        }
53        if (Math.abs(rotateDegree-lrDegree) > 1) {
54            RotateAnimation animation = new RotateAnimation (lrDegree,
55                rotateDegree, Animation.RELATIVE_TO_SELF, 0.5f,
56                Animation.RELATIVE_TO_SELF, 0.5f);
57            animation.setFillAfter (true);
58            imgCompass.startAnimation (animation);
59            lrDegree = rotateDegree;
60        }
61    }
62    @Override
63    public void onAccuracyChanged (Sensor sensor, int accuracy) {
64    }
65 }
```

上述第 3~59 行代码的 onSensorChanged( )方法中分别记录加速度传感器和磁场传感器的值, 然后将这两个值传入 SensorManager 的 getRotationMatrix( )方法中就可以得到一个包含旋转矩阵的 R 数组。得到 R 数组后, 调用 SensorManager 的 getOrientation( )方法计算移动设备的旋转数据。第 15 行代码使用 Math.toDegrees( )方法将弧度转换为角度, 然后将计算出的角度取反, 用于旋转指南针背景图 compass.png。取反后的角度在 90°~180°是西南方向, 180°指向正南; 取反后的角度在 -180°~ -90°是东南方向, -90°指向正东; 取反后的角度在 -90°~ 0°是东北方向, 0°指向正北; 取反后的角度在 0°~ 90°, 90°指向正西。第 53 行代码表示旋转角度差大于 1 时, 使用旋转动画技术旋转指南针背景图, 也就是调用 RotateAnimation( )方法实现动画旋转, 该方法有 6 个参数, 第 1 个参数表示旋转的起始角度; 第 2 个参数表示旋转的终止角度; 第 3~6 个参数表示旋转的中心点坐标。

(3) 注册传感器监听器。

```
1    sm = (SensorManager) getSystemService(Context.SENSOR_SERVICE);
2    aSensor = sm.getDefaultSensor (Sensor.TYPE_ACCELEROMETER);
3    mSensor = sm.getDefaultSensor (Sensor.TYPE_MAGNETIC_FIELD);
4    myListener = new MySensorListener();
5    sm.registerListener (myListener, aSensor, SensorManager.SENSOR_DELAY_UI);
6    sm.registerListener (myListener, mSensor, SensorManager.SENSOR_DELAY_UI);
```

(4) 注销传感器监听器。

```
1    @Override
2    protected void onDestroy() {
3        super.onDestroy();
4        if (sm! = null) {
5            sm.unregisterListener(myListener);
6        }
7    }
```

10.3 百度地图在 Android 中的开发应用

下面通过将百度地图加入 App，在 App 上引入定位功能，在百度地图上添加覆盖物和模式切换等功能的实现过程介绍百度地图在 Android 中的开发应用。

10.3.1 预备知识

1. 获取访问应用程序密钥（AK）

（1）开发者注册。

【百度地图加入App】

打开 http://lbsyun.baidu.com/index.php? title = androidsdk 页面，并按照如图 10.4 所示提示信息进行开发者注册（若已经注册过，则可以忽略此步），输入注册信息后单击"提交"按钮，此时系统会自动发送确认邮件到注册时填写的邮箱。登录邮箱确认后，显示如图 10.5 所示界面。单击"申请密钥"按钮，获取应用程序密钥。单击"创建应用"按钮后，显示如图 10.6 所示界面。

图 10.4 百度地图开放平台开发者注册页面

第10章 传感器与位置服务应用开发

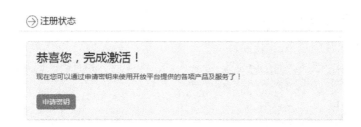

图 10.5 注册成功页面

（2）获取 SHA1。

在图 10.6 所示界面中需要输入应用名称、选择应用类型（本案例是基于 Android 平台的地图开发，选择 Android SDK）及启用服务、输入 SHA1 及包名（包名必须与开发的 App 包名相同）。

图 10.6 创建应用程序界面

在 Android Studio 开发环境中获取 SHA1 的方法如下。

在 Android Studio 开发环境界面右侧依次单击 Gradle→项目名（本案例为 nnutcditu）→Tasks→Android→signingReport，然后单击界面右下方的 Gradle Console，如图 10.7 所示。

将图 10.7 所示的 SHA1 值填入图 10.6 中的对应位置，单击图 10.6 中的"提交"按钮后，显示如图 10.8 所示的界面，界面中的访问应用（AK）列显示的就是该应用程序的 AK。

（3）下载 Android 地图 SDK。

打开 http://lbsyun.baidu.com/index.php?title=androidsdk 页面，单击"产品下载"→"自定义下载"后显示如图 10.9 所示界面。在此界面中开发者可以根据所开发 App 的功能选择相应功能的开发资源，本案例选择了基础定位、基础地图（含室内图）、检索功能、LBS 云检索、计算工具等五项功能的开发资源，接着分别单击"开发包""示例代码"按钮下载开发包和示例代码。

2. 将百度地图加入 App

(1) 创建并配置应用 App 开发环境。

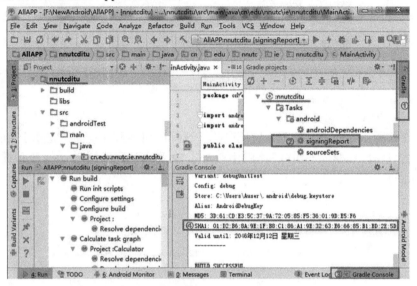

图 10.7　获取 SHA1 界面

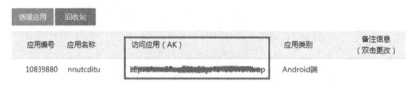

图 10.8　访问应用应用密钥 AK

图 10.9　开发资源下载界面

第10章 传感器与位置服务应用开发

- 按前面介绍的应用程序创建方法创建项目（模块），注意项目的应用程序名和包名都要与图 10.6 中输入的名称一致。
- 将下载的开发包解压后，将 BaiduLBS_AndroidSDK_Lib/libs 目录下的所有内容复制到项目的 libs 目录下（也就是在 Android 应用开发中引入第三方类库）。
- 在 BaiduLBS_Android.jar 文件上右击，选择 Add As Library...，如图 10.10 所示。

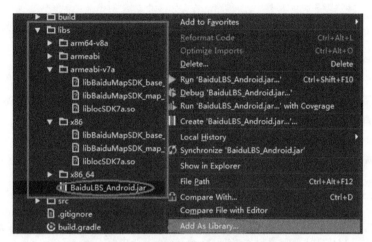

图 10.10　在项目中引入百度地图开发包

（2）在 build.gradle 文件的 android 项添加代码。

```
1       sourceSets {
2           main {
3               jniLibs.srcDirs = ['libs']
4           }
5       }
```

（3）显示地图。

在 AndroidManifest 中添加开发密钥、所需权限等信息。

- 在 application 中添加开发密钥。

```
1   <application>
2       <meta-data
3           android:name="com.baidu.lbsapi.API_KEY"
4           android:value="开发者 key" />
5   </application>
```

上述第 4 行代码的"开发者 key"就是图 10.8 中申请获得的 AK 代码。

- 添加所需权限。

```
1   <uses-permission android:name="android.permission.ACCESS_NETWORK_STATE" />
2   <uses-permission android:name="android.permission.INTERNET" />
3   <uses-permission android:name="com.android.launcher.permission.READ_SETTINGS" />
4   <uses-permission android:name="android.permission.WAKE_LOCK" />
5   <uses-permission android:name="android.permission.CHANGE_WIFI_STATE" />
6   <uses-permission android:name="android.permission.ACCESS_WIFI_STATE" />
7   <uses-permission android:name="android.permission.GET_TASKS" />
```

```
8    <uses-permission android:name="android.permission.WRITE_EXTERNAL_STORAGE" />
9    <uses-permission android: name=" android.permission.WRITE_SETTINGS" />
```

上面的权限主要包括访问网络、WiFi 状态改变、WiFi 状态访问、写 SD 卡等。限于篇幅，读者可以自行查看资料。

● 在布局文件中添加地图控件。

```
1    <com.baidu.mapapi.map.TextureMapView
2        android:id="@+id/bmapView"
3        android:layout_width="fill_parent"
4        android: layout_height="fill_parent"
5    android: clickable=" true" />
```

在应用程序的布局文件（本案例为 main_layout.xml）中添加地图控件 TextureMapView，该地图控件来自 BaiduLBS_Android.jar 包中。

● 在应用程序创建时初始化 SDK 引用的 Context 全局变量。

```
1    public class MainActivity extendsAppCompatActivity{
2        @Override
3        protected void onCreate(Bundle savedInstanceState) {
4            super.onCreate(savedInstanceState);
5            //在使用SDK各组件之前初始化context信息,传入ApplicationContext
6            //注意该方法要在setContentView方法之前实现
7            SDKInitializer.initialize(getApplicationContext());
8            setContentView(R.layout.main_layout);
9        }
10   }
```

● 创建地图 Activity，管理地图生命周期。

```
1    public class MainActivity extends AppCompatActivity {
2        TextureMapView mMapView = null;
3        @Override
4        protected void onCreate(Bundle savedInstanceState) {
5            super.onCreate(savedInstanceState);
6            SDKInitializer.initialize(getApplicationContext());
7            setContentView(R.layout.activity_main);
8            //获取地图控件引用
9            mMapView = (TextureMapView) findViewById (R.id.bmapView);
10       }
11       @Override
12       protected void onDestroy() {
13           super.onDestroy();
14           //在activity执行onDestroy时执行mMapView.onDestroy()，实现地图生命周期管理
15           mMapView.onDestroy();
16       }
17       @Override
18       protected void onResume() {
19           super.onResume();
20           //在activity执行onResume时执行mMapView.onResume()，实现地图生命周期管理
21           mMapView.onResume();
```

第10章 传感器与位置服务应用开发

```
22            }
23        @Override
24        protected void onPause() {
25            super.onPause();
26        //在activity执行onPause时执行mMapView.onPause(),实现地图生命周期管理
27            mMapView.onPause();
28        }
29    }
```

onDestroy( )、onResume( )和onPause( )方法将百度地图mMapView的生命周期与当前显示百度地图的Activity进行绑定。即当Activity销毁时百度地图也同时销毁，以实现性能上的保护，避免Activity已经销毁而百度地图仍然在运行的情况。至此，运行项目就可以显示图10.11所示的地图了。

图10.11 地图显示效果

10.3.2 百度地图应用实现

1. 地图显示界面的优化

（1）去标题栏。

由于Android Studio中的Activity默认是继承于AppCompatActivity类，所以需要修改配置文件中的theme属性值，代码如下。

【切换地图显示类型、地图定位】

```
android:theme ="@style/Theme.AppCompat.Light.NoActionBar"
```

(2) 改变地图的默认显示比例。

默认状态下地图的显示比例为 5km，可以根据用户的需要将地图的显示比例进行调整。例如，将显示比例调整为 500m，可以使用以下代码。

```
1    BaiduMap baiduMap = mMapView.getMap();
2    MapStatusUpdate mapStatusUpdate = MapStatusUpdateFactory.zoomTo(15.0f);
3    baiduMap.setMapStatus(mapStatusUpdate);
```

上述第 2 行代码的 zoomTo（float zoom）方法用于设置地图缩放级别，参数 zoom 为 float 类型的地图缩放级别，其返回值为 MapStatusUpdate 对象。

2. 切换地图显示类型

百度地图中的地图有普通地图、卫星地图和实时交通地图三种类型，使用时可以使用选项菜单、按钮等操作来切换地图类型。本案例使用选项菜单实现，代码如下。

```
1    public boolean onOptionsItemSelected(MenuItem item) {
2        switch (item.getItemId()){
3            case 0:
4                baiduMap.setMapType(BaiduMap.MAP_TYPE_NORMAL);
5                break;
6            case 1:
7                baiduMap.setMapType (BaiduMap.MAP_TYPE_SATELLITE);
8                break;
9            case 2:
10               if (baiduMap.isTrafficEnabled()) {
11                   baiduMap.setTrafficEnabled (false);
12                   item.setTitle (" 实时交通 (off)");
13               } else {
14                   baiduMap.setTrafficEnabled (true);
15                   item.setTitle (" 实时交通 (on)");
16               }
17        }
18   }
```

单击普通地图菜单、卫星地图菜单时就可以直接用代码设置实现切换。而实时交通地图在百度地图中有 off 和 on 两种状态，这就需要使用 if 语句判断当前的状态，然后根据当前的状态进行相应的设置操作。运行上述代码所得地图切换效果如图 10.12 所示。

3. 百度地图定位

百度地图定位实现的功能是当开启 App 后，自动定位到用户当前所在位置并用 Toast 显示出来，当用户拖动地图后，可以使用选项菜单中的"我的位置"重新让地图以用户所在位置为中心显示在屏幕上。

(1) 百度地图定位的关键 API。

● BDLocationListener：是需要重写的百度地图定位监听器类，该类中有一个最重要的方法——onReceiverLocation（BDLocation location），当触发监听的时候系统就会自动调用此方法，此方法可以获得使用者目前位置的所有信息，这些信息存放在 BDLocation 类中，通过 BDLocation 类中的一些方法可以获得需要的信息。例如，getCity( ) 方法可以获得当前的城市、getAddr( ) 方法可以获得当前位置的所有信息，也可以通过 getLatitude( ) 方法和 getLongitude( ) 方法获得当前位置的纬度值和经度值。

● LocationClient：是定位 SDK 的核心类，用于定位服务的客户端，只支持在主线程中运行。

● BDLocation：封装了定位 SDK 的定位结果，在 BDLocationListener 的 onReceive( ) 方法中获取；通过该类可以获取 errorcode、位置坐标、精度半径等信息。BDLocation 的常用方法及功能说明见表 10 - 5。

图 10.12　地图切换效果

表 10 - 5　BDLocation 的常用方法及功能说明

方　法　名	功　能　说　明
int getLocType ( )	获取 error code。61——GPS 定位结果，62——扫描整合定位失败（结果无效），63——网络异常，65——定位缓存的结果，66——离线定位结果，67——离线定位失败，68——网络连接失败，161——网络定位结果，162～167——服务端定位失败，502——key 参数错误，601——key 服务被开发者自己禁用
double getLatitude ( )	获取纬度
double getLongitude ( )	获取经度
boolean hasRadius ( )	判断是否有定位精度半径
float getRadius ( )	获取定位精度半径，单位是米
String getAddrStr ( )	获取反地理编码
String getProvince ( )	获取省份信息
String getCity ( )	获取城市信息
String getDistrict ( )	获取区县信息
float getDirection( )	获得设备方向，范围（0～360），上部正朝向北的方向为 0°方向

● LocationClientOption：该类用来设置定位 SDK 的各种参数，如定位模式、定位时间间隔、坐标系类型等。LocationClientOption 的常用方法及功能说明见表 10 - 6。

表 10-6 LocationClientOption 的常用方法及功能说明

方 法 名	功 能 说 明
setLocationMode（LocationMode mode）	设置定位模式，有 Hight_Accuracy（高精度）、Battery_Saving（低功耗）和 Device_Sensors（仅 GPS 设备）几种模式
setOpenGps(boolean)	设置是否打开 GPS（默认不打开），使用前必须打开硬件的 GPS
setNeedDeviceDirect(boolean)	设置是否需要设备方向信息
setIsNeedAddress(boolean)	设置是否需要地址信息
setAddrType(String)	设置是否返回地址信息（默认无地址信息），all 表示返回地址信息，其他值表示不返回地址信息
setCoorType(String)	设置坐标类型，返回若干种坐标系，包括国测局坐标系（gcj02）、百度坐标系（bd09、bd09ll）。百度移动端地图对外接口中的坐标系默认是 bd09ll
setProdName(String)	设置产品线名称
setScanSpan(int)	设置定时定位的时间间隔，单位为毫秒

（2）百度地图定位的实现。
- 定义变量。

```
1       private TextureMapView mMapView = null;
2       private BaiduMap baiduMap;
3       private LocationClient client;
4       private boolean isFirstin = true;//用于判断用户是不是第 1 次进入 App
5       private Context context;
6       private double myLat;//用于记录纬度值
7       private double myLon;//用于记录经度值
```

因为地图显示界面上用户可以用拖动方式使地图位置发生改变，为了地图位置改变后能够重新回到原来的显示状态，所以使用 myLat、myLon 两个变量记录用户在地图的位置，便于使用"我的位置"菜单让用户的当前位置回到界面中央。
- 定义类实现 BDLocationListener 接口——MyLocListener.java。

BDLocationListener 接口有一个方法——onReceiveLocation()需要实现，即接收异步返回的定位结果，参数是 BDLocation 类型参数。其关键代码如下。

```
1       private class MyLocListener implements BDLocationListener {
2           @Override
3           public void onReceiveLocation(BDLocation location) {
4               MyLocationData data = new MyLocationData.Builder()//
5                   .accuracy(location.getRadius())//精度
6                   .latitude(location.getLatitude())//纬度
7                   .longitude(location.getLongitude())//经度
8                   .build();
9               baiduMap.setMyLocationData(data);
10              if (isFirstin) {
11                  LatLng latLng = new LatLng(location.getLatitude(),
location.getLongitude());//获得位置的纬度值和经度值
```

```
    12                    MapStatusUpdate msu =
MapStatusUpdateFactory.newLatLng(latLng);
    13                    baiduMap.animateMapStatus(msu);//使用动画效果传递地图的位置
    14                    isFirstin = false;
    15                    Toast.makeText(context, location.getAddrStr(),
Toast.LENGTH_LONG).show();
    16                }
    17            }
    18        }
```

上述第 9 行代码后面没有其他语句时，应用程序运行后直接使用系统默认的图标显示在地图上；若此处使用语句自定义图标，则应用程序中的图标会改变为用户自定义的图标，即在应用程序第 1 次进入时在地图中心点显示的图标。定位地图显示效果如图 10.13 所示。

图 10.13 定位地图显示效果

- 自定义方法初始化 LocationClient 和配置定位 SDK 参数。

本案例自定义的方法名为 initLocation()，详细代码如下。

```
    1    void initLocation() {
    2        client = new LocationClient(context); //声明 LocationClient 类
    3        MyLocListener myLocListener = new MyLocListener();
    4        client.registerLocationListener(myLocListener); //注册监听函数
    5        LocationClientOption option = new LocationClientOption();
    6        option.setCoorType("bd09ll");
```

```
7        option.setScanSpan(1000);
8        option.setIsNeedAddress(true);
9        option.setOpenGps(true);
10       client.setLocOption(option);
11   }
```

设置定位参数包括定位模式（高精度定位模式、低功耗定位模式和仅用设备定位模式）、返回坐标类型、是否打开 GPS、是否返回地址信息和位置语义化信息、POI 信息等，通常包括如下代码。

```
1    LocationClientOption option = new LocationClientOption();
2    option.setLocationMode(LocationMode.Hight_Accuracy); //可选, 默认高精度,
设置定位模式, 高精度, 低功耗, 仅设备
3    option.setCoorType ("bd0911"); //可选, 默认 gcj02, 设置返回的定位结果坐标系
4    option.setScanSpan (1000); //可选, 默认 0, 即仅定位一次, 设置发起定位请求的
间隔需要大于等于 1000ms 才是有效的
5    option.setIsNeedAddress (true); //可选, 设置是否需要地址信息, 默认不需要
6    option.setOpenGps (true); //可选, 默认 false, 设置是否使用 GPS
7    option.setLocationNotify (true); //可选, 默认 false, 设置是否当 GPS 有效时
按照每秒 1 次的频率输出 GPS 结果
8    option.setIsNeedLocationDescribe (true); //可选, 默认 false, 设置是否需要
位置语义化结果, 可以在 BDLocation.getLocationDescribe 里得到, 结果类似于"在北京天安
门附近"
9    option.setIsNeedLocationPoiList (true); //可选, 默认 false, 设置是否需要
POI 结果, 可以在 BDLocation.getPoiList 里得到
10   option.setIgnoreKillProcess (false); //可选, 默认 true, 定位 SDK 内部是一
个 SERVICE, 并放到了独立进程, 设置是否在 stop 的时候杀死这个进程, 默认不杀死
11   option.SetIgnoreCacheException (false); //可选, 默认 false, 设置是否收集
CRASH 信息, 默认收集
12   option.setEnableSimulateGps (false); //可选, 默认 false, 设置是否需要过滤
GPS 仿真结果, 默认需要
13   mLocationClient.setLocOption (option);
```

上述的第 1 行代码、第 3～6 行代码和第 13 行代码通常在 App 开发中都是需要的，而其他行的代码可以根据 App 的功能需要进行选择。另外需要读者注意：高精度定位模式会同时使用网络定位和 GPS 定位，优先返回最高精度的定位结果；低功耗定位模式不会使用 GPS，只会使用网络定位（WiFi 和基站定位）；仅用设备定位模式不需要连接网络，只使用 GPS 进行定位，这种模式下不支持室内环境的定位。

● 配置 Service。

在项目的配置文件 androidManifest.xml 中添加如下代码。

```
1        <service
2            android:name ="com.baidu.location.f"
3            android:enabled ="true"
4            android:process =":remote" >
5        </service>
```

● 在 Activity 的 onCreate( )方法中实现相关功能。

```
1        @Override
2        protected void onCreate(Bundle savedInstanceState) {
```

```
3        super.onCreate(savedInstanceState);
4        SDKInitializer.initialize(getApplicationContext());
5        setContentView(R.layout.main_layout);
6        context = this;
7        mMapView = (TextureMapView) this.findViewById (R.id.bmapView);
8        baiduMap = mMapView.getMap();
9        MapStatusUpdate mapStatusUpdate = MapStatusUpdateFactory.zoomTo (15.0f);
10       baiduMap.setMapStatus (mapStatusUpdate);
11       initLocation();
12   }
```

此时运行效果如图10.13所示。但是如果用户现在拖动地图,界面上的地图会随之移动变化,若想回到原来的定位位置就比较难。为了在现有功能的基础上增加一个"我的位置"功能,即单击"我的位置"按钮或菜单,可以让地图回到用户所在位置,需要继续完成下列步骤。

● 在 MyLocListener 类中的 baiduMap.setMyLocationData(data)代码下增加下列两行代码,用于获得用户定位位置的纬度值和经度值并保存在 myLat 和 myLon 中。

```
1    myLat = location.getLatitude();
2    myLon = location.getLongitude();
```

● 在前面建立的选项菜单中增加"我的位置"菜单项,并在监听事件中添加下列代码,即在单击"我的位置"菜单时,把前面保存的纬度值和经度值传递过来后进行重新定位,这样就可以让地图回到原来的位置。

```
1    case 3:
2        LatLng latLng = new LatLng(myLat, myLon);//获得位置的纬度和经度值
3        MapStatusUpdate msu = MapStatusUpdateFactory.newLatLng(latLng);
4        baiduMap.animateMapStatus(msu);//使用动画效果传递地图的位置
5        break;
```

当用户退出 App 时,App 需要关闭地图定位;当用户开启 App 时,App 需要开启地图定位,所以需要重写 Activity 的 onStop()、onStart()方法,代码如下。

```
1    protected void onStart() {
2        super.onStart();
3        baiduMap.setMyLocationEnabled(true);
4        if(! client.isStarted())
5            client.start();//开启定位
6    }
7    protected void onStop() {
8        super.onStop();
9        baiduMap.setMyLocationEnabled(false);
10       client.stop();//关闭定位
11   }
```

4. 百度地图中方向传感器的使用

通常当 Android 终端设备方向改变时,地图中央标注用户位置的图标方向也会跟着改变,下面以此功能的实现过程介绍 Android 终端设备中方向传感器的使用方法。

（1）准备图标。

将带有箭头方向的图标复制到项目的 res/mipmap-hdpi 目录下，本案例中的图标文件名为 navi.png。

（2）自定义定位图标。

在实现 BDLocationListener 接口的 MyLocListener 类中使用下列代码自定义定位图标。

```
1    BitmapDescriptor    iconLocation =
BitmapDescriptorFactory.fromResource(R.mipmap.navi); //初始化图标
2    MyLocationConfiguration config = new MyLocationConfiguration
(MyLocationConfiguration.LocationMode.NORMAL,true,iconLocation);
3    baiduMap.setMyLocationConfiguration(config);
```

上述代码添加在 baiduMap.setMyLocationData（data）语句后面，此时运行程序后自定义的图标就会显示在地图中央（即当前设备所在位置处），运行效果如图 10.14 所示。

图 10.14　自定义图标显示定位位置效果

（3）集成方向传感器。

现在需要与方向传感器结合来实现带方向的定位图标，并在旋转设备的时候，图标自动跟着旋转。

● 定义类实现 SensorEventListener 接口用于监听方向传感器——MyOrientationListener.java。

自定义一个子类 MyOrientationListener 实现 SensorEventListener，并重写 onSensorChanged()方法和 onAccuracyChanged()方法。其中 onAccuracyChanged()方法监听精度改变并不需要，所以只要通过 onSensorChanged()方法监听 x 轴方向的改变就可以满足功能需要。其详细代码如下。

```java
1   public class MyOrientationListener implements SensorEventListener {
2       private SensorManager mSensorManager;
3       private Context mContext;
4       private Sensor mSensor;
5       private float mLastX; //方向传感器有三个坐标,现在只关注 x
6       public MyOrientationListener(Context context) {
7           this.mContext = context;
8       }
9       public void start() { //开始监听
10          mSensorManager = (SensorManager) mContext.getSystemService(Context.SENSOR_SERVICE);
11          if (mSensorManager ! = null) {
12              mSensor = mSensorManager.getDefaultSensor (Sensor.TYPE_ORIENTATION); //获得方向传感器
13          }
14          if (mSensor ! = null) {注册方向传感器
15              mSensorManager.registerListener (this, mSensor, SensorManager.SENSOR_DELAY_UI);
16          }
17      }
18      public void stop() {//结束监听
19          mSensorManager.unregisterListener (this); //注销方向传感器
20      }
21      @Override
22      public void onSensorChanged (SensorEvent event) {
23          if (event.sensor.getType() == Sensor.TYPE_ORIENTATION) {
24              float x = event.values [SensorManager.DATA_X]; //只获取 x 的值
25              if (Math.abs (x - mLastX) > 1.0) {
26                  if (onOrientationListener ! = null) {
27                      onOrientationListener.onOrientationChanged (x);
28                  }
29              }
30              mLastX = x;
31          }
32      }
33      @Override
34      public void onAccuracyChanged (Sensor sensor, int accuracy) {
35      }
36      private OnOrientationListener onOrientationListener;
37      public void setOnOrientationListener (OnOrientationListener onOrientationListener) {
38          this.onOrientationListener = onOrientationListener;
39      }
40      public interface OnOrientationListener {
41          void onOrientationChanged (float x);
42      }
43  }
```

- 在定位中实现方向传感器的监听，在 Activity 中的 onCreate( )方法中添加如下代码。

```
1    myOrientationListener = new MyOrientationListener(this); //实例化监听方
向传感器对象
2    myOrientationListener.setOnOrientationListener(new MyOrientationLis-
tener.OnOrientationListener() {
3        @Override
4        public void onOrientationChanged(float x) {
5            mLastX = x; //将获取的 x 轴方向赋值给全局变量
6        }
7    });
```

- 修改获取 location 信息时的配置——添加旋转角度。

因为要实现方向传感器的监听需要获得设备当前的方向信息,要获得当前的方向信息就需要获得方向传感器的旋转角度,要获得旋转角度就需要修改上面地图定位中自定义的 MyLocListener 类(实现 BDLocationListener 接口)中的 onReceiveLocation( )方法,也就是通过 direction(mLastX)方法获得方向传感器传过来的 x 轴数值,其修改位置如下所示。

```
1    MyLocationData data = new MyLocationData.Builder()//
2        //............
3        .direction(mLastX)//此处设置开发者获取到的方向信息,顺时针 0°~360°,
mLastX 就是获取到的方向传感器传过来的 x 轴数值
4        //............
```

- 启动/停止方向传感器监听。

当 App 启动后要自动启动方向传感器监听,要实现此功能只要在 activity 中的 onStart( )方法中添加如下代码。

```
myOrientationListener.start();
```

当 App 停止后要自动停止方向传感器监听,要实现此功能只要在 activity 中的 onStop( )方法中添加如下代码。

```
myOrientationListener.stop();
```

到此,随着终端设备的方向改变,地图上的图标指向也会随之改变,其运行结果如图 10.14 所示。

5. 百度地图中添加覆盖物

很多包含地图的 App 都可以实现搜索周边的加油站、宾馆等功能,当搜索到需要的加油站、宾馆后,可以看到在对应的位置会出现一些特别的图标,这种图标其实就是在百度地图上添加覆盖物。下面将从百度地图上添加覆盖物、周边搜索相关热点、单击覆盖物弹出信息(如图片、与本地的距离及点赞的数量)等功能的实现来介绍百度地图应用的扩展。

(1) 创建一个用于存储弹出信息的类——Info.java。

该类中包括单击位置处的信息,即纬度、经度、图片 ID、名称、与本地的距离及点赞数量等信息,其关键代码如下。

```
1    public class Infoimplements Serializable {
2        private double lat;//纬度
3        private double lon;//经度
```

```
4      private   int img;//图片 ID
5      private   String name;//名称
6      private   String distance;//与本地的距离
7      private   int zan;//点赞数量
8      public static List <Info> infos = new ArrayList <Info>();
9      static
10     {
11         infos.add(new Info(32.412796,119.932188, R.mipmap.m5, "小旅馆",
"距离 209 米", 1456));
12         infos.add(new Info(32.413466,119.941675, R.mipmap.m6, "洗浴会所",
"距离 897 米", 456));
13         infos.add(new Info(32.410997,119.937183, R.mipmap.m7, "服装城",
"距离 249 米", 100));
14         infos.add(new Info(32.411795, 119.934183, R.mipmap.m8, "饭店",
"距离 679 米", 256));
15     }
16     ……//构造方法
17     ……//get、set 方法
18  }
```

Info 类中的 infos 在实际应用中是从服务器返回的数据部分（通常是 Json 格式的数据，即从服务器返回的纬度、经度、图片、热点名称、与本地的距离及点赞数量）转换成的实体集合，上述代码中模拟了这部分数据。

（2）在地图上添加覆盖物——绑定选项菜单事件。

首先在选项菜单中添加"添加覆盖物"菜单，在单击"添加覆盖物"菜单后，需要清除原来的定位图层，通过迭代的方法将热点图标在对应位置标出。也就是需要在选项菜单中绑定"添加覆盖物"菜单监听事件，实现代码如下。

```
1    case 4:
2        BitmapDescriptor markerMap =
BitmapDescriptorFactory.fromResource(R.mipmap.navi);
3        baiduMap.clear();//清除定位图层
4        LatLng nlatLng = null;//定义经纬度对象
5        OverlayOptions overlayOptions = null;
6        Marker marker = null;//定义放在地图上的 Marker
7        for (Info info : infos) {
8            nlatLng = new LatLng(info.getLat(),info.getLon());
9            overlayOptions = new MarkerOptions().position(nlatLng)
10             .icon(markerMap).zIndex(5);//生成图层信息(位置、图标及显示层数)
11           marker = (Marker)(baiduMap.addOverlay(overlayOptions));//根据
图层信息生成 Marker 对象
12           Bundle bundle = new Bundle();
13           bundle.putSerializable("info", info);
14           marker.setExtraInfo(bundle);//Marker 需要携带对应的信息
15       }
16       MapStatusUpdate lastStatus =
MapStatusUpdateFactory.newLatLng(nlatLng); // 将地图移到最后一个经纬度位置
17       baiduMap.setMapStatus(lastStatus);
18       break;
```

上述第 7~15 行代码使用了迭代的方法在地图上添加了图层（Overlay），然后在返回的 Marker 中设置了热点位置的相关信息（经纬度、名称、图片、与本地的距离及点赞数量等）。运行效果如图 10.15 所示。

图 10.15　在地图上添加热点图标效果

（3）创建显示热点位置详细信息的布局文件——markerLayout.xml。

当用户单击热点 Marker 时，就可以获得该热点位置的详细数据信息并显示。要显示这些信息，就需要为这个信息创建一个布局文件。该布局文件包含 2 个 ImageView、3 个 TextView 分别用于显示热点照片、赞图片、热点名称、与本地的距离及点赞数量，显示效果如图 10.16 所示。布局文件的详细代码如下：

```
1    <RelativeLayout
2        android:background="#6aa66a"
3        android:id="@+id/markinfo"
4        android:layout_width="wrap_content"
5        android:layout_height="wrap_content"
6        android:layout_alignParentBottom="true">
7        <ImageView
8            android:paddingTop="20dp"
9            android:id="@+id/img"
10           android:layout_width="match_parent"
11           android:layout_height="220dp"
12           android:src="@mipmap/m5"/>
13       <LinearLayout
```

```
14            android:layout_width=" wrap_content"
15            android: layout_height=" wrap_content"
16            android: layout_below=" @id/img"
17            android: orientation=" vertical"
18            android: paddingLeft=" 20dp" >
19            <TextView
20                android: id=" @ +id/tvname"
21                android: layout_width=" wrap_content"
22                android: layout_height=" wrap_content"
23                android: text=" 客栈" />
24            <TextView
25                android: id=" @ +id/tvdistance"
26                android: layout_width=" wrap_content"
27                android: layout_height=" wrap_content"
28                android: text=" 距离100 米" />
29        </LinearLayout>
30        <LinearLayout
31            android: layout_width=" wrap_content"
32            android: layout_height=" wrap_content"
33            android: layout_alignParentRight=" true"
34            android: layout_below=" @id/img"
35            android: paddingRight=" 20dp" >
36            <ImageView
37                android: layout_width=" wrap_content"
38                android: layout_height=" wrap_content"
39                android: src=" @mipmap/zan" />
40            <TextView
41                android: id=" @ +id/tvzan"
42                android: layout_width=" wrap_content"
43                android: layout_height=" wrap_content"
44                android: layout_gravity=" center"
45                android: text=" 987" />
46        </LinearLayout>
47    </RelativeLayout>
```

图10.16　热点信息显示效果

(4) 为地图上的 Marker 添加单击事件。

默认情况下,显示的 Marker 热点信息处于隐藏状态,所以需要将 markinfo 的 visibility 属性设置为 gone。当单击 Marker 热点后,将该属性设置为 visible,让热点信息显示出来。其关键代码如下。

```
1    markerLayout = (RelativeLayout) this.findViewById(R.id.markinfo);
2    //设置单击地图 Marker 的监听事件
3    baiduMap.setOnMarkerClickListener(new BaiduMap.OnMarkerClickListener() {
4        @Override
5        public boolean onMarkerClick(Marker marker) {
6            Bundle extraInfo = marker.getExtraInfo();
7            Info info = (Info) extraInfo.getSerializable("info");
8            //以下 4 行代码用于从 Marker 布局中实例化对象
9            ImageView img = (ImageView) markerLayout.findViewById(R.id.img);
10           TextView tvname = (TextView) markerLayout.findViewById(R.id.tvname);
11           TextView tvdistance = (TextView) markerLayout.findViewById(R.id.tvdistance);
12           TextView tvzan = (TextView) markerLayout.findViewById(R.id.tvzan);
13           //以下代码分别为上述 4 个控件设置对应的值
14           img.setImageResource(info.getImg());
15           tvname.setText(info.getName());
16           tvdistance.setText(info.getDistance());
17           tvzan.setText(info.getZan() + "");
18           markerLayout.setVisibility(View.VISIBLE); //将 Marker 层显示
19           return true;
20       }
21   });
```

当单击地图上 Marker 热点之外的位置时,显示的热点信息消失。运行后的效果如图 10.17 所示,其关键代码如下。

```
1    //设置单击 Marker 之外的位置时的监听事件
2    baiduMap.setOnMapClickListener(new BaiduMap.OnMapClickListener() {
3            @Override
4            public void onMapClick(LatLng latLng) {
5                markerLayout.setVisibility(View.GONE);
6            }
7            @Override
8            public boolean onMapPoiClick(MapPoi mapPoi) {
9                return false;
10           }
11   });
```

(5) 覆盖物显示文本信息。

当用户单击热点位置图标(覆盖物)时,能够在覆盖物上显示该热点的名称信息,可以使用 InfoWindow() 方法来封装显示的内容、位置及当单击其他热点时显示信息消失。运行效果如图 10.18 所示。其关键代码如下。

```
1    InfoWindow infoWindow;
2    TextView tv = new TextView(context);
```

```
3      tv.setBackgroundColor(Color.GRAY);
4      tv.setPadding(30, 20, 30, 50);
5      tv.setText(info.getNaame());
6      LatLng latLng = marker.getPosition();//将 Marker 所在位置的经纬度信息转化为
屏幕上的坐标
7      android.graphics.Point point = baiduMap.getProjection().toScreenLocation
(latLng);
8      point.y = point.y - 47;
9      LatLng latLngInfo = baiduMap.getProjection().fromScreenLocation(point);
10     BitmapDescriptor tips = BitmapDescriptorFactory.fromView(tv);
11     //为弹出的 InfoWindows 添加单击事件
12     infoWindow = new InfoWindow(tips, latLngInfo, 10, new OnInfoWindowClick-
Listener() {
13         @Override
14         public void onInfoWindowClick() {
15             baiduMap.hideInfoWindow();//隐藏弹出的消息框
16         }
17     });
18     baiduMap.showInfoWindow(infoWindow);//显示消息框
```

以上代码中实例化 InfoWindow 的构造方法 InfoWindow（BitmapDescriptor，LatLng，int，InfoWindow.OnInfoWindowClickListener）有 4 个参数，其中第 1 个参数表示要显示的组件；第 2 个参数表示要显示的位置；第 3 个参数表示 y 轴偏移量；第 4 个参数表示单击消息框的监听事件。

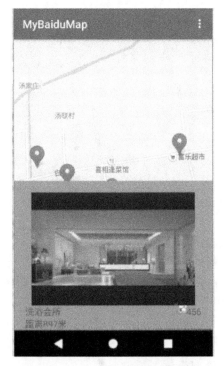

图 10.17　热点详细信息显示效果

图 10.18　覆盖物上显示文本信息效果

至此，实现了百度地图在 Android 中的相关应用，读者可以参阅 ALLAPP 代码包中 ditunnutc 文件夹里的详细代码。另外，读者也可以查阅百度地图的开发文档进行其他扩展功能的开发。

本 章 小 结

近年来，基于传感器和位置的服务发展尤为迅速，涉及商务、医疗、工作和生活的各个方面，为用户提供定位、追踪和敏感区域警告等一系列服务。本章结合实际案例项目的开发过程介绍了 Android 中加速度传感器、磁场传感器、方向传感器和百度地图的应用开发方法，让读者能够结合实际需求开发出更多有趣且有用的应用程序。

习 题

一、选择题

1. 获取传感器管理类对象的方法是（　　）。
   A. getSystemService(SENSOR_SERVICE)
   B. getSystemService(AlARM_SERVICE)
   C. getDefaultSensor(inttype)
   D. getSensorManager( )

2. 在 Android 中，注册陀螺仪传感器时需要使用的传感器的参数是（　　）。
   A. Sensor. TYPE_LIGHT   B. Sensor. TYPE_GYROSCOPE
   C. Sensor. TYPE_ACCELEROMETER D. Sensor. TYPE_TEMPERATURE

3. 下列哪个传感器可以用于制作微博里的"摇一摇"功能（即晃动手机来寻找周围同上微博的人）？（　　）
   A. Sensor. TYPE_ORIENTATION
   B. Sensor. TYPE_PROXIMITY
   C. Sensor. TYPE_ACCELEROMETER
   D. Sensor. TYPE_LIGHT

二、填空题

1. _____定位是通过终端设备内置的硬件与卫星交互来获取当前设备所在的位置的经纬度坐标。

2. _____定位是根据手机、平板电脑等终端设备，利用附近的基站进行三角定位。

3. 用完传感器需要注销传感器监听器，通常需要调用 SensorManager 的_____方法。

4. 百度地图为用户提供了普通地图、卫星地图和_____地图，方便用户进行选择。

5. SensorEventListener 是使用传感器的核心部分，以下两个方法必须实现：_____和 onAccuracyChanged( )方法。

6. _____接口定义了常见的 Provider 状态变化和位置变化的方法。

7. 可以用来辅助 WebView 设置其一些属性和状态的类是_____。

8. 要注册各种传感器需要先获取_____对象。

三、判断题

1. GoogleMap API 中用于显示地图的是 MapView 组件，必须和 MapActivity 配合使用。
（　　）
2. 光线感应传感器主要用于 Android 的 LCD 自动亮度功能。（　　）
3. Android 所有的传感器都归传感器管理器 SensorManager 管理。（　　）
4. 加速度传感器又叫 G-sensor，返回 x 轴、y 轴、z 轴的加速度数值。（　　）
5. WiFi 定位是根据一个固定的 WifiMAC 地址，通过收集到的该 WiFi 热点的位置，然后访问网络上的定位服务器以获得经纬度坐标。只要有 WiFi 的地方可以使用，并且定位精度较高。（　　）
6. 陀螺仪传感器经常被用来计算手机已转动的角度，当手机逆时针旋转时，角速度为正值，顺时针旋转时，角速度为负值。（　　）

【第10章参考答案】